## SEVENTH EDITION

# THE SKILLED HELPER

## A PROBLEM-MANAGEMENT AND OPPORTUNITY-DEVELOPMENT APPROACH TO HELPING

### Gerard Egan

Professor Emeritus
Loyola University of Chicago

**BROOKS/COLE**

**THOMSON LEARNING**

Australia • Canada • Mexico • Singapore • Spain • United Kingdom • United States

**BROOKS/COLE**

**THOMSON LEARNING**

Executive Editor: *Lisa Gebo*
Assistant Editors: *Julie Dillemuth, Alma Dea Michelena*
Marketing Team: *Caroline Concilla, Justine Ferguson,*
and *Tami Strang*
Editorial Assistant: *Sheila Walsh*
Project Editor: *Mary Anne Shahidi*
Production Service & Compositor: *Pre-Press Co., Inc.*

Permissions Editor: *Sue Ewing*
Cover Design: *Andrew Ogus*
Cover Illustration: *Ryuichi Okano/Photonica*
Interior Illustration: *Pre-Press Co., Inc.*
Print Buyer: *Vena Dyer*
Printing and Binding: *Phoenix Color Corp*

*For more information about this or any other Brooks/Cole product, contact:*
BROOKS/COLE
511 Forest Lodge Road
Pacific Grove, CA 93950 USA
www.brookscole.com
1-800-423-0563 (Thomson Learning Academic Resource Center)

Printed in the United States of America

10   9   8   7   6   5   4

**Library of Congress Cataloging-in-Publication Data**

Egan, Gerard.
   The skilled helper: a problem-management and opportunity-development approach to helping / Gerard Egan.— 7th ed.
        p.        cm.
   Includes bibliographical references and index.
   ISBN 0 534-36731-3 (alk. paper)
   1. Counseling. 2. Helping behavior. I. Title.

BF637.C6.E39 2001                           2001025786
158'.3—dc21

# C O N T E N T S

## Part One

### LAYING THE GROUNDWORK    1

CHAPTER 1

## INTRODUCTION TO HELPING    2

FORMAL AND INFORMAL HELPERS: A VERY BRIEF HISTORY    3

WHAT HELPING IS ABOUT    3

POSITIVE PSYCHOLOGY AND HELPING    6

THE TWO PRINCIPAL GOALS OF HELPING    7

DOES HELPING HELP?    9

IS HELPING FOR EVERYONE?    14

WHAT THIS BOOK IS—AND WHAT IT IS NOT    15

MOVING FROM SMART TO WISE: MANAGING THE SHADOW SIDE OF HELPING    16

CHAPTER 2

## OVERVIEW OF THE HELPING MODEL    21

RATIONAL PROBLEM SOLVING AND ITS LIMITATIONS    23

THE SKILLED-HELPER MODEL: A PROBLEM-MANAGEMENT AND OPPORTUNITY-DEVELOPMENT APPROACH TO HELPING    24

THE STAGES AND STEPS OF THE HELPING MODEL    25

STAGE I: "WHAT'S GOING ON?" HELPING CLIENTS CLARIFY THE KEY ISSUES CALLING FOR CHANGE    26

STAGE II: "WHAT SOLUTIONS MAKE SENSE FOR ME?" HELPING CLIENTS DETERMINE OUTCOMES    28

STAGE III: "WHAT DO I HAVE TO DO TO GET WHAT I NEED OR WANT?" HELPING CLIENTS DEVELOP STRATEGIES FOR ACCOMPLISHING GOALS    30

ACTION: "HOW DO I GET RESULTS?" HELPING CLIENTS IMPLEMENT THEIR PLANS    31

"How Are We Doing?" Ongoing Evaluation of the Helping Process    32
Flexibility in Using the Model    33
Brief Therapy and a Hologram Approach to Helping    35
Problem Management and Culture: A Human Universal    36
Using the Model as a "Browser": The Search for Best Practice    37
Understanding and Dealing with the Shadow Side of Helping Models    38

## Chapter 3
### The Helping Relationship: Values in Action    40

The Helping Relationship    42
The Relationship as a Working Alliance    43
Values in Action    44
Respect as the Foundation Value    46
Empathy as a Primary Orientation Value    48
Genuineness as a Professional Value    53
Client Empowerment as an Outcome Value    55
A Working Charter: A Client-Helper Contract    58
Shadow-Side Realities in the Helping Relationship    59

## Part Two
### The Therapeutic Dialogue    63

## Chapter 4
### Introduction to Communication and the Skill of Visibly Tuning in to Clients    64

The Importance of Dialogue in Helping    65
Visibly Tuning in to Clients: The Importance of Empathic Presence    66
The Shadow Side of Communication Skills    70

## Chapter 5
### Active Listening: The Foundation of Understanding    73

Inadequate Listening    75
Empathic Listening    76
Listening to Words: Clients' Stories, Points of View, Decisions, and Intentions or Proposals    77
Listening to Clients' Nonverbal Messages and Modifiers    83
Processing What You Hear: The Thoughtful Search for Meaning    85

Listening to Oneself: The Helper's Internal Conversation    88

The Shadow Side of Listening to Clients    89

CHAPTER 6

## SHARING EMPATHIC HIGHLIGHTS: COMMUNICATING AND CHECKING UNDERSTANDING    93

Responding Skills    95

The Three Dimensions of Responding Skills: Perceptiveness, Know-How, and Assertiveness    95

Sharing Empathic Highlights: Communicating Understanding to Clients    97

The Key Building Blocks of Empathic Highlights    98

Principles for Sharing Highlights    105

Tactics for Communicating Highlights    112

A Caution: The Importance of Empathic Relationships    112

The Shadow Side of Sharing Empathic Highlights    113

CHAPTER 7

## THE ART OF PROBING AND SUMMARIZING    117

Nonverbal and Verbal Prompts    119

Different Forms of Probes    120

Using Questions Effectively    121

Principles in the Use of Probes    122

The Relationship Between Sharing Highlights and Using Probes    129

The Art of Summarizing: Providing Focus and Direction    131

How to Become Proficient in Using Communication Skills    134

Shadow Side Realities of Communication Skills    135

## Part Three

## STAGE I OF THE HELPING MODEL AND ADVANCED COMMUNICATION SKILLS    137

CHAPTER 8

## STEP I-A: "WHAT ARE MY CONCERNS?" HELPING CLIENTS TELL THEIR STORIES    138

An Introduction to Stage I: Identifying and Exploring Problems and Opportunities    139

Step I-A: "What's Going On?"    139

HELPING CLIENTS EXPLORE PROBLEM SITUATIONS AND UNEXPLOITED OPPORTUNITIES    141

STEP I-A AND ACTION    153

IS STEP 1-A ENOUGH?    157

THE SHADOW SIDE OF STEP I-A    158

EVALUATION QUESTIONS FOR STEP I-A    161

CHAPTER 9

# RELUCTANT AND RESISTANT CLIENTS    162

RELUCTANCE: MISGIVINGS ABOUT CHANGE    163

RESISTANCE: REACTING TO COERCION    165

PRINCIPLES FOR MANAGING RELUCTANCE AND RESISTANCE    167

PSYCHOLOGICAL DEFENSES: THE SHADOW SIDE OF RELUCTANCE AND RESISTANCE    171

CHAPTER 10

# STEP I-B: I. THE NATURE OF CHALLENGING    174

CHALLENGING: THE BASIC CONCEPT    176

BLIND SPOTS: THE TARGETS OF CHALLENGING    177

FROM BLIND SPOTS TO NEW PERSPECTIVES    181

THE GOALS OF CHALLENGING    184

APPLICATIONS: FROM BLIND SPOTS TO NEW PERSPECTIVES TO ACTION    184

CHAPTER 11

# STEP I-B: II. SPECIFIC CHALLENGING SKILLS    199

ADVANCED EMPATHIC HIGHLIGHTS: THE MESSAGE BEHIND THE MESSAGE    200

INFORMATION SHARING: FROM NEW PERSPECTIVES TO ACTION    205

HELPER SELF-DISCLOSURE    207

IMMEDIACY: DIRECT, MUTUAL TALK    209

USING SUGGESTIONS AND RECOMMENDATIONS    214

CONFRONTATION    215

ENCOURAGEMENT    216

EVALUATION QUESTIONS FOR STEP I-B: THE USE OF SPECIFIC CHALLENGING SKILLS    217

CHAPTER 12

# STEP I-B: III. THE WISDOM OF CHALLENGING    218

GUIDELINES FOR EFFECTIVE CHALLENGING    219

LINKING CHALLENGE TO ACTION    224

The Shadow Side of Challenging    224

Evaluation Questions For Step I-B: The Process and Wisdom of
Challenging    228

CHAPTER 13

# STEP I-C: LEVERAGE—HELPING CLIENTS WORK ON THE RIGHT THINGS    230

The Economics of Helping    231

Screening: The Initial Search for Leverage    231

Leverage: Working on Issues That Make a Difference    233

Some Principles of Leverage    233

Focus and Leverage: The Lazarus Technique    238

Step I-C and Action    239

The Shadow Side of Step I-C    240

Evaluation Questions for Step I-C    240

Part Four

# STAGE II: HELPING CLIENTS DETERMINE WHAT THEY NEED AND WANT    241

CHAPTER 14

# INTRODUCTION TO STAGE II: "WHAT SOLUTIONS MAKE SENSE FOR ME?" HELPING CLIENTS IDENTIFY, CHOOSE, AND SHAPE GOALS    242

The Three Steps of Stage II    243

Solution-Focused Helping    243

Helping Clients Discover and Use Their Power Through
Goal Setting    249

Helping Clients Become More Effective Decision Makers    251

CHAPTER 15

# STEP II-A: "WHAT DO I NEED AND WANT?" POSSIBILITIES FOR A BETTER FUTURE    260

Possibilities for a Better Future    261

Skills for Identifying Possibilities for a Better Future    263

Cases Featuring Possibilities for a Better Future    270

Evaluation Questions for Step II-A    274

CHAPTER 16

## STEP II-B: "WHAT DO I REALLY WANT?" MOVING FROM POSSIBILITIES TO CHOICES    275

FROM POSSIBILITIES TO CHOICES    276
HELPING CLIENTS SHAPE THEIR GOALS    276
NEEDS VERSUS WANTS    286
EMERGING GOALS    288
ADAPTIVE GOALS    288
THE "REAL-OPTIONS" APPROACH    292
A BIAS FOR ACTION AS A METAGOAL    292
EVALUATION QUESTIONS FOR STEP II-B    293


CHAPTER 17

## STEP II-C: "WHAT AM I WILLING TO PAY FOR WHAT I WANT?" COMMITMENT    294

HELPING CLIENTS COMMIT THEMSELVES TO A BETTER FUTURE    295
GREAT EXPECTATIONS: CLIENT SELF-EFFICACY    301
STAGE II AND ACTION    304
THE SHADOW SIDE OF GOAL SETTING    305
EVALUATION QUESTIONS FOR STEP II-C    307


Part Five

## STAGE III: HELPING CLIENTS DEVELOP STRATEGIES TO ACCOMPLISH THEIR GOALS    309


CHAPTER 18

## STEP III-A: "HOW MANY WAYS ARE THERE TO GET WHAT I NEED AND WANT?" ACTION STRATEGIES    310

INTRODUCTION TO STAGE III    311
MANY DIFFERENT PATHS TO GOALS    313
"WHAT SUPPORT DO I NEED TO WORK FOR WHAT I WANT?"    317
"WHAT WORKING KNOWLEDGE AND SKILLS WILL HELP ME GET WHAT I NEED AND WANT?"    319
LINKING STRATEGIES TO ACTION    320
EVALUATION QUESTIONS FOR STEP III-A    322

CHAPTER 19

## STEP III-B: "WHAT STRATEGIES ARE BEST FOR ME?" BEST-FIT STRATEGIES    323

"WHAT'S BEST FOR ME?" THE CASE OF BUD    324

HELPING CLIENTS CHOOSE BEST-FIT STRATEGIES    325

STRATEGY SAMPLING    327

A BALANCE-SHEET METHOD FOR CHOOSING STRATEGIES    328

LINKING STEP III-B TO ACTION    331

THE SHADOW SIDE OF SELECTING STRATEGIES    331

EVALUATION QUESTIONS FOR STEP III-B    333

CHAPTER 20

## STEP III-C: "WHAT KIND OF PLAN WILL HELP ME GET WHAT I NEED AND WANT?" HELPING CLIENTS MAKE PLANS    334

NO PLAN OF ACTION: THE CASE OF FRANK    335

HOW PLANS ADD VALUE TO CLIENTS' CHANGE PROGRAMS    336

SHAPING THE PLAN: THREE CASES    338

HUMANIZING THE TECHNOLOGY OF CONSTRUCTIVE CHANGE    341

TAILORING READY-MADE PROGRAMS TO CLIENTS' NEEDS    344

EVALUATION QUESTIONS FOR STEP III-C    346

PART SIX

## THE ACTION ARROW: MAKING IT ALL HAPPEN    347

CHAPTER 21

## "HOW DO I MAKE IT ALL HAPPEN?" HELPING CLIENTS GET WHAT THEY WANT AND NEED    348

HELPING CLIENTS BECOME EFFECTIVE TACTICIANS    351

GETTING ALONG WITHOUT A HELPER: DEVELOPING SOCIAL NETWORKS FOR SUPPORTIVE CHALLENGE    359

THE SHADOW SIDE OF IMPLEMENTING CHANGE    363

**REFERENCES**    369
**NAME INDEX**    397
**SUBJECT INDEX**    403

# GERARD EGAN

Gerard Egan, Ph.D., is Professor Emeritus of Organization Development and Psychology in the Center for Organization Development of Loyola University of Chicago. He has written over fifteen books, some in the field of counseling and communication, including *The Skilled Helper, Interpersonal Living,* and, with Michael Cowan, *People in Systems. The Skilled Helper,* translated into both European and Asian languages, is currently the most widely used counseling text in the world.

His other books, dealing with business and management, include *Change Agent Skills in Helping and Human Service Settings; Adding Value: A Systematic Guide to Business-Based Management and Leadership; Working the Shadow Side: A Guide to Positive Behind-the-Scenes Management;* and, with Andrew Bailey, *TalkWorks at Work: How to Become a Better Communicator and Make the Most of Your Career.* The last book has been published by British Telecommunications (now BT) in London.

Through these writings, complemented by extensive consulting, he has created a comprehensive business-based system of management focusing on strategy, operations, structure, human resource management, the managerial role itself, and leadership. The management system includes a framework for initiating and managing change and a framework for managing such "shadow-side" complexities as organizational culture and politics and resistance to change.

He has lectured, consulted, and given workshops in Africa, Asia, Australia, Europe, and both North and South America, and Mexico. In China he has worked with university- and community-based professionals on counselor-training systems. He has benefitted enormously from working with people around the world and has tried to suffuse his writings with the spirit and benefits of his multicultural experiences.

He consults with a variety of companies and institutions worldwide. Now most of the counseling he does takes place within these organizations as part of his fourfold role as consultant, coach, counselor, and confidant. He specializes in working with senior managers, often on a longer term basis, in the areas of board relationships, strategy, business and organization effectiveness, human resource development, management development, leadership, the challenge and redesign of corporate culture, and the design and management of change.

Since this edition of *The Skilled Helper* is seeing the light of day in the second half of 2001, it could well be called the "millennium edition" (the "real" millennium, of course). However, there is nothing particularly magic about a millennium. Indeed, for many people in the world who march to a different calendar, there was no millennium. That said, for many the millennium has a degree of psychological fascination. Some, in awe of the fact that we have not yet destroyed ourselves, see it as time to face up to the grave challenges the world faces. Others see it as a time to celebrate the advances of humankind and envision the creation of a better world for all. On a more modest scale, millennial time is time out. Time to take stock. Time for renewal. And the psychological professions, like all institutions, are in constant need of renewal.

The last decade has presented providers of psychological services with many challenges, some of them two-edged—the marketplace and managed care; diversity and multiculturalism; the "two cultures" of America, one me-first, individualistic, and get-all-that-you-can, the other group-oriented, proud of traditional values, concerned with integrity, accountability, and discipline; new definitions of family; troubled schools; the problems of affluence; the persistence of poverty and discrimination; the aging of the nation—to name a few.

To this mix add a new interest in "positive psychology," a renewed focus on such human realities as opportunities, optimism, hope, caring, virtue, strength, resourcefulness, resilience, dialogue, encouragement, vocation, work ethic, creativity, originality, spirituality, and wisdom. These upbeat realities, translated into projects, programs, methods, skills, and attitudes, can do much for the helping professions, which, like other institutions, are in constant need of reform. The helping professions could use a "second spring" of some sort in order to cope more pragmatically, enthusiastically, vigorously, and effectively with the challenges outlined above.

Positive psychology is not "the answer" since nothing is, but it is infinitely better than some of the squabbling that characterizes "dialogue" in the helping professions. It strikes the right chord. The challenge is to get it out of our textbooks and articles and into the social settings of our lives. There is no grand plan for this. Thousands of individuals will have to make it happen. Then we can begin to codify what works best.

Perhaps positive psychology can bring new life to the practice of prevention in our individual lives, relationships, homes, schools, workplaces, churches, courts, legislatures, and places of entertainment. The problem with prevention is, in a sense, its "hollow" nature. When it works, something doesn't happen. The child doesn't develop a sense of inferiority. The individual doesn't get AIDS. Spousal abuse doesn't take place. Racist attitudes don't develop. But people, including politicians, don't get excited about what doesn't happen. Therefore, if prevention practices are to work, they should not just prevent something bad; they should confer something good. Something tangible. Something useful. Something enjoyable. To be fair, many prevention programs not only prevent, but they confer. But there is still a prevention vacuum and positive psychology can help us fill it. Furthermore, even though every dollar spent on prevention yields four or five in terms of benefits, most of prevention should be based on human capital rather than dollars. Much or most of prevention is free. Our communities, families, schools, and workplaces should be hotbeds of prevention.

Helping encounters should be both remedial and preventive. Clients should leave helping sessions more capable of managing problems and both spotting and developing opportunities. Ideally, they should leave with the ability to communicate more effectively with themselves and others. But in the bigger picture, this is but a drop in the bucket. The real challenge is for our society to find ways of imparting life skills such as dialogue, problem management, and opportunity development to all citizens. When I was a kid, my school made sure that I learned the multiplication tables cold, but there was nothing about problem solving, opportunity spotting, and interpersonal communication. We should know better now. To be fair, in grammar school and high school I learned how to read and to love reading and how to write and love writing. I also learned a great deal about responsibility, accountability, and discipline. For this I am extremely grateful. But now is the time to wed the best of the past with the best of the future.

## Acknowledgments

I would like to thank the following reviewers for their helpful insights and suggestions: Jane Arscott, Athabasca University, Alberta, Canada; Dale Blumen, University of Rhode Island; Stuart W.B. Evans, The University of Melbourne, Australia; Nelica La Gro, University of East London, England; James R. Mahalik, Boston College; Russell D. Miars, Portland State University; and Douglas A. Whyte, Community College of Philadelphia.

*Gerard Egan*

# LAYING THE GROUNDWORK

Although the centerpiece of this book is a problem-management and opportunity-development helping model and the methods and communication skills that make it work, there is some groundwork to be laid. This includes outlining the nature and goals of helping (Chapter 1), providing an overview of the helping process itself (Chapter 2), and describing the helping relationship and the values that should drive it (Chapter 3).

# INTRODUCTION TO HELPING

FORMAL AND INFORMAL HELPERS: A VERY BRIEF HISTORY

WHAT HELPING IS ABOUT

   Clients with Problem Situations, Missed Opportunities, or Unused Potential

      Problem situations

      Missed opportunities or unused potential

POSITIVE PSYCHOLOGY AND HELPING

THE TWO PRINCIPAL GOALS OF HELPING

      The importance of results

      A results-focused case

DOES HELPING HELP?

      What evidence is there for the effectiveness of helping?

      What do clients think?

      What about efficacy studies and treatment manuals?

      Are there good helpers and bad helpers?

      What can we conclude?

IS HELPING FOR EVERYONE?

WHAT THIS BOOK IS—AND WHAT IT IS NOT

      It is a practical model of helping

      It is not the total curriculum

MOVING FROM SMART TO WISE: MANAGING THE SHADOW SIDE OF HELPING

   The Downside: The Messiness of Helping

   The Upside: Common Sense and Wisdom in the Helping Professions

## FORMAL AND INFORMAL HELPERS: A VERY BRIEF HISTORY

Throughout history, there has been a deeply embedded conviction that, under the proper conditions, some people are capable of helping others come to grips with problems in living. This conviction, of course, plays itself out differently in different cultures, but it is still a cross-cultural phenomenon. Today this conviction is often institutionalized in a variety of formal helping professions. Counselors, psychiatrists, psychologists, social workers, and ministers of religion are counted among those whose formal role is to help people manage the distressing problems of life.

There is also a second set of professionals who, although they are not helpers in the formal sense, often deal with people in times of crisis and distress. This group includes organizational consultants, dentists, doctors, lawyers, nurses, probation officers, teachers, managers, supervisors, police officers, and practitioners in other service industries. Although these people are specialists in their own professions, there is still some expectation that they will help those they serve manage a variety of problem situations. For instance, teachers teach English, history, mathematics, and science to students who are growing physically, intellectually, socially, and emotionally, and struggling with developmental tasks and crises. Teachers are, therefore, in a position to help their students, in direct and indirect ways, explore, understand, and deal with the problems of growing up. Managers and supervisors help workers cope with problems related to work performance, career development, interpersonal relationships in the workplace, and a variety of personal problems that affect their ability to do their jobs. This book is addressed directly to the first set of professionals and indirectly to the second.

In addition to these professional helpers are any and all who try to help relatives, friends, acquaintances, strangers (on buses and planes), and themselves come to grips with problems in living. Indeed, only a small fraction of the help provided on any given day comes from helping professionals. Informal helpers—bartenders and hairdressers are often mentioned—abound in the social settings of life. Friends help one another through troubled times. Parents need to manage their own marital problems while helping their children grow and develop. A study has shown that more than a quarter of Americans have at one time or another felt that they were headed for serious psychological trouble. However, of these more sought help from informal social supports (for instance, friends, relatives) than from any other source (Swindle, Heller, Pescosolido, & Kikuzawa, 2000). In the end, of course, all of us must learn how to help ourselves cope with the problems and crises of life.

Since helping is such a common human experience, training in both solving one's own problems and helping others solve theirs should be as common as training in reading, writing, and math. Unfortunately, this is not the case.

## WHAT HELPING IS ABOUT

To determine what helping is about, it is useful to consider (1) why people seek—or are sent to get—help in the first place and (2) what the principal goals of the helping process are.

## Clients with Problem Situations, Missed Opportunities, or Unused Potential

Many clients become clients because, either in their own eyes or in the eyes of others, they are involved in problem situations they are not handling well. Others come because they feel they are not living as fully as they might. Many come because of a mixture of both. Therefore, clients' problem situations or missed opportunities and unused potential constitute the starting points of the helping process.

**Problem situations.** Clients come for help because they have crises, troubles, doubts, difficulties, frustrations, or concerns. These are often generically labeled "problems," but they are not problems in a mathematical sense, because these problems often cause emotional turmoil and have no clear-cut solutions. It is probably better to say that clients come with *problem situations*—that is, with complex and messy *problems in living* that they are not handling well. These problem situations are often poorly defined. Even when problem situations are well defined, clients still don't know how to handle them. Or they feel that they do not have the resources needed to cope with them adequately. Or they have tried solutions that did not work.

Problem situations arise in our interactions with ourselves, with others, and with the social settings, organizations, and institutions of life. Clients—whether they are hounded by self-doubt, tortured by unreasonable fears, grappling with the stress that accompanies serious illness such as cancer, addicted to alcohol or drugs, involved in failing marriages, fired from jobs because they do not have the skills needed in the "new economy," confused in their efforts to adapt to a new culture, suffering from catastrophic loss, jailed because of child abuse, wallowing in midlife crises, lonely and out of community with no family or friends, battered by their spouses, or victimized by racism—all face problem situations that move them to seek help or, in some cases, move others to refer them for help or even send them for help.

People with even devastating problem situations can often, with some help, handle them more effectively. Consider the following example:

Martha, 58, suffered three devastating losses within six months. One of her four sons, who lived in a different city, died suddenly of a stroke. He was only 32. Shortly after, she lost her job in a "downsizing" move stemming from the merger of her employer with another company. Finally, her husband, who had been ill for about two years, died of cancer. Though she was not destitute, her financial condition could not be called comfortable, at least not by middle-class North American standards. Two of her three surviving sons were married with families of their own; one lived in a distant suburb, the other in a different city. The unmarried son, a sales rep for an international company, traveled abroad extensively. After her husband's death, she became agitated, confused, angry, and depressed. She also felt guilty—first, because she believed that she should have done "more" for her husband; second, because she also felt strangely responsible for her son's early death. Finally, she was deathly afraid of becoming a "burden" to her children.

At first, retreating into herself, she refused solace or help from anyone. But eventually, she responded to the gentle persistence of her church minister. He helped her become a member of a "grieving family" instead of a guilty outcast. She began attending a support group at the church. A psychologist who worked at a local university provided some direction for the group. As a result of her interactions within the group, she slowly started to accept help from her sons. She began to realize that she was not alone in her grief but was a member of a family that needed to help one another cope with the sense of loss they were all experiencing. She also began relating with some of the members of the group outside the group sessions. This filled the social void she experienced when her company laid her off. She had an occasional informal chat with the psychologist who provided services for the group. Through contacts within the group, she got another job.

Notice that help came from many quarters. Her newfound solidarity with her family, the church support group, the active concern of the minister, and the informal chats with the psychologist helped Martha enormously. Furthermore, since she had always been a resourceful person, the help she received enabled her to tap into her own ingenuity.

It is important to note that none of this "solved" Martha's problems or made up for the losses she had experienced. Indeed, the goal of helping is not to *solve* problems but to help the troubled person *manage* them more effectively or even transcend them by taking advantage of new possibilities in life.

In a way, the principal focus on problem management is not the problem at all. Jones, Rasmussen, and Moffitt (1997) see problem solving as an opportunity for learning. The problem situation itself is a stimulus to learning. While their book focuses directly on educational systems, the kind of "engaged learning" they discuss can take place in the helper-client relationship. What clients learn in the give-and-take of helping sessions they can apply first to managing the presenting problem, but what they learn often has wider application. When clients learn how to sort out difficult relationships, they can apply their learning both to sorting out other problems in relationships and to preventing problems from arising in the first place.

**Missed opportunities or unused potential.**  Some clients come for help not because they are dogged by problems like those previously mentioned but because they are not as effective as they would like to be. So clients' *missed opportunities* and *unused potential* constitute a second starting point for helping. Most clients have resources they are not using or opportunities they are not developing. People who feel locked in dead-end jobs or bland marriages, who are frustrated because they lack challenging goals, who feel guilty because they are failing to live up to their own values and ideals, who want to do something more constructive with their lives, or who are disappointed with their uneventful interpersonal lives—such clients come to helpers not to manage their problems better but to live more fully.

In these cases, it is a question not of what is going wrong but of what could be better. It has often been suggested that most of us use only a small fraction of our potential. Most of us are capable of dealing much more creatively with ourselves, with our relationships with others, with our work life, and generally, with the ways in which we involve ourselves with the social settings of our lives. Consider the following case:

After ten years as a helper in several mental health centers, Carol was experiencing burnout. In the opening interview with a counselor, she berated herself for not being dedicated enough. Asked when she felt best about herself, she said that it was on those relatively infrequent occasions when she was asked to provide help for other mental health centers that were experiencing problems, having growing pains, or reorganizing. The counselor helped her explore her potential as a consultant to human-service organizations and make a career adjustment. She enrolled in an organization-development program at a local university. Carol stayed in the helping field but with a new focus and a new set of skills.

In this case, the counselor helped the client manage her problems (burnout, guilt) by helping her explore and develop an opportunity (a new career).

## Positive Psychology and Helping

Helping clients identify and develop unused potential and missed opportunities can be called a *positive psychology* goal. As guest editors of the special millennium issue of the *American Psychologist*, Seligman and Csikszentmihalyi (2000) called for a better balance of perspectives in the helping professions. In their minds, too much attention is focused on pathology and too little on positive psychology: "Our message is to remind our field that psychology is not just the study of pathology, weakness, and damage; it is also the study of strength and virtue. Treatment is not just fixing what is broken; it is nurturing what is best" (p. 7). They and their fellow authors discuss such upbeat topics as

- subjective well-being, happiness, hope, optimism (see Chang, 2001);
- a sense of vocation, developing a work ethic;
- interpersonal skills; the capacity for love, forgiveness, civility, nurturance, altruism;
- an appreciation of beauty and art;
- responsibility, self-determination, courage, perseverance, moderation;
- future mindedness, originality, creativity, talent;
- a civic sense;
- spirituality, wisdom.

Seligman and Csikszentmihalyi suggest that counselors improve their services by weaving the spirit of these topics into their interactions with clients. Seeing problem management as life-enhancing learning and treating all encounters with clients as opportunity-development sessions are part of the positive psychology approach.

In 2000, the Templeton Foundation instituted The Templeton Positive Psychology Prize for excellent research in various areas of positive psychology (Azar, 2000). The substantial monetary awards were conferred for research on the beneficial effects of positive emotions, optimism and forward-thinking behavior, the ways positive emotions help people form and maintain relationships, and intellectual precocity and ways to help people realize their exceptional potential. Themes such as these need to find their way into helping sessions. The positive psychology theme is revisited throughout this book.

A note of caution: Positive psychology is not an "everything's going to be all right" approach to life and helping. Richard Lazarus (2000 ) puts it well:

> However, it might be worthwhile to note that the danger posed by accentuating the positive is that if a conditional and properly nuanced position is not adopted, positive psychology could remain at a Pollyanna level. Positive psychology could come to be characterized by simplistic, inspirational, and quasi-religious thinking and the message reduced to "positive affect is good and negative affect is bad." I hope that this ambitious and tantalizing effort truly advances what is known about human adaptation, as it should, and that it will not be just another fad that quickly comes and goes. (p. 670)

The term *positive psychology* appears frequently in this book. You, the reader, should take a critical look at each use; the term should not be trivialized.

## THE TWO PRINCIPAL GOALS OF HELPING

The goals of helping should be based on the needs of clients. Positive psychology offers us an overall foundation or quality-of-life goal for all clients—subjective well-being (SWB), or happiness (Diener, 2000; Myers, 2000; Robbins & Kliewer, 2000). Clients come to helpers because they are unhappy with one or more dimensions of their lives. Diener points out that scientific knowledge of SWB is both possible and desirable but the psychological community does not take it seriously. The SWB research that does exist tends to focus on who or what groups are happy rather than when and why people are happy and what processes lead to SWB. It goes without saying that philosophical concern about happiness is as old as Plato and Aristotle.

More immediately, there are two basic counseling goals—one relating to clients' managing specific problems in living more effectively or developing missed opportunities and unused resources, and the other relating to their general ability to manage problems and develop opportunities in everyday life. Both goals are related to increasing clients' happiness.

> GOAL ONE: *Help clients manage their problems in living more effectively and develop unused resources and missed opportunities more fully.*

Helpers are successful to the degree to which their clients—through client-helper interactions—are better positioned to manage specific problem situations and develop specific unused resources and missed opportunities more effectively. Notice that this stops short of saying that clients actually end up managing problems and developing opportunities better. Although counselors help clients achieve valued outcomes, they do not control those outcomes directly. In the end, clients can choose to live more effectively or not.

Since helping is a two-way, collaborative process, clients, too, have a primary goal. Clients are successful to the degree that they commit themselves to the helping process and capitalize on what they learn from the helping sessions to manage problem situations more effectively and develop opportunities more fully "out there" in their day-to-day lives.

**The importance of results.** A corollary to Goal One is that *helping is about results, outcomes, accomplishments, impact.* For many, this principle is so important that they have developed over the years what is called "solution-focused" therapy (Manthei, 1998; O'Hanlon & Weiner-Davis, 1989; Rowan, O'Hanlon, & O'Hanlon, 1999). Helping is an *-ing* word: It includes a series of activities in which helpers and clients engage. These activities, however, have value only to the degree that they lead to valued outcomes in clients' lives. Ultimately, statements such as "We had a good session," whether spoken by the helper or by the client, must translate into more effective living on the part of the client. If a helper and a client engage in a series of counseling sessions productively, something of value will emerge that makes the sessions worthwhile. Unreasonable fears will disappear or diminish to manageable levels, self-confidence will replace self-doubt, addictions will be conquered, an operation will be faced with a degree of equanimity, a better job will be found, a woman and man will breathe new life into their marriage, a battered wife will find the courage to leave her husband, a man embittered by institutional racism will regain his self-respect and take his rightful place in the community. In a word, helping should make a substantive difference in the life of the client. Helping is about constructive change.

**A results-focused case.** The need for results is seen clearly in the case of a battered woman, Andrea, outlined by Driscoll (1984):

> The mistreatment had caused her to feel that she was worthless even as she developed a secret superiority to those who mistreated her. These attitudes contributed in turn to her continuing passivity and had to be challenged if she was to become assertive about her own rights. Through the helping interactions, she developed a sense of worth and self-confidence. This was the first outcome of the helping process. As she gained confidence, she became more assertive; she realized that she had the right to take stands, and she chose to challenge those who took advantage of her. She stopped merely resenting them and did something about it. The second outcome was a pattern of assertiveness, however tentative in the beginning, that took the place of a pattern of passivity. When her assertive stands were successful, her rights became established, her social relationships improved, and her confidence in herself increased, thus further altering the original self-defeating pattern. This was a third set of outcomes. As she saw herself becoming more and more an "agent" rather than a "patient" in her everyday life, she found it easier to put aside her resentment and the self-limiting satisfactions of the passive-victim role and to continue asserting herself. This constituted a fourth set of outcomes. The activities in which she engaged, either within the helping sessions or in her day-to-day life, were valuable because they led to these valued outcomes. (p. 64)

Andrea needed much more than "good sessions" with a helper. She needed to work for outcomes that made a difference in her life. Today there is another reason for focusing on outcomes. In the United States, many psychological services are offered in managed-care settings. In these settings, more and more third-party payments depend on meaningful treatment plans and the delivery of problem-managing outcomes (Meier & Letsch, 2000). But economics should not force helpers to do what they should be doing anyway in the service of their clients.

> **GOAL TWO: Help clients become better at helping**
> **themselves in their everyday lives.**

Clients often are poor problem solvers, or whatever problem-solving ability they have tends to disappear in times of crisis. What Miller, Galanter, and Pribram (1960) said many years ago is, unfortunately, probably just as true today:

> In ordinary affairs we usually muddle about, doing what is habitual and customary, being slightly puzzled when it sometimes fails to give the intended outcome, but not stopping to worry much about the failures because there are still too many other things still to do. Then circumstances conspire

against us and we find ourselves caught failing where we must succeed—
where we cannot withdraw from the field, or lower our self-imposed standards, or ask for help, or throw a tantrum. Then we may begin to suspect that
we face a problem. . . . An ordinary person almost never approaches a problem systematically and exhaustively unless he or she has been specifically educated to do so. (pp. 171, 174)

Most people in our society are not "educated to do so." And if many clients are poor
at managing problems in living, they are equally poor in identifying and developing
opportunities and resources. We have yet to find ways of making sure that our children develop what most consider to be essential "life skills," such as problem management, opportunity identification and development, sensible decision making,
and the skills of interpersonal relating.

It is no wonder, then, that clients are often poor problem solvers or that whatever problem-solving ability they have tends to disappear in times of crisis. If the
second goal of the helping process is to be achieved—that is, if clients are to go
away better able to manage their problems in living more effectively and develop
opportunities on their own—then sharing some form of the problem-management
and opportunity-development process with them is essential.

So the second goal of helping deals with clients' needs (1) to participate actively in the problem-management process during the helping sessions themselves,
(2) to apply what they learn to managing immediate problems and opportunities,
and (3) to continue to manage their lives more effectively after the period of formal
helping is over. Just as doctors want their patients to learn how to prevent illness
through exercise, good nutritional habits, and the avoidance of toxic substances
and activities, and just as dentists want their patients to engage in preventing cavi-
tites through effective flossing and brushing, so skilled helpers want to see each of
their clients not only managing a specific problem situation more effectively but
also becoming more capable of managing subsequent problems in living more effectively. That is, helping at its best provides clients with tools to become more effective
self-helpers. Therefore, although this book is about a process helpers can use to help
clients, more fundamentally, it is about a problem-management and opportunity
–development process that clients can use to help themselves. This process can help
clients become more effective students of problem management and opportunity
development, better decision makers, and more responsible "agents of change" in
their own lives. Chapter 2 provides an overview of this process.

## DOES HELPING HELP?

This question must sound strange in a book on helping. However, ever since Eysenck
(1952) questioned the usefulness of psychotherapy, there has been an ongoing debate
as to the efficacy of both helping in general and different approaches to helping. Over
the years, some critics have expressed grave doubts about the very legitimacy of the
helping professions themselves, even claiming that helping is a fraudulent process, a
manipulative and malicious enterprise (see, for example, Cowen, 1982; Eysenck,
1984, 1994; Masson, 1988). Masson went so far as to claim that in the United States,
helping is a multibillion-dollar business that does no more than profit from people's

misery. He also maintained that devaluing people is part and parcel of all therapy and that the helper's values and needs are inevitably imposed on the client. Although such criticisms are extreme, they should not be dismissed out of hand. What they say may be true of some forms of helping and of some helpers.

On the other hand, we have yet to come up with an unqualified yes to the question, Does helping help? Good news and bad news abound. Therefore, a few words about the overall effectiveness of helping, how clients themselves feel about the help they have received, and the efficacy of specific treatments can provide the beginning helper with some insight into the helping profession. What follows is a taste of the debate. If you want the full menu, entering such terms as *effectiveness of psychotherapy, psychotherapy outcome studies, customer satisfaction with psychotherapy, treatment efficacy studies,* and *manualized treatments* into an Internet psychology search engine will provide the entire feast.

**What evidence is there for the effectiveness of helping?** There is a long history of outcome research. Hill and Corbett (1993) and Whiston and Sexton (1993) provide reviews that cover 50-year periods. There is a great deal of evidence showing that different kinds of helping, including counseling (Lambert & Cattani-Thompson, 1996), do help many people in many different situations. One sign of the maturing of the helping professions is the fact that the surgeon general of the United States has issued the first-ever report on mental health (see Satcher, 2000 for an executive overview). Even though many would say that the findings stated in this report have been known for some time, four are relevant here: (1) Mental health is fundamental to physical or overall health, (2) mental disorders are real health conditions, (3) the efficacy of mental health treatments is well documented, and (4) a range of treatments exists for most mental disorders. To come to these conclusions, contributors, under the guidance of the Office of the Surgeon General, reviewed thousands of research studies and first-person accounts from individuals who had experienced mental disorders.

Although conventional reviews of groups of helping-outcome studies over the years produced mixed and ambiguous results (see Rossi & Wright, 1984; Schmidt, 1992), an approach called *meta-analysis*—a kind of study of studies and a reinterpretation of their findings—has been a powerful tool in demonstrating that helping helps (see Lipsey & Wilson, 1993; Smith, Glass, & Miller, 1980). In the words of Lipsey and Wilson, "Meta-analytic reviews [of helping outcomes] show a strong, dramatic pattern of positive overall effects that cannot readily be explained as artifacts of meta-analytic technique or generalized placebo effects" (p. 1181). Hundreds of meta-analytical studies have been done over the past 20 years, and although some are admittedly quite crude, they still add up, in the eyes of many, to convincing evidence of the overall efficacy of helping. Further studies of the efficacy of helping are presently underway.

Still doubts persist. For instance, many outcome studies on helping are done in the lab or under lablike conditions. But lab results cannot be automatically compared to clinical-setting results. Real helping does not take place in a lab (see Henggeler, Schoenwald, & Pickrel, 1995; Weisz, Donenberg, Han, & Weiss, 1995). Furthermore, meta-analysis, the major tool used in efficacy studies, has itself come under criticism (Matt & Navarro, 1997).

**What do clients think?**  Some studies ask clients whether they have been helped by counseling and psychotherapy and to what degree (Pekarik and Guidry, 1999). Client satisfaction studies have been widely used over the years and are highly regarded in many different mental-health treatment centers (Lambert, Salzer, & Bickman, 1998). *Consumer Reports* (1994; 1995) published the results of a sophisticated large-scale survey project on client satisfaction with helping. The findings indicated that

- clients believed that they had benefitted very substantially from psychotherapy;
- psychotherapy alone did not differ in effectiveness from psychotherapy plus medication;
- no specific form of helping did better than any other for any particular kind of problem;
- psychiatrists, psychologists, and social workers did not differ in their effectiveness as helpers;
- long-term treatment produced appreciably better results than did short-term treatment;
- clients whose choice of helper or length of therapy was limited by insurance or managed-care systems did not benefit as much as clients without those restrictions.

The study dealt with the responses of real clients to questions about themselves, their helpers, processes used, and benefits received (see Seligman, 1995, for a discussion and critique of this study). Case closed? Hardly.

The *Consumer Reports* study has received a great deal of criticism on both theoretical and methodological grounds (Brock, Green, & Reich, 1998; Jacobson & Christensen, 1996). It has also been demonstrated that client satisfaction does not always mean that problems are being managed and opportunities developed (Pekarik & Guidry, 1999; Pekarik & Wolff, 1996): "Although often considered a traditional outcome measure, there are only low-to-moderate correlations between satisfaction and other measure of outcome. When traditional adjustment measures are clearly distinguished from satisfaction items, these correlations are especially low" (Pekarik & Guidry, p. 474).

Clients might be satisfied for a variety of reasons—for instance, because they like their counselors or because they feel less stressed. This does not mean that, in terms used in this book, they are managing problem situations or developing opportunities any better. The ideal, of course, is that clients feel satisfied because they have changed their lives for the better (Ankuta & Abeles, 1993).

**What about efficacy studies and treatment manuals?**  Efficacy studies focus on the usefulness of a specific helping methodology for a particular kind of problem—for instance, a cognitive therapy approach with clients suffering from panic disorders or some form of behavior therapy with clients suffering from phobias. In these studies, comparisons in outcome effectiveness and efficiency are made between two different treatments for the same disorder or between clients who receive a particular treatment and clients who do not. Efficacy studies are carried out under carefully controlled conditions. For instance, Seligman (1995) proposed eight rigorous conditions for the ideal efficacy study (see also Luborsky, 1993). Because of the rigor demanded, these studies are expensive and time-consuming. There are hundreds of efficacy studies, many of them very well designed, that demonstrate the efficacy of a particular therapy for a particular psychological disorder.

Over the years, these studies have led to *empirically supported treatments* (ESTs) or *empirically validated treatments* (EVTs) (Nathan, 1998; Waehler, Kalodner, Wampold, & Lichtenberg, 2000). When a treatment proves to be robust, it becomes *manualized*; that is, a manual on how to deliver the treatment for a particular kind of client is written for practitioners. Manuals have been developed for a wide range of human problems, including anxiety, phobias, depression, personality disorders, post-traumatic stress disorder, substance abuse, panic, and borderline personality disorder, to name a few. According to Waehler and his associates, the forces driving the creation of manualized treatments include managed care's push to maximize the value of all treatments, including mental health treatments, the challenge coming from biological psychiatry to justify empirically the efficacy of psychosocial approaches to helping, and the attempt of psychology itself to answer the question: "What treatment, by whom, is most effective for this individual with that specific problem under which set of circumstances?" (Paul, 1967).

But there is also a great deal of controversy over manualized treatments. On the upside, a study of 47 therapists (Najavits, Weiss, Shaw, & Dierberger, 2000) found very positive views of treatment manuals, extensive use, and few concerns. Toward the other end of the scale, Garfield (1996) responded to guidelines on the use of empirically validated treatments issued by the Division of Clinical Psychology of the American Psychological Association (Division 12 Task Force, 1995). He expressed a range of concerns that have since been echoed and amplified by other researchers and practitioners. Some common concerns include: the language of the task force report was too strong, some of its recommendations were premature, manuals often idealize and thus distort psychotherapy, the research base underpinning some manuals is questionable, clients are messy and don't come to therapy with neatly categorized problems, studies don't factor in the competence (or the lack thereof) of the therapists involved, manuals ignore the role of the therapist as a model of adult living, the place of art and clinical judgment (see Soldz & McCullough, 2000; Waddington, 1997) is demeaned or ignored, and manual treatments are often highly specialized, cumbersome, time consuming, expensive, and difficult to use.

Others have responded to these concerns and see a bright future for manualized treatments (Nathan, 1998; Wilson, 1998). Two journals provide, in special sections, a range of articles on empirically supported therapies and manualized treatments: Kazdin (1996) introduces a series of articles on empirically validated treatments in *Clinical Psychology: Science and Practice* (Fall), and Kendall (1998) introduces a special section of the *Journal of Consulting and Clinical Psychology* (February) on empirically supported therapies.

That said, many clients have problem situations rather than specific problems such as phobias. Or they have both. Consider the case of Martha outlined earlier in the chapter. Her problems would not be amenable to manualized treatments. Furthermore, manuals focus on problems and not the development of opportunities, which should be one of the principal forms of helping. Carol, whose anxiety and depression were described earlier, probably would not have identified and developed a new career opportunity if she had received manualized treatments. Still, if you intend to become a professional helper, keep your eye on developments in man-ualized treatments. They are not going to disappear (Foxhall, 2000).

**Are there good helpers and bad helpers?** Yes. There is indeed some evidence that therapy sometimes not only does not help but also actually makes things worse. That is, some helping leads to negative outcomes (see Mohr, 1995, and Strupp, Hadley, & Gomes-Schwartz, 1977, for reviews of the negative-outcome literature). Research shows that some of the factors associated with negative outcomes in helping are associated with clients, others with helpers. Sometimes clients who have severe in-terpersonal problems and severe symptomatology, who are poorly motivated, or who expect helping to be painless become more dysfunctional through therapy. Helpers who underestimate the severity of clients' problems, experience interpersonal diffi-culties with clients, use poor techniques, overuse any technique, or disagree with clients over helping methodology can make things worse rather than better.

Finally, some helpers are incompetent. And even the competent and commit-ted have their lapses. As noted by Luborsky and his associates (1986):

- There are considerable differences between therapists in their average success rates.
- There is considerable variability in outcome within the caseload of individual therapists.
- Variations in success rates typically have more to do with the therapist than with the type of treatment.

Although helping can and often does work, there is plenty of evidence that ineffec-tive helping also abounds. Helping is a potentially powerful process that is all too easy to mismanage. It is no secret that because of inept helpers, some clients get worse as a result of treatment. Helping is not neutral; it is "for better or for worse." Ellis (1984) claimed that inept helpers are either ineffective or inefficient. Even though the ineffi-cient may ultimately help their clients, they use "methods that are often distinctly in-ept and that consequently lead these clients to achieve weak and unlasting results, frequently at the expense of enormous amounts of wasted time and money" (p. 24). Since studies on the efficacy of counseling and psychotherapy do not usually make a distinction between high-level and low-level helpers, and because the research on de-terioration effects in therapy suggests that there is a large number of low-level or in-adequate helpers, the negative results found in many studies are predictable.

**What can we conclude?** Common sense tells us that some forms of helping actually help in the hands of good helpers. Personally, I have no doubt that in the hands of skilled and socially intelligent helpers, helping can do a great deal of good. Norman Kagan (1973) long ago suggested that the basic issue confronting the helping profes-sions is not validity—that is, whether helping helps or not—but reliability: "Not, can counseling and psychotherapy work, but does it work consistently? Not, can we edu-cate people who are able to help others, but can we develop methods which will in-crease the likelihood that most of our graduates will become as effective mental health workers as only a rare few do?" (p. 44). The question then is not "Does helping work?" but rather "How and under what conditions does it work?" The answer to the first question is easier than the answer to the second (see Bergin & Garfield, 1994).

You are encouraged to acquaint yourself with the ongoing debate concerning the efficacy of helping. Study of this debate is meant not to discourage you but to help you

- appreciate the complexity of the helping process;
- acquaint yourself with the issues involved in evaluating the outcomes of helping;
- appreciate that, poorly done, helping can actually harm others;
- become reasonably cautious as a helper;
- find incentives for becoming a high-level helper, learning and using practical models, methods, skills, and guidelines for helping.

Until we have better answers to the questions posed here, *caveat emptor*—that is, "Let the buyer [the client] beware." The problem is that the typical client is not aware of the issues that are being discussed here. Often clients' problems are so pressing that they just want help. Therefore, "Let the practitioner beware." Become competent. Don't overpromise. Remain professionally self-critical. Keep your eye on results—that is, problem-managing and opportunity-developing outcomes for clients. Strive continually to become a better helper. Because helping in some generic sense "works," do not assume that it will always work with everyone. Finally, do not confuse difficult cases with impossible cases.

## IS HELPING FOR EVERYONE?

To say that in the main, helping works does not mean that it is for everyone. Helpers see only a fraction of the population with problems. Some people don't go to helpers because they don't know how to access them. Others feel that they are not ready for change. Some don't acknowledge that they have problems. Some have "aversive expectations" of mental health services (Kushner & Sher, 1989) and stay clear. Most people muddle through without professional help, picking up help wherever they find it. I often show a popular series of videotaped counseling sessions in which the client receives help from a number of helpers, each with a different approach to helping. Once the course participants have seen the tapes, I ask, "Does this client need counseling?" They all practically yell, "Yes!" Then I ask, "Well, if this client needs counseling, how many people in the world need counseling?" The question proves to be very sobering because the client in question is struggling with issues we all struggle with. She was one of what some would call "the worried well." Just because a person might benefit from counseling does not mean that he or she "needs" counseling. Furthermore, many clients can "get better" in a variety of ways without help.

Working with clients who, for whatever reason, don't want to grapple with their problem situations or develop their unused resources is a waste of time and money. Of course, it doesn't hurt to try, but the helper should know when to quit. In addition, clients with certain kinds of problems seem to be beyond help, at least at this stage of the development of the helping professions. Knowing when to help is important, as is knowing how much to help. Even though studies show that the longer a client is in therapy, the greater the benefit, spending a great deal of time in therapy is not always feasible. Helping is an expensive proposition, both monetarily and psychologically. Even when it is "free," someone is paying for it through tax dollars, insurance premiums, or free-will offerings. Therefore, without rushing nature or your clients, every session should be focused and to the point. Do not assume that you have a client for life.

## WHAT THIS BOOK IS—AND WHAT IT IS NOT

"Beware the person of one book," we are told. For some people the central message of a book becomes a cause. Religious books such as the Bible or the Koran are examples. Although causes empower people, we should beware when a person believes that all he or she needs to know is in one book. Such a person remains closed to new ideas and growth. Certainly all truth about helping cannot be found in one book. Therefore, *The Skilled Helper* is not and cannot be "all you need to know about helping." So what is it, and what is it not?

**It is a practical model of helping.** Because this book cannot do everything, it is important to state what it is meant to do. Its purpose is to provide helpers—whether novices or those with experience—with a practical framework or model of helping and some methods and skills that make the model work. It is designed to enable helpers to engage in activities that will help their clients manage their lives more effectively. Helpers can also train their clients to use the model to manage their lives better on their own.

**It is not the total curriculum.** This book is part of the helping curriculum—an extremely important part—but it is not the whole. Clients are the "customers" of helpers and have every right to expect the best of service from them. Beyond a helping model and the skills that make it work, what kind of training enables helpers to "deliver the goods" to their clients? A practical curriculum is one that enables helpers to understand and work with their clients in the service of problem management and opportunity development. The curriculum includes both working knowledge and skills. *Working knowledge* is the translation of theory and research into the kind of applied understandings that enable helpers to work with clients. *Skill* refers to the actual ability to deliver services.

In addition to a working model of helping, a fuller curriculum for training professional helpers might include the following:

- A working knowledge of *applied developmental psychology*—how people develop or create their lives across the life span and the impact of environmental factors such as culture and socioeconomic status on development.

- An understanding of the principles of *cognitive psychology* as applied to helping, because the way people think and construct their worlds has a great deal to do with both getting into and getting out of trouble.

- An understanding of the *dynamics of the helping professions themselves* as they are currently practiced in our society together with the challenges they face.

- An understanding of clients as *psychosomatic beings* and the interaction between physical and psychological states.

- The ability to apply *the principles of human behavior*—what we know about incentives, rewards, and punishment—to the helping process, because wrestling with problem situations and undeveloped opportunities always involves incentives and rewards.

- *Abnormal psychology*—a systematic understanding of the ways in which individuals get into psychological trouble.

- An understanding of the *diversity* of age, race, ethnicity, religion, sexual orientation, culture, social standing, economic standing, and the like among clients.

- An understanding of the ways in which people act when they are in social settings—that is, the *people-in-systems framework* (Egan & Cowan, 1979)—together with an understanding of clients in context—that is, an understanding of clients in the social settings of their lives.

- An understanding of the needs and problems of *special populations* with which one works, such as the physically challenged, substance abusers, the homeless.

- *Applied personality theory*, because this area of psychology helps us understand in very practical ways what makes people "tick" and many of the ways in which individuals differ from one another.

In the end, there is no such thing as the perfect professional curriculum to which all helpers should subscribe. Although much of this curriculum, summarized in Figure 1-1, is touched on indirectly in the pages that follow, especially through the many examples offered, this book is not a substitute for such a curriculum. However, the fact that paraprofessional and informal helpers do not go through such a rigorous curriculum does not mean that they cannot be effective helpers. Indeed, studies have demonstrated that paraprofessional helpers can be as effective as professional helpers and in some cases even more helpful (Durlak, 1979; Hattie, Sharpley, & Rogers, 1984).

## Moving from Smart to Wise: Managing the Shadow Side of Helping

This book outlines a model of helping that is rational, linear, and systematic. What good is that, you well might ask, in a world that is often irrational, nonlinear, and chaotic? One answer is that rational models help clients bring much-needed discipline and order into their chaotic lives. Effective helpers do not apologize for using such models. But they also make sure that their humanity permeates their models.

More than intelligence is needed to apply the model well—smart is not smart enough. The helper who understands and uses the model together with the skills and techniques that make it work might well be smart, but he or she must also be wise. Effective helpers understand the limitations not only of helping models but also of helpers, the helping profession, clients, and the environments that affect the helping process. One dimension of wisdom is the ability to understand and manage these limitations, which in sum constitute what I call the arational dimensions, or the "shadow side," of life. The shadow side of helping can be defined as

*all those things that adversely affect the helping relationship, process,*
*outcomes, and impact in substantive ways but that are not*
*identified and explored by helper or client*
*or even the profession itself.*

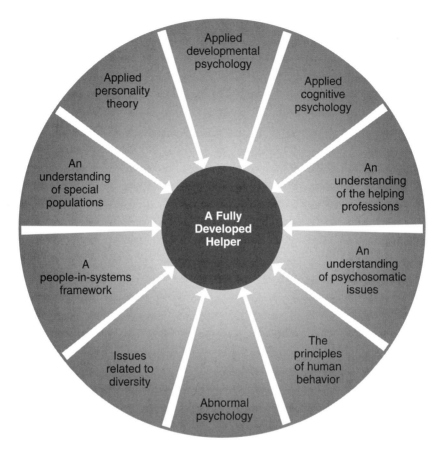

FIGURE 1-1
A Sample Curriculum for Helpers

Helping models are flawed; helpers are sometimes selfish, lazy, and even predatory, and they are prone to burnout. Clients are sometimes selfish, lazy, and predatory even in the helping relationship.

Indeed, if the world were completely rational, we would run out of clients. Not to worry, however, since many clients cause their own problems. People knowingly head down paths that lead to trouble. Life is not a straight road; often it is more like a maze. Many times life seems to be a contradictory process in which good and evil, the comic and the tragic, cowardice and heroism are inextricably intermingled. This book describes and illustrates the helping relationship and process in very positive terms—as they should be, not as they always are.

## The Downside: The Messiness of Helping

We have already seen a cardinal example of the messiness of helping: the ongoing debate as to its efficacy and what makes for both effectiveness and efficiency. These are shadow-side issues in that they are hidden from clients and ignored by many practitioners. And there are plenty of other examples. Nevertheless, the ability to understand how helper, client, the relationship, and the helping process itself can go wrong is the first step toward managing the shadow side.

For instance, helpers' motives are not always as pure as they are portrayed in this book. Some helpers are not very committed, even though they are in a profession in which success demands a high degree of commitment. Some helpers are competent, others are not. Incompetent helpers pass themselves off as professionals. Finally, although, like other people, helpers get into trouble, often they don't use the tools of their trade to get themselves out of trouble—"Don't do as I do; do as I say."

Clients often play games with themselves, their helpers, and the helping process. Helpers sometimes seduce their clients, and clients seduce their helpers—not necessarily sexually. Both helper and client pursue hidden agendas, and the helping relationship itself ends up as a conspiracy to do nothing. Clients, aided and abetted by their helpers, work on the wrong issues. Helpers fail to keep up with developments in the profession. They end up using helping methods that are not likely to benefit their clients. Helping is continued even though it is going nowhere. The list goes on.

Managing the shadow side is an exercise in integrity, social intelligence, and competence, not cynicism. Clearly, not all these shadow-side scenarios happen all the time; to assume that they do would be cynical. The shadow side is not usually an exercise in ill will. Few helpers set out to seduce their clients. Few clients set out to seduce their helpers. Helpers don't realize that they are incompetent. Clients don't realize that they are playing games. Most clients most of the time are well-intentioned, and most helpers do their best to put their own concerns aside to help their clients as best they can. But this does not mean that any given helper is always giving his or her best. Both clients and helpers have their temptations. Clients, as we shall see, have their blind spots that keep them mired down in their problems, but helpers also have theirs.

All human endeavors have their shadow side. Companies and institutions are plagued with internal politics and are often guided by covert or vaguely understood beliefs, values, and norms that do not serve the best interests of the business, its employees, or its shareholders. If helpers don't know what's in the shadows, they are naive. If they believe that shadow-side realities win out more often than not, they are cynical. To avoid both naïveté and cynicism, helpers should pursue a course of upbeat and compassionate realism.

## The Upside: Common Sense and Wisdom in the Helping Professions

Much of the stuff in the shadows contributes to the downside of helping, but the shadow side also has its upside (see Kottler, 1992, 1993, 1997). Although helping is sometimes referred to as an art, the emphasis in the journals is on theory and research. In a way, the emphasis is on "smart" rather than "wise." Some writers and researchers, however, have begun to focus on such things as wisdom, sagacity, street smarts, practical intelligence, and common sense in helping (for instance, Baltes & Staudinger, 2000; Hanna, 1994; Hanna & Ottens, 1995; Schmidt & Hunter, 1993; Sternberg, 1990; Sternberg, Wagner, Williams, & Horvath, 1995). This is part of the positive-psychology approach to helping.

Take self-control—a prerequisite for clients if they are to manage problems and develop opportunities. While latter-day research has helped us understand more fully the dynamics of self-control, in truth, most of the techniques used to facilitate self-control have not been derived from formal theory and research. When it comes to self-control, the Bible, Koran, Talmud, and other books of wisdom contain much of what has been discovered through research studies (see Karoly, 1995, p. 273). Whereas "smart" helpers see the client as a "clinical entity subject to the templates and tools of the psychotherapeutic trade," wise helpers see each client as a "vital, dynamic personage" (Hanna, 1994, p. 132) who needs to be helped to take advantage of the wisdom and common sense that is already within themselves, the history of their societies, their cultures, their environments, and their helpers.

Helpers need to be wise, and part of their job is to impart some of their wisdom, however indirectly, to their clients. Baltes and Staudinger (2000) define wisdom as "an expertise in the conduct and meaning of life" or "an expert knowledge system concerning the fundamental pragmatics of life" (pp. 124, 122). What is it that characterizes wisdom? Here are some possibilities (see Sternberg, 1990, 1998):

- Self-knowledge and maturity.
- Knowledge of life's obligations and goals.
- An understanding of cultural conditioning.
- The guts to admit mistakes and the sense to learn from them.
- A psychological and a human understanding of others; insight into human interactions.
- The ability to "see through" situations; the ability to understand the meaning of events.
- Tolerance for ambiguity and the ability to work with it.
- Being comfortable with messy and ill-structured cases.
- An understanding of the messiness of human beings.
- Openness to events that don't fit comfortably into logical or traditional categories.
- The ability to frame a problem so that it is workable; the ability to reframe information.
- Avoidance of stereotypes.
- Holistic thinking; open-mindedness; open-endedness; contextual thinking.
- "Meta-thinking," or the ability to think about thinking and become aware about being aware.
- The ability to see relationships among diverse factors; the ability to spot flaws in reasoning; intuition; the ability to synthesize.
- The refusal to let experience become a liability through the creation of blind spots.
- The ability to take the long view of problems.
- The ability to blend seemingly antithetical helping roles—being one who cares and understands while also being one who challenges and "frustrates" (see Levin & Shepherd, 1974).
- An understanding of the spiritual dimensions of life.

Wisdom is about excellence in living. As such, it focuses on knowing "how" (the procedural dimension) rather than merely knowing "what" (the factual dimension).

Wisdom, then, is critical in helping others. Such things as wisdom and common sense can be considered part of the upside shadows because they have not received a great deal of attention in the helping literature and they do not form part of the curriculum in helper training programs. Perhaps this is beginning to change.

Even the downside of the shadow side of helping can provide benefits. Consider an analogy. The shadow side of helping is a kind of "noise" in the system. But scientists have discovered that sometimes a small amount of noise in a system, called *stochastic resonance*, makes the system more sensitive and efficient (*Economist*, 1995). For instance, in the helping professions, noise in the guise of the debate around what makes helping both effective and efficient can ultimately benefit clients.

Despite the downside of the shadow side of helping, the tone of this book is unabashedly upbeat. However, helpers-to-be must not ignore the less palatable dimensions of the helping professions, including the less palatable dimensions of themselves. Therefore, throughout this book, some of the common shadow-side realities that plague client, helper, and the profession itself are noted at the service of managing them. This book is by no means a treatise on the shadow side of helping. Rather, its intent is to get helpers to begin to think about the shadow side of the profession. Wise helpers are idealistic without being naive. They also know the difference between realism and cynicism and opt for the former. They see the journey "from smart to wise" as a never-ending one.

# OVERVIEW OF THE HELPING MODEL

RATIONAL PROBLEM SOLVING AND ITS LIMITATIONS

THE SKILLED-HELPER MODEL: A PROBLEM-MANAGEMENT AND OPPORTUNITY-DEVELOPMENT APPROACH TO HELPING

THE STAGES AND STEPS OF THE HELPING MODEL

STAGE I: "WHAT'S GOING ON?" HELPING CLIENTS CLARIFY THE KEY ISSUES CALLING FOR CHANGE

The Three Steps of Stage I

Step I-A: Help clients tell their *stories*

Step I-B: Help clients break through *blind spots* that prevent them from seeing themselves, their problem situations, and their unexplored opportunities as they really are

Step I-C: Help clients choose the *right* problems and/or opportunities to work on

STAGE II: "WHAT SOLUTIONS MAKE SENSE FOR ME?" HELPING CLIENTS DETERMINE OUTCOMES

The Three Steps of Stage II

Step II-A: Help clients use their imaginations to spell out *possibilities for a better future*

Step II-B: Help clients choose *realistic and challenging goals* that are real solutions to the key problems and unexplored opportunities identified in Stage I

Step II-C: Help clients find the *incentives* that will help them *commit* themselves to their change agendas

STAGE III: "WHAT DO I HAVE TO DO TO GET WHAT I NEED OR WANT?" HELPING CLIENTS DEVELOP STRATEGIES FOR ACCOMPLISHING GOALS

The Three Steps of Stage III

Step III-A: Possible actions: Help clients see that there are many different ways of achieving goals

Step III-B: Help clients choose best-fit strategies

Step III-C: Help clients craft a plan

ACTION: "HOW DO I GET RESULTS?" HELPING CLIENTS IMPLEMENT THEIR PLANS

"HOW ARE WE DOING?" ONGOING EVALUATION OF THE HELPING PROCESS

FLEXIBILITY IN USING THE MODEL

BRIEF THERAPY AND A HOLOGRAM APPROACH TO HELPING

PROBLEM MANAGEMENT AND CULTURE: A HUMAN UNIVERSAL

USING THE MODEL AS A "BROWSER": THE SEARCH FOR BEST PRACTICE
    Eclecticism
    Problem management as underlying process
    The "browser" approach

UNDERSTANDING AND DEALING WITH THE SHADOW SIDE OF HELPING MODELS
    No model
    Fads
    Rigid applications of helping models
    Virtuosity

# RATIONAL PROBLEM SOLVING AND
# ITS LIMITATIONS

The problem-solving process is often described as a more or less straightforward, natural, and rational process of decision making. For instance, Yankelovich (1992) offered a seven-step process. Applied to helping, it looks something like this:

- **Initial awareness.** First, the client becomes aware of an issue or a set of issues. For instance, a couple, after a number of disputes over household finances, develop a vague awareness of dissatisfaction with the relationship itself.

- **Urgency.** Second, a sense of urgency develops, especially as the underlying problem situation—the dissatisfaction with the relationship itself—becomes more distressing. Even small annoyances are now seen in the light of overall dissatisfaction.

- **Initial search for remedies.** Third, the client begins to look for remedies. However implicitly or perfunctorily, the client explores different strategies for managing the problem situation. For instance, a client in a difficult marriage begins thinking about complaining openly to her partner or friends, separating, getting a divorce, instituting subtle acts of revenge, having an affair, going to a marriage counselor, seeing a minister, unilaterally withdrawing from the relationship in some way, and so forth. The client may try out one more of these remedies without evaluating their cost or consequences.

- **Estimation of costs.** Fourth, the costs of pursuing different remedies begin to become apparent. Someone in a troubled relationship might say to herself: "Being open and honest hasn't really worked. If I continue to put my cards on the table, I'll have to go through the agony of confrontation, denial, argument, counter accusations, and who knows what else." Or he might say, "Simply withdrawing from the relationship in small ways has been painful. What would I do if I were to go out on my own?" Or, "What would happen to the kids?" At this point, the client often backs away from dealing with the problem situation directly because there is no cost-free or painless way of dealing with it.

- **Deliberation.** Fifth, since the problem situation does not go away, it is impossible to retreat completely. At this point, a more serious weighing of choices takes place. For instance, the costs of confronting the situation are weighed against the costs of merely withdrawing. Often, a kind of dialogue goes on in the client's mind between steps 4 and 5; for example, "I might have to go through the agony of a separation for the kids' sake. Maybe time apart is what we need."

- **Rational decision.** Sixth, an intellectual decision is made to accept some choice and pursue a certain course of action: "I'm going bring all of this up with my spouse and suggest we see a marriage counselor." Or, "I'm going to get on with my life, find other things to do, and let the marriage go where it will."

- **Rational-emotional decision.** However, a merely intellectual decision is often not enough to drive action. So the heart joins the head, as it were, in the decision. One spouse might finally say, "I've had enough of this! I'm leaving. It won't be comfortable, but it's better than living like this." The other might say, "It is unfair to both of us to go on like this, and it's certainly not good for the kids," and this drives the decision to seek help, even if it means going alone. Decisions driven by emotion and convictions are more likely to be translated into action.

Four things should be noted. First, these steps, however logically sequenced on paper, are often jumbled and intermingled in real-life problem-management situations. Second, this natural process can be derailed at almost any point along the way. For instance, uncontrolled emotions can spill out and make a bad situation worse. Or the costs of managing the problem seem too high and so the process itself is put on the back burner. Third, decision making in difficult situations is seldom as rational as this process suggests. Indeed, decision making can be viewed as a journey as complex as life itself (Scott, 2000). Fourth, this natural process often lacks a method for turning decisions into solution-focused action.

The problem-management and opportunity-development process outlined in this chapter and developed in the rest of the book borrows from this natural process, complements it with other steps and techniques, suggests ways of helping clients turn decisions into action, focuses on solutions, that is, life-enhancing outcomes, provides ways of challenging backsliding, and, at its best, speeds up the entire process.

## THE SKILLED HELPER MODEL: A PROBLEM-MANAGEMENT AND OPPORTUNITY-DEVELOPMENT APPROACH TO HELPING

Common sense suggests that problem-solving models, techniques, and skills are important for all of us, since all of us must grapple daily with problems in living of greater or lesser severity. Ask parents whether problem-management skills are important for their children, and they would say "Certainly." But ask where and how their children pick up these skills, and they hem and haw: "Sometimes at home, but not always." Parents don't always see themselves as paragons of effective problem solving: "Maybe at school," some would wonder. Yet review the curricula of our primary, secondary, and tertiary schools, and you will find little about problem solving that focuses on problems in living. Some say that formal courses in problem-solving skills are not found in our schools because such skills can only be learned through experience. To a certain extent, that's true. However, if problem-management skills are so important, we might wonder why society leaves the acquisition of these skills to chance. A problem-solving mentality should be second nature to us. The world may be the laboratory for problem solving, but the skills needed to optimize learning in this lab should be taught. They are too important to be left to chance.

Let's move to the helping professions. Institute an Internet search and you will soon discover that there are dozens, if not hundreds, of models or approaches to helping, all of them claiming a high degree of success. Library shelves are filled with books on counseling and psychotherapy—books dealing with theory, research, and practice. Back in the 1980s, it was estimated that there were between 250 and 400 different approaches to helping (see Herink, 1980; Karasu, 1986). Some of the approaches discussed then have, of course, fallen by the wayside, but many more have been added since. All are proposed with equal seriousness and their proponents say that they work. In the face of all this diversity, helpers, especially beginning helpers, need a basic, practical, working model of helping.

Since all approaches must eventually help clients manage problems and develop unused resources, the model of choice outlined in these pages is a flexible, humanistic, broadly based problem-management and opportunity-development model—a model that is straightforward without ignoring the complexities of clients' lives or of the helping process itself. Indeed, since the problem-management and opportunity-development process outlined in this book is embedded in almost all approaches to helping, this model provides an excellent foundation for any "brand" of helping you eventually choose. This book provides the basics for them all. Finally, a problem-management model in counseling and therapy has the advantage of the vast amount of research that has been done on the problem-solving process itself. The model, techniques, and skills outlined in this book tap that research base.

## THE STAGES AND STEPS OF THE HELPING MODEL

All worthwhile helping frameworks, models, or processes ultimately help clients ask and answer for themselves four fundamental questions:

- **What's going on?** What are the problems, issues, concerns, or undeveloped opportunities I should be working on?
- **What do I need or want?** What do I want my life to look like? What changes would make me happier?
- **What do I have to do to get what I need or want?** What plan will get me where I want to go?
- **How do I get results**? How do I turn planning and goal setting into solutions, results, outcomes, or accomplishments? How do I get going and keep going?

These four questions—turned into three logical "stages" and an implementation phase in Figure 2-1—provide the basic framework for the helping process. The term *stage* is in quotation marks because it has sequential overtones that are somewhat misleading. Each stage is a set of tasks around a theme that helps clients move forward in managing problems and developing opportunities. The theme of Stage I is problem/opportunity clarification and ownership. Stage II is about goal setting and commitment to goals. Stage III is about strategies for accomplishing goals. In practice, the three stages overlap and interact with one another as clients struggle to manage problems and develop opportunities. And, as we shall see, helping, like life itself, is not as logical as the models used to describe it.

This chapter presents an extended example to bring this process to life. The case, though real, has been disguised and simplified. It is not a session-by-session presentation. Rather, it illustrates ways in which one client was helped to ask and answer the four fundamental questions just outlined. The client, Carlos, is voluntary, verbal, and for the most part, cooperative. In actual practice, cases do not always flow as easily as this one. The simplification of the case, however, will help you see the main features of the helping process in action.

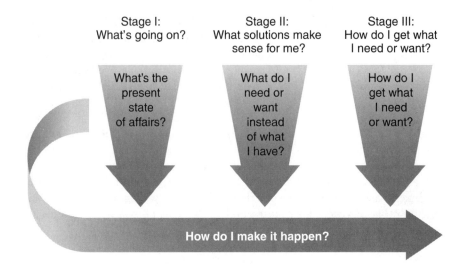

Stage I:
What's going on?

Stage II:
What solutions make
sense for me?

Stage III:
How do I get what
I need or want?

What's the
present
state
of affairs?

What do I
need or
want
instead
of what
I have?

How do I
get what
I need
or want?

**How do I make it happen?**

FIGURE 2-1
The Skilled Helper Model

## STAGE I: "WHAT'S GOING ON?"
## HELPING CLIENTS CLARIFY THE KEY ISSUES
## CALLING FOR CHANGE

The present state spells out the range of difficulties the client is facing. What are the problems, issues, concerns, and undeveloped opportunities with which Carlos needs to grapple? Here is a thumbnail sketch:

Carlos, a man in his mid-twenties, has been working for a consulting firm for about a year. He is well educated, though unlike some of his fellow workers, his undergraduate and MBA degrees are not from the "best" schools. He is bright, though practical rather than academic intelligence is his strong suit. He can be quite personable when he wants to be. But he doesn't always want to be. He comes from a family with traditional Hispanic cultural values. He can speak Spanish fairly well but prefers not to, even at home.

Trouble started when one of his colleagues, a consultant some 20 years older than Carlos, buttonholed him after a meeting one day and said: "I've seen you in action a few times. You know, you're your own worst enemy. You're headed for a fall and don't even know it. If I were you, I'd get some help." With that he walked away, and Carlos had no further interaction with him.

Though Carlos was first bothered by the incident, he sloughed it off. However, a few weeks later, while talking with one of his colleagues who had been at the firm for about three years, Carlos recounted the incident in a joking way. But his colleague didn't laugh. He merely said in a lighthearted manner, "Well, you never know, Carlos, there just might be something there."

Through the company, he has access to a couple of "developmental counselors." He is somewhat put off when he finds out that both of them are women, but he makes an appointment with one.

So here is a young man who has received a couple of shots across his bow. While he certainly doesn't think that he needs help, he is unsettled enough to be willing to talk with someone.

## The Three Steps of Stage I

As we shall see in the chapters that follow, each "stage" is divided into three "steps." Like the stages themselves, the steps are not steps in a mechanistic, "first do this, then do this" sense. And like the stages, the steps are interactive. In Stage I, they are activities that help clients develop answers to two questions: What's going on in my life? What should I work on?

**Step I-A: Help clients tell their *stories*.** Though helping is ultimately about solutions, some review of the problem or missed opportunity is called for. So Elena, Carlos's counselor, helps him tell his story. Through their dialogue, she helps him review what is happening in the workplace. Elena knows that if she can help Carlos get an undistorted picture of himself, his problems, and his unused opportunities, he will have a better chance of doing something about them. Her overall goal is to help Carlos manage the interpersonal dimensions of his work life better. His interpersonal style is both problem and opportunity.

Carlos gets over his initial reluctance and brings up a number of things that bother him. He feels that he is being discriminated against at work. He doesn't feel that he's on the fast track, and he thinks that he should be. People at work don't understand and appreciate him. Though he lives at home because of financial reasons, he feels distant from his family. "We don't have that much in common anymore. They don't want to be mainstream." He and his girlfriend are currently at odds. At one point he says, "I think I've moved beyond her anyway."

**Step I-B: Help clients break through *blind spots* that prevent them from seeing themselves, their problem situations, and their unexplored opportunities as they really are.** Counselors add great value when they can help their clients identify significant blind spots related to their problems and unexplored opportunities. Effectively challenged, blind spots yield to new perspectives that help clients think more realistically about problems, opportunities, and solutions.

It does not take Elena long to realize that Carlos has some significant blind spots. For instance, he tends to blame others for his problems: "They are keeping me back." And he does not realize how self-centered he is. He gets angry when others stand in his way or don't cater to his needs and wants. Yet he is quite insensitive to anybody else's needs. His arrogant style rubs both colleagues and customers the wrong way. Without being brutal, Elena helps him see himself as others see him. She points out that being Hispanic and coming from the "wrong" school have little to do with his colleagues' hostility toward him. Elena, Hispanic herself, understands both Carlos's struggles and his excuses.

**Step I-C: Help clients choose the *right* problems and/or opportunities to work on.** If clients have a range of issues, help them gain *leverage* by working on issues that will make a difference. If a client wants to work only on trivialities or does not want to work at all, then it might be better to defer counseling.

Carlos becomes more cooperative as it becomes clear to him that Elena has both his interests and those of the firm in mind. He sees her as "solid"—decent, businesslike, and not overly "psychological." He comes to realize that, although he has a number of concerns, he had better work on his interpersonal communication style and his relationships with both colleagues and clients if he wants to advance in his career. He also needs to do something about the "victim" mentality he has developed. With Elena's help, he realizes that the flip side of his communication style problem is an enormous opportunity. He quickly sees that becoming a better communicator and relationship builder will help him in every social setting in life. Because they are in a work setting and time is limited, Elena does not push him on the problems he's having at home or in his social life outside work. However, she suspects that the changes he makes at work will also apply outside.

## STAGE II: "WHAT SOLUTIONS MAKE SENSE FOR ME?" HELPING CLIENTS DETERMINE OUTCOMES

In Stage II, counselors help clients explore and choose possibilities for a better future—a future in which clients can manage key problem situations and develop key opportunities. Helpers ask, "What do you want this future to look like?" The clients' answers constitute their "change agendas." Stage II focuses on outcomes.

Elena helps Carlos ask himself such questions as: What do I want? What do I need, whether I currently want it or not? What would my business life look like if it were more tolerable or—even better—more engaging and fulfilling?

Unfortunately, some approaches to problem solving or management skip Stage II. They move from the "What's wrong?" stage (Stage I) to a "What do I do about it?" stage (Stage III). As we shall see, however, helping clients discover what they want has a profound impact on the entire helping process.

### The Three Steps of Stage II

Stage II also has three steps—that is, three ways of helping clients answer as creatively as possible the question "What do I need or want?"

**Step II-A: Help clients use their imaginations to spell out *possibilities for a better future.*** This often helps clients move beyond the problem-and-misery mind-set they bring with them and develop a sense of hope. Brainstorming possibilities for a better future can also help clients understand their problem situations better: "Now that I am beginning to know what I want, I can see my problems and unused opportunities more clearly."

Elena helps Carlos brainstorm goals that would help him repair some damaged relationships with both colleagues and clients and help him do something about his victim mentality. Carlos declares that he needs to become a "better communicator." Elena, pointing out that "becoming a better communicator" is a rather vague aspiration, asks him, "What do some of the good communicators you know look like?" Carlos comes up with a range of possibilities. "I'd give great presentations like Jeff." "I'd be a good problem solver like Sharon." "Tony listens a lot better than I do." "I'm not patient at all, but I like it when Abigail is patient with me." "Roger seems to have good relationships with everyone." And so forth. All of these become possibilities for a better communication style. She also enables him to explore further possibilities by asking him what a "repaired relationship" would look like both from his perspective and from the perspective of customers and colleagues he might have alienated.

**Step II-B: Help clients choose *realistic and challenging goals* that are real solutions to the key problems and unexplored opportunities identified in Stage I.** Possibilities need to be turned into goals because helping is about solutions and outcomes. A client's goals, then, constitute his or her *agenda for change*. If goals are to be pursued and accomplished, they should be clear, related to the problems and unexplored opportunities the client has chosen to work on, substantive, realistic, prudent, sustainable, flexible, consistent with the client's values, and set in a reasonable time frame. Effective counselors help clients "shape" their agendas to meet these requirements.

Elena helps Carlos sort through some of the possibilities he has come up with. It becomes clear that changes in his interpersonal communication style would help him manage some problems and develop some opportunities at the same time. If he were to communicate well—in terms of both communication skills and the values that support relationship-building communication—he would come across more effectively and repair damaged relationships. But he needs more than skills. He needs to change his self-centered and "poor me" attitude. Some of the possibilities Carlos comes up with in his brainstorming session with Elena can be put aside for the present. For instance, things like "becoming a terrific presenter" can wait.

Becoming a better communicator coupled with upbeat interpersonal relationship values and attitudes is a substantial package. It includes Carlos's becoming good at the give and take of dialogue and the skills that make it work. These skills include visibly tuning in to others, active listening, thoughtfully processing what he hears, demonstrating understanding of the key points others are making, getting his own points across clearly, drawing others in to the conversation, and the like. It also includes embracing the values that make conversations serve relationships—mutual respect, social sensitivity, emotional control, and collaboration.

**Step II-C: Help clients find the *incentives* that will help them *commit* themselves to their change agendas.** The question clients must ask themselves is: "What am I willing to pay for what I need and want?" Without strong commitment, change agendas end up as no more than some "nice ideas." This does not mean that clients are not sincere in setting goals. Rather, once they leave the counseling session, they run into the demands of everyday life. The goals they set for themselves, however useful, face a great deal of competition. Counselors provide an important service when they help clients test their commitment to the better future embedded in the goals they choose.

Becoming a better communicator at the service of repairing and building relationships is hard work. Elena helps Carlos review the incentives he has for engaging in such work. One very strong one is this: He has to. His current interpersonal communication style will probably get him fired and prevent him from being successful in the future. Developing the values that should permeate dialogue is even harder work.

It means undoing bad habits developed over years. Undoing bad habits is difficult even when a client is committed to doing so. Elena does not dwell on Carlos's bad habits. Rather, she believes that embracing good habits—like showing interest in others and checking his understanding of what they have to say—will drive out his bad habits. If Carlos does all this work, the upside is enormous. Because communication is at the heart of everything he does, better communication skills and values will serve him well in every dimension of his life. Elena does not find it difficult to help Carlos appreciate this attractive package of incentives. Positive psychology wins. In Carlos's case, developing opportunities is the main way of managing problems.

## Stage III: "What Do I Have to Do to Get What I Need or Want?" Helping Clients Develop Strategies for Accomplishing Goals

Stage III defines the actions that clients should take to translate goals into problem-managing accomplishments. Stage II answers the question: "How do I get there?" It is about identifying and choosing action strategies and plans.

### The Three Steps of Stage III

Stage III, too, has three steps that, in practice, intermingle with one another and with the steps of the other stages.

**Step III-A: Possible actions: Help clients see that there are many different ways of achieving their goals.** Stimulating clients to think of different ways of achieving their goals is usually an excellent investment of time. That said, clients should not leap into action. Hasty and disorganized action is often self-defeating. Complaints such as "I tried this and it didn't work. Then I tried that and it didn't work either!" is often a sign of poor planning rather than of the impossibility of the task.

Elena helps Carlos explore different ways of becoming the competent communicator he wants to be so he can establish a better self-image and develop and foster satisfying work relationships with both colleagues and clients. To acquire the skills he needs, Carlos can read books, take courses at local colleges, attend courses for professionals, get a tutor or coach, or come up with his own approach to developing the skills and attitudes he needs. Elena helps Carlos brainstorm the possibilities. She also points out where he can get more information, but she then lets him do his homework.

**Step III-B: Help clients choose best-fit strategies.** While Step III-A provides clients with a pool of possible strategies, Step III-B helps clients choose the action strategies that best fit their talents, resources, style, temperament, environment, and timetable.

With Elena's help, Carlos makes some choices. He chooses to attend an interpersonal communication program for working professionals. Although more expensive than university-based courses, the program offers greater flexibility and fits better with Carlos's rather hectic travel schedule. Second, Elena helps Carlos see that life is his lab. That is, every conversation is part of the program—an opportunity to practice the skills he will be learning and demonstrate the attitudes that foster relationship building. To Carlos's credit, he points out that he needs to find a way of monitoring the degree that the values of effective dialogue are permeating his conversations. He decides to get a peer coach—a colleague that has both an excellent communication style and positive relationships with colleagues and clients. He finds a colleague who fits the bill and who is willing to help. He says to his colleague: "Be honest with me. Tell me the way it is."

**Step III-C: Help clients craft a plan.** Help clients organize the actions they need to take to accomplish their goals. Plans are simply maps clients use to get where they want to go. A plan can be quite simple. Indeed, overly sophisticated plans are often self-defeating.

Carlos's plan is straightforward. He will begin the interpersonal communication course within two weeks. He will use every conversation with colleagues and clients as his lab. At the end of each working day he will review the conversations he has had in terms of both skillfulness and values. He creates a checklist for himself that includes such questions as "How effectively did I listen? How clearly did I get my points across? How respectful and collaborative was I? How did I handle sensitive conversations such as those with colleagues and clients who had been turned off by my interpersonal style?" Time and schedules willing, he will meet with his peer coach once a week and with Elena once a month.

## ACTION: "How Do I Get Results?" HELPING CLIENTS IMPLEMENT THEIR PLANS

All three stages of the helping model sit on the "action" arrow, indicating that clients need to act in their own behalf right from the beginning of the helping process. Stages I, II, and III are about planning for change, not constructive change itself. Planning is not action. Talking about problems and opportunities, discussing goals, and figuring out strategies for accomplishing goals is just so much "blah, blah, blah" without goal-accomplishing action. There is nothing magic about change; it is hard work. But, as we shall see in subsequent chapters, each stage and step of the process can promote problem-managing and opportunity-developing action at the very beginning.

Carlos, like most clients, runs into a number of obstacles as he tries to implement his plan. His travel schedule keeps conflicting with even the flexible communication skills program in which he has enrolled. Because he finds the program very useful, he decides to use a tutor from the program to bring him up to speed whenever he cannot fit a session into his travel schedule. Although the company is paying for the program, he has to pay the tutor out of his own pocket. But it's worth it. He also finds that he is not very consistent in using the skills he is learning in the "lab of life." His progress is much slower that he thought it would be. Sometimes he loses heart. He often fails to do the evening self-evaluation sessions. Getting together with his peer coach proves to be almost impossible. And, while he is doing better with clients, he doesn't see much "repair" going on in his relationships with his colleagues. So in one of his discussions with Elena, he reviews the ups and downs of his program and discusses his discouragement with how slowly some of his disenchanted colleagues are warming up to him.

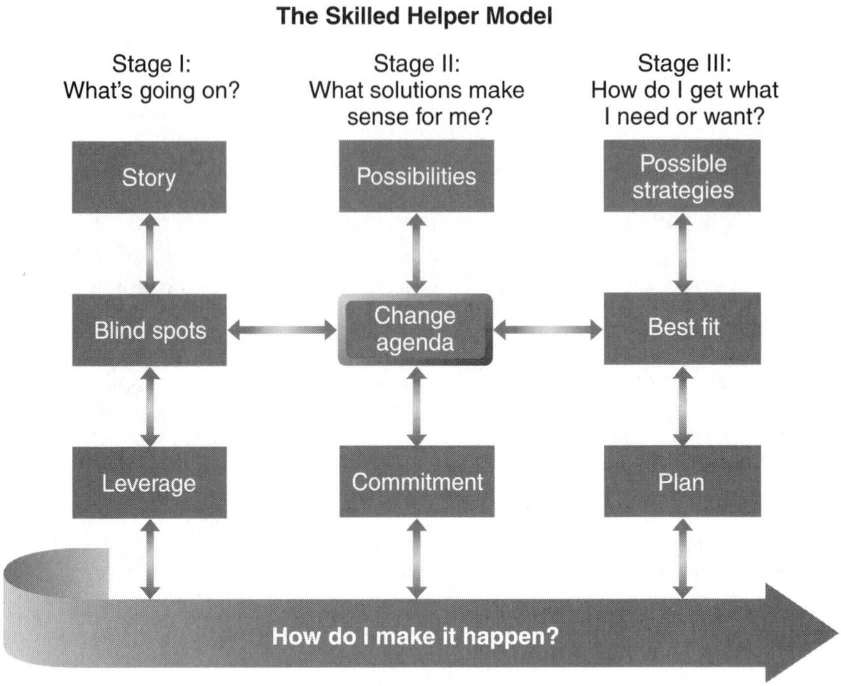

**The Skilled Helper Model**

FIGURE 2-2
**The Helping Model Showing Interactive Stages and Steps**

Following his discussion with Elena, Carlos decides to reset the program. First, he agrees to see her for a half hour every other week. This provides him with an incentive to keep to the program. He wants to give her a good report. Second, he discusses his "bad attitude." Even though she said that the whole program would be a lot of work, he didn't realize just how much work. He finds that he can't do the program well without changing some basic attitudes and habits. It's not just a skills program. It cuts much deeper. He has to recommit himself to a more fundamental attitudinal change. New attitudes look fine on paper, but developing them is another story. So Carlos moves on—two small steps forward and a half step backward. He resets his schedule so he can get together with his peer coach. His sessions with both Elena and his peer coach help, and he does make progress.

Figure 2-2 presents the full model in all its stages and steps and their relationship to the action arrow. It includes two-way arrows between both stages and steps to suggest the kind of flexibility needed to make the process work.

## "How Are We Doing?" Ongoing Evaluation of the Helping Process

In the light of the "Does helping help?" discussion in Chapter 1, how do helpers using the problem-management and opportunity-development framework evaluate what is happening with each client? By making each case a "mini-experiment" in itself. Psychological research has a long history of creating what are called N=1 designs, both to evaluate practice and to conduct research (Blampied, 2000; Hilliard, 1993; Lundervold & Belwood, 2000; Persons, 1991; Valsiner, 1986). What's the value of knowing that helping "works" in general if we do not know that it is working in this case?

In many helping models, evaluation is presented as the last step in the model. However, if evaluation occurs only at the end, it is too late. As Mash and Hunsley (1993) noted, early detection of what is going wrong in the helping process can prevent failure. They claimed that an early-detection framework should be theory based, ongoing, practical, and sensitive to whatever new perspectives might emerge from the helping process. The problem-management and opportunity-development model outlined in this chapter fills the bill. It is a tool to check progress throughout the helping process. As we shall see, it provides criteria for helper effectiveness, for client participation, and for assessing outcomes. In later chapters, questions designed to help with the evaluation process will be provided for each step.

Elena, knowing that helpers and clients need to collaborate in this ongoing evaluation process, works with Carlos in using the helping model as the evaluation framework. Carlos comes to appreciate Elena's skill in using the model. Once Carlos takes ongoing evaluation seriously, he begins to make progress. He gets feedback from the self-evaluation process that he begins to use more frequently, from the observations of his peer coach, and from his sessions with Elena. The ultimate feedback comes from goal accomplishment. In what ways and to what degree is he becoming a more competent communicator? How effectively is he repairing relationships with clients and colleagues? To what degree is he shedding his self-centered approach to relationships that contributes to his "others are out to get me" attitude?

## FLEXIBILITY IN USING THE MODEL

There are many reasons why you need to use the helping model flexibly. The main one is this: Helping is for the client. Clients' needs take precedence over any model. That said, a number of points about flexibility need to be made.

First, clients start and proceed differently. Any stage or step of the helping process can be the entry point. For instance, Client A might start with something that he tried to do to solve a problem but that did not work: "I threatened to quit if they didn't give me a leave of absence, but it backfired. They told me to leave." The starting point is a failed strategy. Client B might start with what she believes she wants but does not have: "I need a boyfriend who will take me as I am. Joe keeps trying to redo me." Stage II is her entry point. Client C might start with the roots of his problem situation: "I don't think I've ever gotten over being abused by my uncle." Stage I is his entry point. Client D might announce that she really has no problems but is still vaguely dissatisfied with her life: "I don't know. Everyone tells me I've got a great life, but something's missing." The implication here is that she has not been seizing the kind of opportunities that could make her happy. Opportunity rather than problem is her starting point.

Second, clients engage in each stage and step of the model differently. Consider clients' stories, for example. Some clients spill out their stories all at once. Others "leak" bits and pieces of their stories throughout the helping process. Still others tell only those parts that put them in a good light. Most clients talk about problems rather than opportunities. Because clients do not always present all their problems at once in neat packages, it is impossible to work through Stage I completely before moving on to Stages II and III and launching into action. It is not even advisable to do so. Some clients don't understand their problems until they begin talking about what they want but don't have. Some clients need to engage in some kind of remedial action before they can adequately define the problem situation; that is, action sometimes precedes understanding. If some supposedly problem-solving action is not successful, then the counselor helps the client learn from it and return to the tasks of clarifying the problem or opportunity and then setting some realistic goals. Take the case of Woody.

> Woody, a sophomore in college, came to the student counseling services with a variety of interpersonal and somatic complaints. He felt attracted to a number of women on campus but did very little to become involved with them. After exploring this issue briefly, he said to the counselor, "Well, I just have to go out and do it." Two months later he returned and said that his experiment had been a disaster. He had gone out with a few women, but the chemistry never seemed right. Then he did meet someone he liked quite a bit. They went out a couple of times, but the third time he called, she said that she didn't want to see him anymore. When he asked why, she muttered vaguely about his being too preoccupied with himself and ended the conversation. He felt so miserable he returned to the counseling center. He and the counselor took another look at his social life. This time, however, he had some experiences to probe. He wanted to explore this "chemistry thing" and his reaction to being described as "too preoccupied with himself."

Woody put into practice Weick's (1979) dictum that chaotic action is sometimes preferable to orderly inactivity. Once he acted, he learned a few things about himself. Some of these learnings proved to be painful, but he now had a better chance of examining his interpersonal style much more concretely.

Third, since the stages and steps of the model intermingle, helpers will often find themselves moving back and forth in the model. Often two or more steps or even two stages of the process merge into one another. For instance, clients can name parts of a problem situation, set goals, and develop strategies to achieve them in the same session. New and more substantial concerns arise while goals are being set, and the process moves back to an earlier, exploratory stage. Helping is seldom a linear event. One client, in discussing a troubled relationship with a friend, said something like this:

> Every time I try to be nice to her, she throws it back in my face. So who says being more considerate is the answer? Maybe my problem is that I'm a wimp, not the self-centered jerk she makes me out to be. Maybe I'm being a wimp with you and you're letting me do it. Maybe it's time for me to start looking out for my own interests—you know, my own agenda—rather than trying to make myself fit into everyone else's plans. I need to take a closer look at the person I want to be in my relationships with others.

In these few sentences the client mentions a failed action strategy, questions a previously set goal, hints at a new problem, suggests a difficulty with the helping relationship itself, offers, at least generically, a different approach to managing his problem, and recasts the problem as an opportunity to develop a more solid interpersonal style. Your challenge is to make sense of clients' entry points and guide them through whatever stage or step will help them move toward problem-managing and opportunity-developing action.

Flexibility, of course, is not mere randomness or chaos. Focus and direction in helping are also essential. Letting clients wander around in the morass of problem situations under the guise of flexibility leads nowhere. The structure of the helping model is the very foundation for flexibility; it is the underlying "system" that keeps helping from being a set of random events. A helping model is like a map that informs you, at any given moment, "where you are" with a client and what kinds of interventions would be most useful. In the map metaphor, the stages and steps of the model are orientation devices. At its best, it is a *shared* map that helps clients participate more fully in the helping process. They, too, need to know where they are going.

## BRIEF THERAPY AND A HOLOGRAM APPROACH TO HELPING

How long does therapy take? That's like asking, How long is a piece of string? It depends. I have known people who were in therapy most of their lives. Others need only one session to reset some area of their lives. Over the past decade, there has been growing interest in "brief therapies" or what Nick and Janet Cummings (2000) call "time-sensitive psychotherapies."

> There is a growing body of evidence that time-sensitive psychotherapies are effective with a large number of patients, perhaps as many as 85% of all those seen in the usual practice (Austad, 1996; Budman & Steenbarger, 1997; Cummings, Pallak, & Cummings, 1996; Hoyt, 1995b), but it is not advocated that all psychotherapy be brief psychotherapy. Even I. P. Miller (1996b), an outspoken critic of short-term therapy, begrudgingly concedes that there is a place for time-sensitive psychotherapies and that psychotherapists of the future must be trained in both short-term and long-term interventions to know when to use one or the other. (p. 44)

They advocate "clinically determined" therapy rather than therapy determined by the economic needs of either clinicians or managed-care enterprises. Effective and efficient helping (Cummings, Budman, & Thomas, 1998) rather than the length of helping should remain center stage.

Although the whole helping process can take days, weeks, months, or even years, it can also take place, literally, within minutes. Consider the following scenario:

> Lara, a social worker in a tough urban neighborhood, gets a call from a local minister who says that a teenager who fears for his life because of neighborhood gang activity needs to see someone. She sees the boy, listens to his concerns, feels his anxiety. While the minister thought that the boy was under some immediate threat, Lara realizes that his anxiety stems from a recent shooting that had nothing to do with him directly. She talks with him to get some idea what his daily comings and goings are like. It is soon clear that he is not the kind of person who courts trouble. She then talks him through the kinds of prudent things he needs to do to avoid trouble "in a neighborhood like ours" and in their dialogue points out that he is already doing most of those things. She gently challenges his false perception that the recent shooting somehow involved him directly. All of this helps allay his fears. But he discovers there are a couple of things he needs to do differently like having "good" friends and not letting himself be a lone target on the streets. She also points out that he is more likely to be hit by lightning than a stray bullet. The boy calms down as he begins to realize that in many ways his life is in his own hands.

Lara listens to the boy's immediate concerns and elicits from him a picture of what his daily life in the community looks like (Stage I). Realizing that he needs some relief from his anxiety (Stage II), she "talks him down." Through their dialogue, the boy learns some things he can do to ensure his safety (Stage III). Lara helps him dispel a couple of blind spots—that is, that he is an immediate target of gang activity and that he is in imminent danger of being hit by a stray bullet. So even if you have only one meeting with a client, it would be a mistake to assume that only Stage I things could happen. Preston, Varzos, and Liebert (2000) have written a manual for clients who are in or are considering going into brief therapy.

Helping can be "lean and mean" and still be fully human. A colleague of mine experimented, quite successfully, with shortening the counseling "hour." He arrived at the point where he would begin a session by saying, "We have five minutes together. Let's see what we can get done." He was very respectful, and it was amazing how much he and his clients could get done in a short time. It is true that the helping industry, driven by politics and the financial dynamics of managed care, is focusing more and more on results-oriented, brief psychotherapy. But even if that were not the case, helpers would still owe their clients value for money. Helping that achieves only partial results may, at times, be the best that we can do.

Helpers would do well to develop a "whole process" mentality about helping. Any part of any stage or step can be invoked at any time in any session if it proves beneficial for the client. Think of the helping process as a hologram—that laser-generated three-dimensional image that seems to float in space. In a hologram, the whole is found in each of its parts. The helping model is more like a hologram than a tool kit. It works best when the whole is found in each of its parts. The hologram is at the center of effective brief therapy.

## PROBLEM MANAGEMENT AND CULTURE: A HUMAN UNIVERSAL

Given the diversity of clients, helping models should be vehicles of personal growth, not cultural domination (see Chapter 3 for a discussion of client diversity). The advantage of problem-management and opportunity-development models of helping is that they are easily recognized across the world. Problem solving seems to be what McCrae and Costa (1997) call a "human universal." Many years ago, before presenting an earlier version of the helping process outlined in this book to some 300 college students and faculty members in Tanzania, I said, "All I can do is present to you the helping process I teach and use. You have to decide whether it makes sense in your own culture." After the presentation, the audience made two comments: First, the communication skills used in the helping process would have to be modified somewhat to fit their culture. Second, the problem-management helping process itself was very useful.

Since then, this scene has been repeated—in conferences and training events I and others have presented—over and over again on every continent. The model presented here spells out, in a flexible, step-by-step fashion, the way human beings think about constructive change. The reason this process crosses cultures so easily is that its logic is embedded in human consciousness. People don't really need to learn the framework of the model because, in essence, they already know it. It is, to use Orlinsky and Howard's (1987) term, a "generic" model of helping. Of course, the process as outlined in these pages together with the skills and techniques that make it work still has to be adapted both to different cultural settings and to different individuals within those settings. This demands cultural sensitivity on the part of helpers.

## Using the Model as a "Browser": The Search for Best Practice

The claim in this book is that problem management and opportunity development comprise one of the principal processes—perhaps *the* principal process—underlying all successful counseling and psychotherapy. What are novice helpers to do in the face of the bewildering array of models and methods available to them? Even though there is only a handful of "major brands" of psychotherapy (see Capauzzi & Gross, 1999; Corey, 1996; Gilliand & James, 1997; Prochaska & Norcross, 1998; Sharf, 1999; Wachtel & Messer, 1997), choices must be made.

**Eclecticism.** Many experienced helpers, even when they choose one specific school or approach to helping, often borrow methods and techniques from other approaches. Other helpers, without declaring allegiance to any particular school, stitch together their own approaches to helping. This borrowing and stitching is called *eclecticism* (Jensen, Bergin, & Greaves, 1990; Lazarus, Beutler, & Norcross, 1992; Prochaska & Norcross, 1998). In one study, some 40% of helpers said that eclecticism was their primary approach to helping (Milan, Montgomery, & Rogers, 1994). Effective eclecticism, however, must be more than a random borrowing of ideas and techniques from here and there. There must be some integrating framework to give coherence to the entire process; that is, to be effective, eclecticism must be systematic.

**Problem management as underlying process.** When any school, model, or eclectic mixture is successful, it is because it helps clients (1) identify and explore problem situations and missed opportunities, (2) determine what they need and want, (3) discover ways of getting what they need and want, and (4) translate what they learn into problem-managing action. That is, the problem-management and opportunity-development process outlined here underlies or is embedded in all approaches to helping, since all approaches deal with constructive client change. Therefore, the helping model described in these pages, together with the basic skills and techniques that make it work, is a very useful starting point for novices, no matter what school or approach or eclectic system they may ultimately choose or devise.

**The "browser" approach.** The helping model in this book can also be used as a tool—a "browser," to use an Internet term—for mining, organizing, and evaluating concepts and techniques that work for clients, no matter what their origin.

 • **Mining.** First, helpers can use the problem-management model to mine any given school or approach, "digging out" whatever is useful without having to accept everything that is offered. The stages and steps of the model serve as tools for identifying methods and techniques that will serve the needs of clients.

- **Organizing.** Second, since the problem-management model is organized by stages and steps, it can be used to organize the methods and techniques that have been mined from the rich literature on helping. For instance, a number of contemporary therapies have elaborated excellent techniques for helping clients identify blind spots and develop new perspectives on the problem situations they face. As we shall see, these techniques can be organized in Step I-B, the "turning blind spots into new perspectives" step of the problem-management model.

- **Evaluating.** Since the problem-management model is pragmatic and focuses on outcomes of helping, it can be used to evaluate the vast number of helping techniques that are constantly being devised. The model enables helpers to ask in what way a technique or method contributes to the "bottom line," that is, to outcomes that serve the needs of clients.

The problem-management and opportunity-development model can serve these functions because it is an open-systems model, not a closed school. Although it takes a stand on how counselors can help their clients, it is open to being corroborated, complemented, and challenged by any other approach, model, or school of helping. The needs of clients, not the egos of model builders, must remain central to the helping process. Our clients deserve "best practice," whatever its source.

## UNDERSTANDING AND DEALING WITH THE SHADOW SIDE OF HELPING MODELS

Besides the broad shadow-side themes mentioned in Chapter 1, there are a number of shadow-side pitfalls in the use of any helping model.

**No model.** Some helpers "wing it." They have no consistent, integrated model that has a track record of benefitting clients. Professional training programs often offer a wide variety of approaches to helping drawn from the "major brands" on offer. If helpers-to-be leave such programs knowing a great deal about different approaches but lacking an integrated approach for themselves, then they need to develop one quickly.

**Fads.** The helping professions are not immune to fads. A fad is an insight or a technique that would have some merit were it to be integrated into some overriding model or framework of helping. Instead it is marketed on its own as the central, if not the only meaningful, intervention needed. A fad need not be something new; it can be the "rediscovery" of a truth or a technique that has not found its proper place in the helping tool kit. Helpers become enamored of these ideas and techniques for a while and then abandon them. There will always be "hot topics" in helping. Note them and integrate them into a comprehensive approach to your clients. Many new approaches to helping make outrageous claims. Don't ignore them, but take the claims with a grain of salt and test the approach.

**Rigid applications of helping models.** Some helpers buy into a model early on and then ignore subsequent challenges or alterations to the model. They stop being learners. The "purity" of the model becomes more important than the needs of clients. Other helpers, especially beginners, apply a useful helping model too rigidly. They drag clients in a linear way through the model even though that is not what clients need. All of this adds up to excessive control. Effective models effectively used are liberating rather than controlling

**Virtuosity.** Another pitfall that some helpers fall into is becoming a virtuoso, specializing in certain techniques and skills—exploring the past, assessment, goal setting, probing, challenging, and the like. Helpers who specialize not only run the risk of ignoring client needs but also often are not very effective even in their chosen specialties. For example, the counselor whose specialty is challenging clients is often an ineffective challenger. The reason is obvious: Challenge must be based on understanding. Challenge is just part of the picture, part of the hologram.

The antidote to these shadow-side tendencies is simple: Helpers need to become radically client centered. Client-centered helping means that the needs of the client, not the models and methods of the helper, constitute the starting point and guide for helping. Therefore, flexibility is essential. In the end, helping is about solutions, results, outcomes, and impact rather than process. The values that drive client-centered helping are reviewed in Chapter 3.

# THE HELPING RELATIONSHIP: VALUES IN ACTION

**THE HELPING RELATIONSHIP**
>   The relationship itself
>   The relationship as a means to an end
>   The relationship and outcomes

**THE RELATIONSHIP AS A WORKING ALLIANCE**
>   The collaborative nature of helping
>   The relationship as a forum for relearning
>   Relationship flexibility

**VALUES IN ACTION**
>   Putting Values into the Broader Context of Culture
>   The Pragmatics of Values

**RESPECT AS THE FOUNDATION VALUE**
>   Do no harm
>   Be competent and committed
>   Make it clear that you are "for" the client
>   Assume the client's goodwill
>   Do not rush to judgment
>   Keep the client's agenda in focus

**EMPATHY AS A PRIMARY ORIENTATION VALUE**
>   The Nature of Empathy
>       A rich term
>       A key helping value
>   Empathy—Understanding Clients as They Are: Diversity and Multiculturalism
>       Understand diversity
>       Challenge whatever blind spots you may have
>       Tailor your interventions in a diversity-sensitive way
>       Work with individuals
>   Guidelines for Integrating Diversity and Multiculturalism into Counseling

GENUINENESS AS A PROFESSIONAL VALUE
    Do not overemphasize the helping role
    Be spontaneous
    Avoid defensiveness
    Be open

CLIENT EMPOWERMENT AS AN OUTCOME VALUE
  Helping as a Social-Influence Process
  Norms for Empowerment and Self-Responsibility
    Start with the premise that clients can change if they choose
    Do not see clients as victims
    Share the helping process with clients
    Help clients see counseling sessions as work sessions
    Become a consultant to clients
    Accept helping as a natural, two-way influence process
    Focus on learning instead of helping
    Do not see clients as overly fragile

A WORKING CHARTER: A CLIENT-HELPER CONTRACT

SHADOW-SIDE REALITIES IN THE HELPING RELATIONSHIP
    Ethical flaws
    Human tendencies in both helpers and clients
    Trouble in the relationship itself
    Vague and violated values
    Failure to share the helping process
    Flawed contracts
    Warring professionals

## THE HELPING RELATIONSHIP

Although theoreticians, researchers, and practitioners alike—not to mention clients—agree that the relationship between client and helper is important, there are significant differences as to how this relationship should be characterized and played out in the helping process (Gaston, et al., 1995; Hill, 1994; Sexton & Whiston, 1994; Weinberger, 1995). Some stress the *relationship* itself (see Bailey, Wood, & Nava, 1992; Kahn, 1990; Kelly, 1994, 1997; Patterson, 1985); others highlight the *work* that is done through the relationship (see Reandeau & Wampold, 1991); still others focus on the *outcomes* to be achieved through the relationship (see Horvath & Symonds, 1991).

**The relationship itself.** Patterson (1985) makes the relationship itself central to helping. He claimed that counseling or psychotherapy does not merely *involve* an interpersonal relationship; rather, it *is* an interpersonal relationship. Kelly (1994, 1997), in offering a humanistic model of counseling integration, argues that all counseling is distinctively human and fundamentally relational. Some traditional schools of psychotherapy indirectly emphasize the centrality of the helping relationship. For instance, in psychoanalytic or psychodynamic approaches, *transference*—the complex and often unconscious interpersonal dynamics between helper and client that are rooted in the client's and even the helper's past—is central (Gelso, Hill, Mohr, Rochlen, & Zack, 1999; Gelso, Kivlighan, Wise, Jones, & Friedman, 1997; Hill & Williams, 2000). Resolving these often murky dynamics is seen as intrinsic to successful therapeutic outcomes. Schneider (1999), in discussing the treatment manuals mentioned in Chapter 1, claims that clients deserve the kind of relationship with their helpers through which human meaning, purpose, and values can be explored.

In a different mode, Carl Rogers (1951, 1957), one of the great pioneers in the field of counseling, emphasizes the quality of the relationship in representing the humanistic-experiential approach to helping (see Kelly, 1994, 1997). Rogers claims that the unconditional positive regard, accurate empathy, and genuineness offered by the helper and perceived by the client are both necessary and often sufficient for therapeutic progress. Through this highly empathic relationship, counselors help clients understand themselves, liberate their resources, and manage their lives more effectively. Rogers's work spawned the widely discussed client-centered approach to helping (Rogers, 1965). Unlike psychodynamic approaches, however, the empathic helping relationship is considered a facilitative condition, not a "problem" in itself to be explored and resolved.

**The relationship as a means to an end.** Others see the helping relationship as very important but still a means to an end. In this view, a good relationship is practical because it enables client and counselor to do the work called for by whatever helping process is used. The relationship is instrumental in achieving the goals of the helping process. Practitioners using cognitive and behavioral approaches to helping, although sensitive to relationship issues (Arnkoff, 1995), tend toward the means-to-an-end view. Overstressing the relationship is a mistake because it obscures the ultimate goal of helping a client manage a particular problem better. This goal cannot be achieved if the relationship is poor; but if too much emphasis is placed on the relationship itself, both client and helper can be distracted from the real work to be done.

**The relationship and outcomes.** Finally, some emphasize outcomes over both means and relationship. Practitioners from solution-focused approaches to helping (see de Shazer, 1985, 1994; Manthei, 1998; O'Hanlon & Weiner-Davis, 1989; Rowan, O'Hanlon, & O'Hanlon, 1999) tend to focus not on relationships as ends or means—though, if pushed, they would categorize relationships as means—but on what clients need to do right away to begin to remedy the problem situations they face. In their eyes, spending a great deal of time exploring the exact character of the problem and its roots is a waste of time. Helping tends to be time limited. Therefore, "Let's get working on this right away" is part of the pragmatics of solution-focused helping.

## THE RELATIONSHIP AS A
## WORKING ALLIANCE

The term *working alliance*, first coined by Greenson (1967) and now used by advocates of different schools of helping, can be used to bring together the best of the relationship-in-itself, relationship-as-means, and solution-focused approaches. Bordin (1979) defines the working alliance as the collaboration between the client and the helper based on their agreement on the goals and tasks of counseling. Although there is, predictably, considerable disagreement among practitioners as to what the critical dimensions of the working alliance are, how it operates, and what results it produces (see Hill & Williams, 2000; Horvath, 2000; Weinberger, 1995), it is relatively simple to outline what it means in the context of the problem-management and opportunity-development process.

**The collaborative nature of helping.** In the working alliance, helpers and clients are collaborators. Helping is not something that helpers do to clients; rather, it is a process that helpers and clients work through together. Helpers do not "cure" their patients. Both have work to do in the problem-management and opportunity-development stages and steps, and both have responsibilities related to outcomes. Outcomes depend on the competence and motivation of the helper, on the competence and motivation of the client, and on the quality of their interactions. Helping is a two-person team effort in which helpers need to do their part and clients theirs. If either party refuses to play or plays incompetently, the entire enterprise can fail.

**The relationship as a forum for relearning.** Even though helpers don't cure their clients, the relationship itself can be therapeutic. In the working alliance, the relationship itself is often a forum or vehicle for social-emotional relearning (Mallinckrodt, 1996). Effective helpers model attitudes and behavior that help clients challenge and change their own attitudes and behavior. It is as if a client were to say to himself (though not in so many words), "She [the helper] obviously cares for and trusts me, so perhaps it is all right for me to care for and trust myself." Or, "He takes the risk of challenging me, so what's so bad about challenge when it's done well?" Or, "I came here frightened to death by relationships, and now I'm experiencing a nonexploitative relationship that I cherish." Furthermore, protected by the safety of the helping relationship, clients can experiment with different behaviors during the sessions themselves. The shy person can speak up, the reclusive person can open up, the aggressive person can back off, the overly sensitive person can ask to be challenged, and so forth.

These learnings can then be transferred to other social settings. It is as if a client might say to himself, "He [the helper] listens to me so carefully and makes sure that he understands my point of view even when he thinks I should reconsider it. My relationships outside would be a lot different if I were to do the same." Or, "I do a lot of stuff in the sessions that would make anyone angry. But she doesn't let herself become a victim of emotions, either her own or mine. And her self-control doesn't diminish her humanity at all. That would make a big difference in my life." The relearning dynamic, however subtle or covert, is often powerful. In sum, needed changes in both attitudes and behavior often take place within the sessions themselves through the relationship.

**Relationship flexibility.** The idea that one kind of perfect relationship or alliance fits all clients is a myth. Different clients have different needs, and those needs are best met through different kinds of relationships and different modulations within the same relationship. One client might work best with a helper who expresses a great deal of warmth, whereas another might work best with a helper who is more objective and businesslike. Some clients come to counseling with a fear of intimacy and can be put off if helpers, right from the beginning, communicate a great deal of empathy and warmth. Once these clients learn to trust their helpers, stronger interventions can be used. Effective helpers use a mix of styles, skills, and techniques tailored to the kind of relationship that is right for each client (Lazarus, 1993; Mahrer, 1993). And they remain themselves while they do so.

We should neither underestimate nor overestimate the importance of the helping relationship. Helpers would do well to stay in touch with what the relationship means to each client, no matter what the literature says. It certainly does contribute to outcomes, but in the end it is one among a number of key variables (Albano, 2000).

## VALUES IN ACTION

One of the best ways to characterize a helping relationship is through the values that should permeate and drive it. The relationship is the vehicle through which values come alive. Values, expressed concretely through working-alliance behaviors, play a critical role in the helping process (Bergin, 1991; Beutler & Bergan, 1991; Kerr & Erb, 1991; Norcross & Wogan, 1987; Vachon & Agresti, 1992). Since it has become increasingly clear that helpers' values influence clients' values over the course of the helping process, it is essential to build a value orientation into the process itself.

### Putting Values into the Broader
### Context of Personal Culture

Values are central to culture, but culture is a wider reality. The bigger picture—the one that applies to societies and various subgroupings such as associations and organizations—is, briefly, this: Shared beliefs and assumptions interact with shared values and produce shared norms that drive shared patterns of behavior. However, counselors don't deal with societies as such but with individuals. So let's apply this basic culture framework to an individual. It goes something like this:

- Over the course of life, individuals develop *assumptions* and *beliefs* about themselves, other people, and the world around them. For instance, Isaiah, a client suffering from posttraumatic stress disorder stemming from being brutally attacked and witnessing gang activity in his neighborhood, has come to believe that the world is a heartless place.
- *Values*, what people prize, are picked up or inculcated along the path of life. Isaiah, because of a number of ups and downs in his life, has come to value or prize personal security.
- Assumptions and beliefs, interacting with values, generate *norms*—the "dos and don'ts" we carry around inside ourselves. For Isaiah, one of these is, "Don't trust people. You'll get hurt."
- These norms drive *patterns of behavior*, and these patterns of behavior constitute, as it were, the *bottom line* of personal or individual culture—"the way I live my life." For Isaiah, this means not taking chances with people. He's a loner.

Effective helpers come to understand the personal cultures of clients and the impact these individual cultures have both in everyday life and in helping sessions. Of course, since no individual is an island, personal cultures do not develop in a vacuum. The beliefs, values, and norms people develop are greatly influenced by their environments.

Culture is usually not applied to individuals but rather to societies, institutions, companies, professions, groups, and families. *Shared* assumptions and beliefs and interaction with *shared* values produce *shared* norms that drive *shared* patterns of behavior. That said, individuals within any given culture can and often do differ widely in their personal cultures. Even though individuals are deeply influenced by both biological and cultural inheritance, over the life span, influenced by both their social environments and their inner lives, they pick and choose their interests, values, and activities, thus creating their own personal cultures (Massimini & Delle Fave, 2000). For instance, Sally has many of the cultural characteristics of the Smith family, but she is still quite different from her parents, brothers, and sisters and from those in the subculture to which she belongs.

Since patterns of behavior constitute the "bottom line" of culture, a popular definition of societal and institutional culture is "the way we do things here." This definition applied to the helper is "the way I do helping." Inevitably, the helper's personal culture interacts with the client's, for better or for worse. The way helpers together with their clients "do" helping constitutes the culture of helping. The focus in this chapter, directly, is on the values of helping and the norms they generate. Indirectly, this entire book is about the culture of helping—that is, the beliefs, values, and norms that can and should drive the helping process.

## The Pragmatics of Values

Values are not just ideals. They are also a set of practical criteria for making decisions. As such, they are drivers of behavior. For instance, a helper might say to himself or herself during a session with a difficult client something like this:

The arrogant, I'm-always-right attitude of this client needs to be challenged. How I challenge her is important, since I don't want to damage our relationship. I value genuineness and openness. Therefore, I can challenge her by describing her behavior and the impact it has on me and might have on others, and I can do so respectfully, that is, without belittling her.

Working values enable the helper to make decisions on how to proceed. Helpers without a set of working values are adrift. Helpers who don't have an explicit set of values have an implicit or "default" set that may or may not serve the helping process.

Helping-related values, like your other values, cannot be handed to you on a platter. Much less can they be shoved down your throat. Therefore, this chapter is meant to stimulate your thinking about the values that should drive helping. In the final analysis, as you sit with your clients, only those beliefs, values, and norms that you have made your own will make a difference in your helping behavior. Therefore, you need to be proactive in your search for the beliefs, values, and norms that will govern your interactions with your clients.

This does not mean that you will invent a set of values different from everyone else's. Tradition is an important part of value formation, and we all learn from the rich tradition of the helping professions. In this chapter, four major values from the tradition of the helping professions—respect, empathy, genuineness, and client empowerment—are translated into a set of norms. Respect is the *foundation* value; empathy is the value that *orients* helpers in their interactions with clients; genuineness is the "what you see is what you get" *professional* value; client empowerment is the value that drives *outcomes*. They serve as a starting point for your reflection on the values that should drive the helping process. Don't just swallow them. Analyze, reflect on, and debate them.

## RESPECT AS THE FOUNDATION VALUE

Respect for clients is the foundation on which all helping interventions are built. Respect is such a fundamental concept that, like most such concepts, it eludes definition. The word comes from a Latin root that includes the idea of seeing or viewing. Indeed, respect is a particular way of viewing oneself and others. Respect, if it is to make a difference, cannot remain just an attitude or a way of viewing others. Here are some norms that flow from the interaction between a belief in the dignity of the person and the value of respect.

**Do no harm.**  This is the first rule of the physician and the first rule of the helper. Yet some helpers do harm either because they are unprincipled or because they are incompetent. Helping is not a neutral process—it is for better or worse. In a world in which such things as child abuse, wife battering, and exploitation of workers is much more common than we care to think, it is important to emphasize a nonmanipulative and nonexploitative approach to clients. Studies show that some instructors exploit trainees both sexually and in other ways and that some helpers do the same with their clients. Such behavior obviously breaches the code of ethics espoused by all the helping professions.

**Be competent and committed.**  Become good at whatever model of helping you use. Learn the basic problem-management and opportunity-development framework outlined in this book and fine-tune the skills that make it work. There is no place for the "caring incompetent" in the helping professions. It would be great to say that everyone who graduates from some kind of helping training program is not only competent but also increases his or her competence over his or her career. But that is not the case.

**Make it clear that you are "for" the client.** The way you act with clients will tell them a great deal about your attitude toward them. Your manner should indicate that you are "for" each of your clients, that you care for each in a down-to-earth, nonsentimental way. It is as if you are saying to the client attitudinally and behaviorally, "Working with you is worth my time and energy." Respect is both gracious and tough minded. Being for the client is not the same as taking the client's side or acting as the client's advocate. *Being for* means taking the client's point of view seriously even when it needs to be challenged. Respect often involves helping clients place demands on themselves. "Tough love" in no way excludes appropriate warmth toward clients.

**Assume the client's goodwill.** Work on the assumption that clients want to work at living more effectively, at least until that assumption is proved false. The reluctance and resistance of some clients, particularly involuntary clients, is not necessarily evidence of ill will. When you respect your clients, you are willing to enter their world to understand their reluctance and to help them work through it.

**Do not rush to judgment.** You are not there to judge clients or to shove your values down their throats. You are there to help them identify, explore, and review and challenge the consequences of the values they have adopted. Suppose a client, during the first session, says somewhat arrogantly, "When I'm dealing with other people, I say whatever I want, when I want. If others don't like it, well, that's their problem. My first obligation is to myself, being the person I am." A helper, irked by the client's attitude, might respond judgmentally by saying, "You've just put your finger on the core of your problem! How can you expect to get along with people with this kind of self-centered philosophy?" However, another counselor, taking a different approach, might respond, "So being yourself is one of your top priorities and being totally frank is, for you, part of that picture." The first counselor rushes to judgment. The second neither judges nor condones; at this point, she tries to understand the client's point of view and lets him know that she understands—even if she realizes that his point of view needs to be reviewed and challenged later.

**Keep the client's agenda in focus.** Helpers should pursue their clients' agendas, not their own. Here are three examples of helpers who lost clients because of lack of appreciation of the clients' agendas: One helper recalled, painfully, that he lost a client because he had become too preoccupied with his theories of depression rather than the client's painful depressive episodes. Another helper, who dismissed as trivial a client's grief over her pet's death, was dumbfounded and crushed when the client made an attempt on her own life. Later, the client related her "gesture," in part, to the loss of her pet. A third helper, a white male who prided himself on his multicultural focus in counseling, went for counseling himself when a Hispanic client quit therapy, saying, perhaps somewhat unfairly, as he was leaving, "I don't think you're interested in me. You're more interested in Anglo-Hispanic politics."

# Empathy as a Primary Orientation Value

Empathy, though a rich concept in the helping professions, has been a confusing one (see Bohart & Greenberg, 1997, and Duan & Hill, 1996, for overviews). Different theoreticians and researchers have defined it different ways. Some have seen it as a *personality trait*, a disposition to feel what other people feel or to understand others "from the inside," as it were. In this view, some people are by nature more empathic than others. Others have seen empathy, not as a personality trait, but as a situation-specific *state* of feeling for and understanding of another person's experiences. The implication is that helpers can learn how to bring about this state in themselves because it is so useful in the counseling process. Still others, building on the state approach, have focused on empathy as a *process* with stages. For instance, Barrett-Lennard (1981) identifies three phases: empathic resonance, expressed empathy, and received empathy; and Carl Rogers (1975) talks about sensing a client's inner world and communicating that sensing. Finally, Egan (1998) focuses on empathy as an *interpersonal communication skill*. Skilled helpers work hard at understanding their clients and then communicate this understanding to help clients understand themselves, their problem situations, their unused resources and opportunities, and their feelings more fully so they can then manage them more effectively.

## The Nature of Empathy

This edition of *The Skilled Helper* clarifies and simplifies the approach to empathy. In this chapter, empathy is seen as a basic *value* that informs and drives *all* helping behavior. Empathy as a communication skill is renamed, discussed, and illustrated in Chapter 6.

**A rich term.** A number of authors look at empathy from a value point of view and talk about the behaviors that flow from it. Sometimes their language is almost lyrical. For instance, Kohut (1978) states, "Empathy, the accepting, confirming, and understanding human echo evoked by the self, is a psychological nutrient without which human life, as we know and cherish it, could not be sustained" (p. 705). To Kohut, empathy is a value, a philosophy, or a cause with almost religious overtones. With empathy apparently in rather short supply, it might be safer to say that life is fuller because of mutual empathy. Covey (1989), naming empathic communication one of the "seven habits of highly effective people," says that empathy provides those with whom we are interacting with "psychological air" that helps them breathe more freely in their relationships. Goleman (1995, 1998) puts empathy at the heart of emotional intelligence; it is the individual's "social radar" through which he or she senses others' feelings and perspectives and takes an active interest in their concerns.

Rogers (1980) talks passionately about basic empathic listening—being with and understanding the other—calling it "an unappreciated way of being" (p. 137). He uses the word *unappreciated* because, in his view, few people in the general population have developed this "deep listening" ability, and even so-called expert helpers do not give it the attention it deserves. Here is his description of basic empathic listening:

It means entering the private perceptual world of the other and becoming thoroughly at home in it. It involves being sensitive, moment by moment, to the changing felt meanings which flow in this other person, to the fear or rage or tenderness or confusion or whatever that he or she is experiencing. It means temporarily living in the other's life, moving about in it delicately without making judgments. (p. 142)

Such empathic listening is selfless because helpers must put aside their own concerns to be fully with their clients.

**A key helping value.**  Empathy as a value is a radical commitment on the part of helpers to understand clients as fully as possible in three different ways. First, empathy is a commitment to work at understanding each client from *his or her point of view* together with the feelings surrounding this point of view and to communicate this understanding whenever it is deemed helpful. Second, empathy is a commitment to understand individuals in and through the *context* of their lives. The social settings, both large and small, in which they have developed and currently "live and move and have their being" provide routes to understanding. Third, empathy is a commitment to understand the *dissonance* between the client's point of view and reality. But, as Goleman (1995, 1998) notes, there is nothing passive about empathy. Empathic helpers respectfully communicate these three kinds of understanding to their clients and generally take an active interest in their concerns.

## Empathy—Understanding Clients as They Are: Diversity and Multiculturalism

Although dealing knowledgeably and sensitively with diversity—and that particular form of diversity called multiculturalism—is part of both respect and empathy, it is given special attention here because of the current emphasis on diversity in both the workplace and the helping professions. There has been an explosion of literature on diversity and multiculturalism over the past few years (see Axelson, 1999; Bernstein, 1994; Cuellar & Paniagua, 2000; Das, 1995; Herman & Kempen, 1998; Hogan-Garcia, 1999; Ivey & Ivey, 1999; Ivey, Ivey, & Simek-Morgan, 1997; Lee, 1997; Okun, Fried & Okun, 1999; Patterson, 1996; Pedersen, 1994, 1997; Ponterotto, Fuertes, & Chen, 2000; Richards & Bergin, 2000; Sue, Carter, Casas, & Fouad, 1998; Sue, Ivey, and Pedersen, 1996; Sue & Sue, 1999; Weinrach & Thomas, 1996; to name but a very few). There is both an upside and a downside to this avalanche. One upside is that helpers are forced to take another look at the blind spots they may have about diversity and culture and to take another look at the world in which we live. One downside is that multiculturalism has become in many ways a fad, if not an industry.

Let's start with an example. Sue, a midwestern American, is married to Lee, an immigrant from Singapore. They are having problems. Many clients come to helpers because they are having difficulties in their relationships with others or because relationship difficulties are part of a larger problem situation. Therefore, understanding clients' different approaches to developing and sustaining relationships is important. Guisinger and Blatt (1994) put this in a broader multicultural perspective.

Western psychologies have traditionally given greater importance to self-development than to interpersonal relatedness, stressing the development of autonomy, independence, and identity as central factors in the mature personality. In contrast, women, many minority groups, and non-Western societies have generally placed greater emphasis on issues of relatedness. (p. 104)

Sue is deep into the development of autonomy, independence, and her identity as a successful working woman. Lee runs a small, successful Web site development business. Guisinger and Blatt go on to point out that both interpersonal relatedness and self-definition are essential for maturity. Helping Sue and Lee, individuals from different cultures, achieve balance depends on understanding what the "right balance" means in any given culture.

As with the "Does helping help?" debate, it is important to come to grips with the debate surrounding diversity and multiculturalism. Some of the literature on diversity and multiculturalism is informative and challenging; some is ridiculous and infuriating. Many professionals have pointed out that both the differences among and the needs of minority groups—whether race, ethnicity, disability, or some other kind of difference is at issue—and the contributions such groups make to society have been systematically ignored or misunderstood. This is an important social problem that has implications for the helping professions. However, the relationship of counseling to social movements is confusing and difficult.

**Understand diversity.** While clients have in common their humanity, they differ from one another in a whole host of ways: accent, age, attractiveness, color, developmental stage, abilities, disabilities, economic status, education, ethnicity, fitness, gender, group culture, health, national origin, occupation, personal culture, personality variables, politics, problem type, religion, sexual orientation, social status, to name some of the major categories. We differ from one another in hundreds of ways. And who is to say which differences are key? This presents several challenges for helpers. For one, it is essential that helpers understand clients and their problem situations contextually. For instance, a life-threatening illness might be one kind of reality for a 20-year-old and quite a different reality for an 80-year-old. We know that homelessness is a complex phenomenon. A homeless client with a history of drug abuse who has dropped out of graduate school is far different from a drifter who hates shelters for the homeless and resists every effort to get him to go to them.

It is true that helpers can, over time, come to understand a great deal about the characteristics of the populations with whom they work; for instance, they can and should understand the different developmental tasks and challenges that take place over the life span, and if they work with the elderly, they should grow in their understanding of the challenges, needs, problems, and opportunities of the aged. Still, it is impossible to know everything about every population. This impossibility becomes even more dramatic if the combinations and permutations of characteristics are taken into consideration. How could an African American, middle-class, highly educated, younger, urban, Episcopalian, female psychologist possibly understand a poor, unemployed, homeless, middle-aged, uneducated, lapsed-Catholic male, born of migrant workers: a Mexican father and a Polish mother? Indeed, how can anybody fully understand anybody else? If the legitimate principles relating to diversity were to be pushed too far, no one would be able to understand and help anybody else.

**Challenge whatever blind spots you may have.** Since helpers often differ from their clients in many ways, they need to challenge themselves to avoid diversity-related blind spots that can lead to inept interactions and interventions during the helping process. For instance, a physically attractive and extroverted helper might have blind spots with regard to the social flexibility and self-esteem of a physically unattractive and introverted client. Much of the literature on diversity and multi-culturalism targets such blind spots. Counselors would do well to become aware of their own cultural values and biases. They should also make every effort to understand the world views of their clients. Helpers with diversity blind spots are handicapped. Helpers should, as a matter of course, become aware of the key ways in which they differ from their clients and take special care to be sensitive to those differences.

**Tailor your interventions in a diversity-sensitive way.** Both this self-knowledge and this practical understanding of diversity need to be translated into appropriate interventions. The way a Hispanic helper challenges a Hispanic client may be inappropriate for a white client and vice versa. The way a younger helper shares his own experience with a younger client might be inappropriate for a client who is older and vice versa. Client self-disclosure, especially more intimate disclosure, might be relatively easy for a person from one culture, say North American culture, but very difficult for a client from another culture, say Asian or British. In this case, interventions that call for intimate self-disclosure may be seen as inappropriate by such a client. Even though a client may be from a culture that is more open to self-disclosure, he or she may be frightened to death by self-disclosure. Therefore, with clients who come from a culture that has a different perception of self-disclosure or with any client who finds self-disclosure difficult, it might make more sense, after an initial discussion of the problem situation in broad terms, to move to what the client wants that he or she currently does not have (that is, Stage II) rather than to the more intimate details of the problem situation. Once the helping relationship is on firmer ground, the client can move to the work he or she sees as more intimate or demanding. Although the helping model outlined here is a "human universal," helpers need to apply its stages and steps with sensitivity.

**Work with individuals.** The diversity principle is clear: The more helpers understand the broad characteristics, needs, and behaviors of the populations with whom they work—African Americans, Caucasian Americans, diabetics, the elderly, drug addicts, the homeless, you name it—the better positioned they are to adapt these broad parameters and the counseling process itself to the individuals with whom they work. But, whereas diversity focuses on differences both between and within groups—cultures and subcultures, if you will—helpers interact with clients as individuals. Your clients are individuals, not cultures, subcultures, or groups. Remember that category traits can destroy understanding as well as facilitate it.

Of course, individuals often have group characteristics, but they do not come as members of a homogeneous group, because there are no homogeneous groups. One of the principal learnings of social psychology is this: There are as many differences, and sometimes more, within groups as between groups (see Weinrach & Thomas, 1996, pp. 473–474). This middle-class black male is this individual. This poor Asian woman is this person. In a very real sense, a conversation between identical twins is a cross-cultural event because they are different individuals with differences in personal assumptions, beliefs, values, norms, and patterns of behavior. Genetics and group culture account for commonalities among individuals, but personhood and personal cultures emphasize each person's uniqueness. Focusing excessively on what makes this client different can be just as injurious as ignoring differences.

Finally, valuing diversity is not the same as espousing a splintered, antagonistic society in which one's group membership is more important than one's humanity. On the other hand, valuing individuality is not the same as espousing a "society of one"—radical individualism being the ultimate form of diversity. Moving to a "society of one" makes counseling and other forms of human interaction impossible. Take Sean, a client you are seeing for the second time. He is very bright, well spoken, gay, Hispanic, poorly educated, lower-middle-class, slight of build, indifferent to his Catholic heritage, a churchgoer, underemployed, good-looking, honest, and at sea because he feels "defeated." At his age, 26, life should be opening up, but he feels that it is closing down. He feels trapped. Understand this individual in any way you can, but work with Sean.

## Guidelines for Integrating Diversity and Multiculturalism into Counseling

Since it is impossible to lay down rules for every possible case in which diversity is an issue, some broad guidelines are called for. The norms just outlined can serve as guidelines, but ultimately, you have to come to grips with diversity and pull together your own set of guidelines. Here is one set, drawn from an article by Weinrach and Thomas (1996, pp. 475–476) but reworded and reworked a bit:

- Place the needs of the client above all other considerations.
- Identify and focus on whatever frame of reference, self-definition, or belief system is central to any given client, with consideration for, but not limited to, issues of diversity.
- Select counseling interventions on the basis of the client's agenda. Do not impose a social or political agenda on the counseling relationship.
- Make sure that your own values do not adversely affect a client's best interests.
- Avoid cultural stereotyping. Do not overgeneralize. Recognize that within-group differences are often more extensive than between-group differences.
- Do not define diversity narrowly. This client's concern about being unattractive deserves the helper's engagement just as much as that client's concern about racial intolerance.
- Provide opportunities for practitioners to be trained in the working knowledge and skills associated with diversity-sensitive counseling.

- Subject the assumptions, models, and techniques of diversity-sensitive counseling to the same scrutiny as other aspects of the counseling profession.
- Create an environment that supports professional tolerance.

The fact that not all practitioners would agree with this package highlights the importance of your coming to grips not only with diversity but with the whole range of value questions that permeate helping.

In one way, the diversity and multiculturalism debate does a disservice to the helping professions (see Weinrach & Thomas, 1998 for a fine, balanced critique). Many helpers feel scolded—often by their peers who have taken it on themselves to speak for others, on the assumption, perhaps, that others cannot speak for themselves. When it comes to clients, the very best helpers have always been learners. They instinctively know that they are different in many ways from their clients and they know that these differences can get in the way. They instinctively know that they cannot know everything about everyone but don't find that fact self-defeating. They strive to understand the world from each client's perspective. But they don't apologize for who they are. Why should they? Cultural understanding—or understanding of any form of diversity—is a two-way street. The principles of cultural understanding apply to everyone. I must understand you in context, but it's your job to understand me in context. If helping is to be a collaborative event, mutual understanding must be part of the game.

## GENUINENESS AS A PROFESSIONAL VALUE

Like respect, helper genuineness refers to both a set of attitudes and a set of counselor behaviors. Some writers call genuineness "congruence." Genuine people are at home with themselves and therefore can comfortably be themselves in all their interactions. Being genuine has both positive and negative implications; it means doing some things and not doing others.

**Do not overemphasize the helping role.**  Genuine helpers do not take refuge in the role of counselor. Ideally, relating at deeper levels to others and helping are part of their lifestyles, not roles they put on or take off at will. This keeps them far away from being patronizing and condescending. Years ago, Gibb (1968, 1978) suggested ways of being "role free." He said that helpers should learn how to

- express directly to another whatever they are presently experiencing;
- communicate without distorting their own messages;
- listen to others without distorting the messages they hear;
- reveal their true motivation in the process of communicating their messages;
- be spontaneous and free in their communications with others, rather than using habitual and planned strategies;
- respond immediately to another's need or state instead of waiting for the "right" time or giving themselves enough time to come up with the "right" response;
- manifest their vulnerabilities and, in general, the "stuff" of their inner lives;

- live in and communicate about the here and now;
- strive for interdependence rather than dependence or counterdependence in their relationships with their clients;
- learn how to enjoy psychological closeness;
- be concrete in their communications;
- be willing to commit themselves to others.

By this, Gibb did not mean that helpers should be "free spirits," inflicting themselves on others. Indeed, free-spirit helpers can even be dangerous. Being role free is not license. Freedom from role means that counselors should not use the role or facade of counselor to protect themselves, to substitute for competence, or to fool clients in other ways.

**Be spontaneous.** Many of the behaviors suggested by Gibb are ways of being spontaneous. Effective helpers, while being tactful as part of their respect for others, do not constantly weigh what they say to clients. They do not put a number of filters between their inner lives and what they express to others. On the other hand, being genuine does not mean verbalizing every thought to clients.

**Avoid defensiveness.** Genuine helpers are nondefensive. They know their own strengths and deficits and are presumably trying to live mature, meaningful lives. When clients express negative attitudes toward them, they examine the behavior that might cause clients to think negatively, try to understand the clients' points of view, and continue to work with them. Consider the following example:

CLIENT: I don't think I'm really getting anything out of these sessions at all. I still feel drained all the time. Why should I waste my time coming here?

HELPER A: If you were honest with yourself, you'd see that you are the one wasting time. Change is hard, and you keep putting it off.

\* \* \* \*

HELPER B: Well, that's your decision.

Helpers A and B are both defensive, though in different ways. The client is more likely to react to their defensiveness than move forward.

HELPER C: So from where you're sitting, there's no payoff for being here. Just a lot of dreary work and nothing to show for it.

Helper C centers on the experience of the client, with a view to "resetting the system" and helping her explore her responsibility for making the helping process work. Since genuine helpers are at home with themselves, they can allow themselves to examine negative criticism honestly. Helper C, for instance, would be the most likely of the three to ask himself or herself whether he or she is contributing to the apparent stalemate.

**Be open.** Genuine helpers are capable of deeper levels of self-disclosure even within the helping relationship. They do not see self-disclosure as an end in itself, but they feel free to reveal themselves, even in deeper ways, when and if it is appropriate. Being open also means that the helper has no hidden agendas: "What you see is what you get."

# CLIENT EMPOWERMENT AS AN
# OUTCOME VALUE

The second goal of helping, outlined in Chapter 1, deals with empowerment—that is, helping clients identify, develop, and use resources that will make them more effective agents of change both within the helping sessions themselves and in their everyday lives (Strong, Yoder, & Corcoran, 1995). The opposite of empowerment is dependency (Abramson, Cloud, Keese, & Keese, 1994; Bornstein & Bowen, 1995), deference (Rennie, 1994), and oppression (McWhirter, 1996). Because clients often experience helpers as relatively powerful people, and because even the most egalitarian and client-centered of helpers do influence clients, it is necessary to come to terms with social influence in the helping process.

## Helping as a Social-Influence Process

People influence one another every day in every social setting of life. Smith and Mackie (2000) consider it one of eight basic principles needed to understand human behavior. William Crano (2000) suggests that "social influence research has been, and remains, the defining hallmark of social psychology" (p. 68). Parents influence each other and their kids. In turn, they are influenced by their kids. Teachers influence students, and students influence teachers. Bosses influence subordinates and vice versa. Team leaders influence team members, and members influence both one another and the leader. The world is abuzz with social influence. It could not be otherwise. However, social influence is a form of power, and power too often leads to manipulation and oppression.

It is not surprising, then, that helping as a social-influence process has received a fair amount of attention in the helping literature (see Dorn, 1986; Heppner & Claiborn, 1989; Heppner & Frazier, 1992; Houser, Feldman, Williams, & Fierstien, 1998; Hoyt, 1996; McCarthy & Frieze, 1999; McNeill & Stolenberg, 1989; Strong, 1968, 1991; Tracey, 1991). Helpers can influence clients without robbing them of self-responsibility. Even better, they can exercise their trade in such a way that clients are, to use a bit of current business jargon, "empowered" rather than oppressed, both in the helping sessions themselves and in the social settings of everyday life. With empowerment, of course, comes increased self-responsibility.

Imagine a continuum. At one end lies "directing clients' lives" and at the other "leaving clients completely to their own devices." Somewhere along that continuum is "helping clients make their own decisions and act on them." Most forms of helper influence will fall somewhere in between the extremes. Preventing a client from jumping off a bridge moves, understandably, to the controlling end of the continuum. On the other hand, simply accepting and in no way challenging a client's decision to put off dealing with a troubled relationship because he or she is "not ready" moves toward the other end. As Hare-Mustin and Marecek (1986) note, there is a tension between the right of clients to determine their own way of managing their lives and the therapist's obligation to help them live more effectively.

## Norms for Empowerment and Self-Responsibility

Helpers don't self-righteously "empower" clients. That would be patronizing and condescending. In a classic work, Freire (1970) warns helpers against making helping itself just one more form of oppression for those who are already oppressed. Effective counselors help clients discover, develop, and use the untapped power within themselves. Here, then, is a range of empowerment-based norms, some adapted from the work of Farrelly and Brandsma (1974).

**Start with the premise that clients can change if they choose.** Clients have more resources for managing problems in living and developing opportunities than they—or sometimes their helpers—assume. The helper's basic attitude should be that clients have the resources both to participate collaboratively in the helping process and to manage their lives more effectively. These resources may be blocked in a variety of ways or simply unused. The counselor's job is to help clients identify, free, and cultivate these resources. The counselor also helps clients assess their resources realistically so that their aspirations do not outstrip their resources.

**Do not see clients as victims.** Even when clients have been victimized by institutions or individuals, don't see them as helpless victims. The cult of victimhood is already growing too fast in society. Even if victimizing circumstances have diminished a client's degree of freedom—the abused spouse's inability to leave a deadly relationship—work with the freedom that is left.

Don't be fooled by appearances. One counselor trainer in a meeting with his colleagues dismissed a reserved, self-deprecating trainee with the words, "She'll never make it. She's more like a client than a trainee." Fortunately, his colleagues did not work from the same assumption. The woman went on to become one of the program's best students. She was accepted as an intern at a prestigious mental-health center and was hired by the center after graduation.

**Share the helping process with clients.** Clients, like helpers, can benefit from maps of the helping process. Clients should not have to buy "a pig in a poke." Helping should not be a "black box" for them. Clients have a right to know what they are getting into (Heinssen, 1994; Heinssen, Levendusky, & Hunter, 1995; Hunter, 1995; Manthei & Miller, 2000). How to clue clients into the helping process is another matter. Helpers can simply explain what helping is all about. A simple pamphlet outlining the stages and steps of the helping process can be of great help, provided that it is in language that clients can readily understand. Just what kind of detail will help will differ from client to client. Obviously, clients should not be overwhelmed by distracting detail from the beginning. Nor should highly distressed clients be told to contain their anxiety until helpers teach them the helping model. Rather, the details of the model can be shared over a number of sessions. There is no one right way. In my opinion, however, clients should be told as much about the model as they can assimilate.

**Help clients see counseling sessions as work sessions.** Helping is about client-enhancing change. Therefore, counseling sessions deal with exploring the need for change, the kind of change needed, creating programs of constructive change, engaging in change "pilot projects," and finding ways of dealing with obstacles to change. This is work, pure and simple. The search for and implementation of solutions can be arduous, even agonizing, but it can also be deeply satisfying, even exhilarating. Helping clients develop the "work ethic" that makes them partners in the helping process can be one of the helper's most formidable challenges. Some helpers go so far as to cancel counseling sessions until clients are "ready to work." Helping clients discover incentives to work is, of course, less dramatic and hard work in itself.

**Become a consultant to clients.** Helpers can see themselves as consultants hired by clients to help them face problems in living more effectively. Consultants in the business world adopt a variety of roles. They listen, observe, collect data, report observations, teach, train, coach, provide support, challenge, advise, offer suggestions, and even become advocates for certain positions. But the responsibility for running businesses remains with those who hire consultants. Therefore, even though some of the activities of consultants can be seen as quite challenging, the decisions are still made by managers. Consulting, then, is a social-influence process, but it is a collaborative one that does not rob managers of the responsibilities that belong to them. In this respect, it is a useful analogy to helping. The best clients, like the best managers, learn how to use their consultants to add value in managing problems and developing opportunities.

**Accept helping as a natural, two-way influence process.** Tyler, Pargament, and Gatz (1983) move a step beyond the consultant role to what they called the "resource collaborator role." Seeing both helper and client as people with defects, these researchers focus on the give-and-take that should characterize the helping process. In their view, either client or helper can approach the other to originate the helping process. The two have equal status in defining the terms of the relationship, in originating actions within it, and in evaluating both outcomes and the relationship itself. In the best case, positive change occurs in both parties.

Helping is a two-way street. Clients and therapists change one another in the helping process. Even a cursory glance at helping reveals that clients can affect helpers in many ways. For instance, Wei-Lian has to correct Timothy, his counselor, a number of times when Timothy tries to share his understanding of what Wei-Lian has said. For instance, at one point, when Timothy says, "So you don't like the way your father forces his opinions on you," Wei-Lian replies, "No, my father is my father and I must always respect him. I need to listen to his wisdom." The problem is that Timothy has been inadvertently basing some of his responses on his own cultural assumptions rather than on Wei-Lian's. When Timothy finally realizes what he is doing, he says to Wei-Lian, "When I talk with you, I need to be more of a learner. I'm coming to realize that Chinese culture is quite different from mine. I need your help."

**Focus on learning instead of helping.** Although many see helping as an education process, it is probably better characterized as a learning process. Effective counseling helps clients get on a learning track. Both the helping sessions themselves and the time between sessions involve learning, unlearning, and relearning. Howell (1982) gives us a good description of learning when he says that "learning is incorporated into living to the extent that viable options are increased" (p. 14). In the helping process, learning takes place when options that add value to life are opened up, seized, and acted on. If the collaboration between helpers and clients is successful, clients learn in very practical ways. They have more "degrees of freedom" in their lives as they open up options and take advantage of them. This is precisely what counseling helped Carlos do (see Chapter 2). He unlearned, learned, relearned, and acted on his learnings.

**Do not see clients as overly fragile.** Neither pampering nor brutalizing clients serves their best interests. However, many clients are less fragile than helpers make them out to be. Helpers who constantly see clients as fragile may well be acting in a self-protective way. Driscoll (1984) notes that too many helpers shy away from doing much more than listening early in the helping process. The natural deference many clients display early in the helping process (Rennie, 1994)—including their fear of criticizing the therapist, understanding the therapist's frame of reference, meeting the perceived expectations of the therapist, and showing indebtedness to the therapist—can send the wrong message to helpers. Clients early on may be fearful of making some kind of irretrievable error. This does not mean that they are fragile. Reasonable caution from helpers is appropriate, but it is easy to become overly cautious. Driscoll suggests that helpers intervene more right from the beginning—for instance, by reasonably challenging the way clients think and act and by getting them to begin to outline what they want and are willing to work for.

## A WORKING CHARTER:<br>A CLIENT-HELPER CONTRACT

Both implicit and explicit contracts govern the transactions that take place between people in a wide variety of situations, including marriage (where some but by no means all of the provisions of the contract are explicit) and friendship (where the provisions are usually implicit). For helping to be a collaborative venture, both parties must understand their responsibilities. Perhaps the term *working charter* is better than *contract*. It avoids the legal implications of the latter term and connotes a cooperative venture.

To achieve these objectives, the working charter should include, generically, the issues that are covered in Chapters 1 through 3: (a) the nature and goals of the helping process, (b) an overview of the helping model together with the techniques to be used and a sense of the flexibility built into the process, (c) how this process will help the client achieve her or his goals, (d) relevant information about yourself and your background, (e) how the relationship is to be structured and the kinds of responsibilities both you and the client will have, (f) the values that will drive the helping process, and (g) procedural issues. Procedural issues are the nuts and bolts of the helping process, such things as where sessions will be held and how long they will last. Procedural limitations should also be discussed—for instance, how free the client is to contact the helper between sessions: "Ordinarily we won't contact each other between sessions, unless we prearrange it for a particular purpose." However, key ground rules should not come as a surprise to clients. Manthei and Miller (2000) have written a practical book for clients on the elements of a working charter. Charters also work with the seriously mentally ill (Heinssen, Levendusky, & Hunter, 1995).

The working charter need not be too detailed, nor should it be rigid. The question is, How much structure will help this client at this time? The helper needs to provide structure for the relationship and the work to be done without frightening or overwhelming the client. Ideally, the working charter is an instrument that makes the client more informed about the process, more collaborative with the helper, and more proactive in managing his or her problems. At its best, a working charter can help the client and the helper develop realistic mutual expectations, give the client a flavor of the mechanics of the helping process, diminish initial client anxiety and reluctance, provide a sense of direction, and enhance the client's freedom of choice.

## SHADOW-SIDE REALITIES IN THE HELPING RELATIONSHIP

There are common flaws in the working alliance that remain in the shadows, either because they are not dealt with effectively by the helping professions themselves or because individual helpers are inept at addressing them with clients.

**Ethical flaws.** Little has been said about ethics in the helping process so far, not because it is not important but because it is so important. There is a vast amount of literature on ethical responsibilities in the helping professions (see Bersoff, 1995; Canter, Bennett, Jones, & Nagy, 1994; Claiborn, Berberoglu, Nerison, & Somberg, 1994; Corey, Corey, & Callanan, 1997; Cottone & Claus, 2000; Fisher & Younggren, 1997; Keith-Spiegel, 1994, Loman, 1998). There is also a growing literature on ways in which helpers violate their ethical responsibilities. Since this area is too vast and too important to be given summary treatment here, helpers-to-be are urged to make this part of their professional development program.

**Human tendencies in both helpers and clients.** Neither helpers nor clients are usually heroic figures. They are human beings with all-too-human tendencies. For instance, helpers find clients attractive or unattractive; there is nothing wrong with this. However, they must be able to manage closeness in therapy in a way that furthers the helping process (Schwartz, 1993). They must deal with both positive and negative feelings toward clients lest they end up doing silly things. They may have to fight the tendency to be less challenging with attractive clients or not to listen carefully to unattractive clients. Clients, too, have their tendencies. Some have unrealistic expectations of counseling (Tinsley, Bowman, & Barich, 1993), while others trip over their own distorted views of their helpers. In such cases, helpers have to manage both expectations and the relationship. Very often these human tendencies on the part of both client and helper are not center stage in awareness. Rather they constitute a subtext within the relationship. Unskilled helpers can get caught up in both their own and their clients' games, causing the working alliance to break down. Skilled helpers, on the other hand, understand the shadow sides of both themselves and their clients and manage them. Tools that helpers need to challenge themselves and their clients—for instance, the skill of immediacy—are discussed in Chapters 10, 11, and 12.

**Trouble in the relationship itself.** The helping relationship might be flawed from the beginning. That is, the fit or chemistry between helper and client might not be right. But, for a variety of reasons, it is not easy for a helper to say, "I don't think I'm the one for you." On the other hand, high-level helpers can work with a wide variety of clients. They create their own chemistry. They make the relationship work.

One coach/counselor in a work setting was asked to work with a very bright manager whose interpersonal style left much to be desired. But the relationship was troubled from the start. Early on it become clear to the coach that his client expected him to "say good things" about him to senior managers. The client also had a tendency to play "mind games" with the coach, saying things like, "I wonder what's going on in your mind right now. I bet you're thinking things about me that you're not telling me." Managing expectations and managing the relationship proved to be hard work. However, the coach knew enough about the company to realize that "style" was an issue for the senior team. Because the client was bright and innovative, promotion was a distinct possibility, but because of his style, promotion was probably "his to lose." The coach remained respectful and empathetic but challenged "the crap." This shocked the client because he had always been able to "win" in his encounters with subordinates and peers. He stopped playing games and eventually realized that becoming better at interpersonal relations had only an upside.

Even if the relationship starts off on the right foot, it can deteriorate (Arnkoff, 2000; Omer, 2000). In fact, some deterioration is normal. Kivlighan and Shaugnessy (2000) talk about the "tear-and-repair" phenomenon. Many therapeutic relationships start well, get into trouble, and then recover. Experienced helpers are not surprised by this. However, some helping relationships get caught up in what Binder and Strupp (1997) call "negative process." They suggest that the ability of therapists to establish and maintain a good alliance has been overestimated. Hostile interchanges between helpers and clients are common in all treatment models. When impasses and ruptures in the relationship take place, ineffective helpers get bogged down. Many helpers and clients lack both the skill and the will for repair (Watson & Greenberg, 2000). Factors associated with relationship breakdowns include "a client history of interpersonal problems, a lack of agreement between therapists and clients about the tasks and goals of therapy, interference in the therapy by others, transference, possible therapist mistakes, and therapist personal issues" (Hill, Nutt-Williams, Heaton, Thompson, & Rhodes, 1996, p. 207). If impasses and ruptures are not addressed, premature termination often takes place. When this happens, helpers predictably blame clients: "She wasn't ready," "He didn't want to work," "She was impossible," and so forth. Such helpers fail to create the right chemistry.

**Vague and violated values.** Helpers do not always have a clear idea of what their values are, or the values they say they hold—that is, their espoused values—do not always coincide with their actions. Values too often remain "good ideas" and are not translated into specific norms that drive helping behavior. For instance, even though helpers value self-responsibility in their clients, they see them as helpless, make decisions for them, and direct rather than guide. Often they do so out of frustration. Expediency leads them to compromise their values and then rationalize their compromises. "I blew up at a client today, but he really deserved it. It probably did more good than my unappreciated patience." I bet.

**Failure to share the helping process.** When it comes to sharing the helping process itself, some counselors are reluctant to let clients know what the process is all about. Of course, helpers who "fly by the seat of their pants" can't tell clients what it's all about because they don't know what it's all about themselves. Still others seem to think that knowledge of helping processes is secret or sacred or dangerous and should not be communicated to clients, even though there is no evidence to support such beliefs (Dauser, Hedstrom, & Croteau, 1995; Somberg, Stone, & Claiborn, 1993; Sullivan, Martin, & Handelsman, 1993; Winborn, 1977).

**Flawed contracts.** There is an extensive shadow side to both explicit and implicit contracts. Even when a contract is written, the contracting parties interpret some of its provisions differently. Over time they forget what they contracted to and differences become more pronounced. These differences are seldom discussed. In counseling, the client-helper contract has traditionally been implicit, even though the need for more explicit structure has been discussed for years (Proctor & Rosen, 1983). Because of this, the expectations of clients may differ from the expectations of their helpers (Benbenishty & Schul, 1987). Implicit contracts are not enough, but they still abound (Handelsman & Galvin, 1988; Weinrach, 1989; Woody, 1991).

**Warring professionals.** There are not just debates but also conflicts close to internecine wars in the helping professions. For instance, the debate on the "correct" approach to diversity and multiculturalism brings out some of the best and some of the worst in the helping community. Accusations, however subtle or blatant, of cultural imperialism on the one side and "political correctness" on the other fly back and forth. The debate on whether or how the helping professions should take political stands or engage in social engineering generates, as has been noted, more heat than light. No significant article is published about any significant dimension of counseling without a barrage of often testy replies. What happened to learning from one another and integration? The search for the truth gives way at times to the need to be right. It is not always clear how all of this serves the needs of clients. Indeed, clients are often enough left out of the debate. Just as many businesses today are reinventing themselves by starting with their customers and markets, so the helping professions should continually reinvent themselves by looking at helping through the eyes of clients.

# THE THERAPEUTIC DIALOGUE

In Part Two, the basic communication skills needed to be an effective helper are reviewed and illustrated. These skills are integrated under the rubric of the therapeutic dialogue. There are less high sounding names than therapeutic dialogue—the helping dialogue, the problem-management dialogue, the opportunity-development dialogue. But dialogue is at the heart of the communication between helper and client.

Chapter 4 includes an overview of both interpersonal communication and dialogue together with the basic skill often called "attending" but now called, more pragmatically, "visibly tuning in" to clients. This skill focuses on the helper's empathic presence to the client. Chapter 5 outlines the skill of active listening. Helpers visibly tune in, not only to demonstrate their solidarity with their clients but also to understand what their clients are saying both directly and indirectly. Finally, Chapter 6 deals with the skill helpers need both to check out and to share their understanding with clients. This skill, called "basic empathy" in previous editions, has been renamed "sharing empathic highlights" to distinguish it from empathy as a value that should permeate all helping skills.

# INTRODUCTION TO COMMUNICATION AND THE SKILL OF VISIBLY TUNING IN TO CLIENTS

THE IMPORTANCE OF DIALOGUE IN HELPING

VISIBLY TUNING IN TO CLIENTS: THE IMPORTANCE OF EMPATHIC PRESENCE

   Nonverbal Behavior as a Channel of Communication

   Helpers' Nonverbal Behavior

   The Skill of Visibly Tuning in to Clients

      S: Face the client *Squarely*

      O: Adopt an *Open* posture

      L: Remember that it is possible at times to *Lean* toward the other

      E: Maintain good *Eye* contact

      R: Try to be relatively *Relaxed* or natural in these behaviors

THE SHADOW SIDE OF COMMUNICATION SKILLS

## THE IMPORTANCE OF DIALOGUE IN HELPING

Conversations between helpers and their clients should be a therapeutic or helping *dialogue*. Interpersonal communication competence means not only being good at the individual communication skills outlined in this and following chapters but also marshaling them at the service of dialogue. There are four requirements for true dialogue (Egan, in press):

- *Turn taking.* Dialogue is interactive. You talk, then I talk. In counseling, this means that, generally speaking, monologues on the part of either client or helper don't add value. On the other hand, turn taking opens up the possibility for mutual learning. Helpers learn about their clients and base their interventions on what they come to understand through the give-and-take of the dialogue. Clients come to understand themselves and their concerns more fully and learn how to face up to the challenge their problems and opportunities present.

- *Connecting.* What each person says in the conversation should be connected in some way to what the other person has said. The helper's comments should be connected to the client's remarks and, ideally, vice versa. Helper and client need to engage each other if their working alliance is to be productive.

- *Mutual influencing.* Each party in a dialogue should be open to being influenced by what the other person says. This echoes the social-influence dimension of counseling discussed in Chapter 3. Helpers certainly influence their clients, and the best helpers learn from and are influenced by their clients. Therefore, counselors need to be open-minded and help their clients to be open to new learning.

- *Cocreating outcomes.* Good dialogue leads to outcomes that benefit both parties. As we have seen, counseling is about results, accomplishments, outcomes. The job of the counselor is neither to tell clients what to do nor merely to leave them to their own devices. The counselor should act as a catalyst for the kind of problem-managing dialogue that helps clients find their own answers. In true dialogue, neither party should know exactly what the outcome will be. If the counselor knows what he or she is going to tell a client, or if the client has already decided what he or she is going to say and do, the two of them may well have a conversation, but it is probably not a dialogue. Only clients can change themselves, but because of the helping dialogue, these changes will have the mark of effective helpers on them.

Dialogue is beginning to be discussed in mental health (Corrigan, Lickey, Schmook, Virgil, & Juricek, 1999) and other human service settings such as medicine (Hellstroem, 1998). Dialogue is essential because helping is a collaborative endeavor (Roberts, 1998). It is through dialogue that helpers act as catalysts for change. It is through dialogue that clients give expression to their responsibility and accountability for change in their lives. Bugas and Silberschatz (2000) even see clients as "coaches" for their helpers. Clients should "prompt, instruct, and educate" their helpers to key aspects of themselves and their plans to accomplish treatment goals. If helpers and clients alike are to avoid the kind of "negative process" mentioned in Chapter 3, they must engage in transparent dialogue.

While individual communication skills are a necessary part of communication competence, dialogue is the integrating mechanism. Individual skills are the building blocks for effective dialogue. The first skill—called attending or, more colloquially, visibly tuning in—is discussed and illustrated in this chapter. Chapter 5 focuses on active listening. Empathy in the form of sharing empathic highlights is the topic in Chapter 6. Finally, probing and summarizing, the last of the basic communication skills, are dealt with in Chapter 7. The advanced communication skills needed to challenge clients or help them challenge themselves are outlined and illustrated in Chapters 10, 11, and 12. All these skills serve every stage and every step of the helping process.

These skills are not special skills peculiar to helping. Rather, they are extensions of the kinds of skills all of us need in our everyday interpersonal transactions. Ideally, helpers-to-be would enter training programs with this basic set of interpersonal communication skills in place, and training would simply help them adapt the skills to the helping process. Unfortunately, that is often not the case. Indeed, some of the problems clients have either focus on or are complicated by a lack of interpersonal communication skills.

Of course, effective helpers weave these communication skills together seamlessly in their interactions with clients. Communication skills need to become "second nature" if they are to serve the helping process and clients' needs. Bob Carkhuff (1987), Allen Ivey (Ivey & Ivey, 1999), and Carl Rogers (1951, 1957, 1965) were trailblazers in developing and humanizing communication skills and integrating them into the helping process. Their influence is seen throughout this book. In this text, these communication skills are described, outlined, and illustrated. The manual that accompanies this text, *Exercises in Helping Skills*, provides more extensive practice in all of them.

## Visibly Tuning in to Clients: The Importance of Empathic Presence

At some of the more dramatic moments of life, simply being with another person is extremely important. If a friend of yours is in the hospital, just your being there can make a difference, even if conversation is impossible. Similarly, being with a friend who has just lost his wife can be very comforting to him, even if little is said. Your empathic presence is comforting. Most people appreciate it when others pay attention to them. By the same token, being ignored is often painful: The averted face is too often a sign of the averted heart. Given how sensitive most of us are to others' attention or inattention, it is paradoxical how insensitive we can be at times about paying attention to others.

Helping and other deep interpersonal transactions demand a certain robustness or intensity of presence. Attending, that is, visibly tuning in to others, contributes to this presence. Visibly tuning in as an expression of empathy tells clients that you are with them, and it puts you in a position to listen carefully to their concerns. Your attention can be manifested in both physical and psychological ways. Let's start by briefly exploring nonverbal behavior as a channel of communication.

## Nonverbal Behavior as a Channel of Communication

Over the years, both researchers and practitioners have come to appreciate the importance of nonverbal behavior in counseling (Andersen, 1999; Ekman, 1992, 1993; Ekman & Friesen, 1975; Ekman & Rosenberg, 1998; Grace, Kivlighan, & Kunce, 1995; Hicksen & Stacks, 1993; Highlen & Hill, 1984; Knapp & Hall, 1996; McCroskey, 1993; Mehrabian, 1971, 1972, 1981; Norton, 1983; Richmond & McCroskey, 2000; Russell, 1995; Russell, Fernandez-Dols, & Mandler, 1997; for a wealth of information about nonverbal behavior see the following Internet site: http://www3.usal.es/~nonverbal/introduction.htm). Highlen and Hill suggest that nonverbal behaviors regulate conversations, communicate emotions, modify verbal messages, provide important messages about the helping relationship, give insights into self-perceptions, and provide clues that clients (or counselors) are not saying what they are thinking. This area has taken on even more importance because of the multicultural nature of helping.

The face and body are extremely communicative. We know from experience that even when people are together in silence, the atmosphere can be filled with messages. Sometimes the facial expressions, bodily motions, voice quality, and physiological responses of clients communicate more than their words do.

Studies in nonverbal behavior should not be overinterpreted. Taken togther, however, they highlight the importance of nonverbal behavior in the communication process. The following factors, on the part of both helpers and clients, play an important role in the therapeutic dialogue:

- *Bodily behavior,* such as posture, body movements, and gestures.
- *Eye behavior,* such as eye contact, staring, and eye movement.
- *Facial expressions,* such as smiles, frowns, raised eyebrows, and twisted lips.
- *Voice-related behavior,* such as tone of voice, pitch, volume, intensity, inflection, spacing of words, emphases, pauses, silences, and fluency.
- *Observable autonomic physiological responses,* such as quickened breathing, blushing, paleness, and pupil dilation.
- *Physical characteristics,* such as fitness, height, weight, and complexion.
- *Space;* that is, how close or far a person chooses to be during a conversation.
- *General appearance,* such as grooming and dress.

People constantly "speak" to one another through their nonverbal behavior. Effective helpers learn this "language" and how to use it effectively in their interactions with their clients. They also learn how to "read" relevant messages embedded in the nonverbal behavior of their clients.

## Helpers' Nonverbal Behavior

Before you begin interpreting the nonverbal behavior of your clients (discussed in Chapter 5), take a look at yourself. You speak to your clients through all the nonverbal categories outlined in the previous section. At times, your nonverbal behavior is as important as or even more important than your words. Your nonverbal behavior influences clients for better or worse. In your nonverbal behavior, clients read cues that indicate the quality of your presence to them. Attentive presence can invite or encourage them to trust you, open up, and explore the significant dimensions of their problem situations. Half-hearted presence can promote distrust and lead to clients' reluctance to reveal themselves to you. Clients might misinterpret your nonverbal behavior. For instance, you might be comfortable with the space between you and your client, but it is too close for the client. Or remaining silent might in your mind mean giving a client time to think, but the client might feel embarrassed. Part of listening, then, is being sensitive to clients' reactions to your nonverbal behavior.

Effective helpers are mindful of the stream of nonverbal messages they send to clients. Reading your own bodily reactions is an important first step. For instance, if you feel your muscles tensing as the client talks to you, you can say to yourself, "I'm getting anxious here. What's going on? And what nonverbal messages indicating my discomfort am I sending to the client?" Again, you probably would not use these words. Rather you read the signals your body is sending you without letting them distract you from your client.

You can also use your body to censor instinctive or impulsive messages that you feel are inappropriate. For instance, if the client says something that instinctively angers you, you can control the external expression of the anger (for instance, a sour look) to give yourself time to reflect. Such self-control is not phony because your respect for your client takes precedence over your instinctive reactions. Not dumping your annoyance or anger on your clients through nonverbal behavior is not the same as denying it. Becoming aware of it is the first step in dealing with it.

In a more positive vein, you can "punctuate" what you say with nonverbal messages. For instance, Denise is especially attentive when Jennie talks about actions she could take to do something about her problem situation. She leans forward, nods, and says "uh-huh." She uses nonverbal behavior to reinforce Jennie's intention to act.

On the other hand, don't become preoccupied with your body and the qualities of your voice as a source of communication. Rather, learn to use your body instinctively as a means of communication. Being aware of and at home with nonverbal communication can reflect an inner peace with yourself, with the helping process, and with your clients. Your nonverbal behavior should enhance rather than stand in the way of your working alliance with your clients.

Although you can learn the skills of visibly tuning in, they will be phony if they are not driven by the attitudes and values, such as respect and empathy, discussed in Chapter 3. Your mind set—what's in your heart—is as important as your visible presence. If you are not actively interested in the welfare of a client, or if you resent working with a client, subtle or not-so-subtle nonverbal clues will color your behavior. I once mentioned to a doctor my concerns about an invasive diagnostic procedure he intended to use. The doctor said the right words to reassure me, but his physical presence and the way he rushed his words said, "I've heard this dozens of times. I really don't have time for your concerns. Let's get on with this." His words were right but the real communication was in the nonverbal messages that accompanied his words.

## The Skill of Visibly Tuning in to Clients

You can use certain key nonverbal skills to visibly tune in to clients. These skills can be summarized in the acronym SOLER. Since communication skills are particularly sensitive to cultural differences, care should be taken in adapting what follows to different cultures. This is only a framework:

**S: Face the client *Squarely*;** that is, adopt a posture that indicates involvement. In North American culture, facing another person squarely is often considered a basic posture of involvement. It usually says, "I'm here with you; I'm available to you." Turning your body away from another person while you talk to him or her can lessen your degree of contact with that person. Even when people are seated in a circle, they usually try in some way to turn toward the individuals to whom they are speaking. The word *squarely* here should not be taken too literally. *Squarely* is not a military term. The point is that your bodily orientation should convey the message that you are involved with the client. If, for any reason, facing the person squarely is too threatening, then an angled position may be more helpful. The point is not inches and angles but the quality of your presence. Your body sends out messages whether you like it or not. Make them congruent with what you are trying to do.

**O: Adopt an *Open* posture.** Crossed arms and crossed legs can be signs of lessened involvement with or availability to others. An open posture can be a sign that you're open to the client and to what he or she has to say. In North American culture, an open posture is generally seen as a nondefensive posture. Again, the word *open* can be taken literally or metaphorically. If your legs are crossed, this does not mean that you are not involved with the client. But it is important to ask yourself, "To what degree does my present posture communicate openness and availability to the client?" If you are empathic and open-minded, let your posture mirror what is in your heart.

**L: Remember that it is possible at times to *Lean* toward the other.** Watch two people in a restaurant who are intimately engaged in conversation. Very often they are both leaning forward over the table as a natural sign of their involvement. The main thing is to remember that the upper part of your body is on a hinge. It can move toward a person and back away. In North American culture, a slight inclination toward a person is often seen as saying, "I'm with you, I'm interested in you and in what you have to say." Leaning back (the severest form of which is a slouch) can be a way of saying, "I'm not entirely with you" or "I'm bored." Leaning too far forward, however, or doing so too soon, may frighten a client. It can be seen as a way of placing a demand on the other for some kind of closeness or intimacy. In a wider sense, the word *lean* can refer to a kind of bodily flexibility or responsiveness that enhances your communication with a client. And bodily flexibility can mirror mental flexibility.

**E: Maintain good *Eye* contact.** In North American culture, fairly steady eye contact is not unnatural for people deep in conversation. It is not the same as staring. Again, watch two people deep in conversation. You may be amazed at the amount of direct eye contact. Maintaining good eye contact with a client is another way of saying, "I'm with you; I'm interested; I want to hear what you have to say." Obviously, this principle is not violated if you occasionally look away. Indeed, you have to if you don't want to stare. But if you catch yourself looking away frequently, your behavior might give you a hint about some kind of reluctance to be with this person or to get involved with him or her. Or it might say something about your own discomfort. In other cultures, however, too much eye contact, especially with some-one in a position of authority, is out of order. I have learned much about the cul-tural meaning of eye contact from my Asian students and clients.

**R: Try to be relatively *Relaxed* or natural in these behaviors.** Being relaxed means two things: First, it means not fidgeting nervously or engaging in distracting facial expressions; the client might wonder what's making you nervous. Second, it means becoming comfortable with using your body as a vehicle of personal contact and expression. Your being natural in the use of these skills helps put the client at ease.

A counselor trained in the skilled helper model was teaching counseling to vi-sually impaired students in the Royal National College for the Blind. Most of her clients were visually impaired. However, she wrote this about SOLER:

> In counseling students who are blind or visually impaired, eye contact has little or no relevance. However, attention on voice direction is extremely important, and people with a visual impairment will tell you how insulted they feel when sighted people are talking to them while looking somewhere else.
>
> I teach SOLER as part of listening and attending skills and can adapt each letter of the acronym [to my visually impaired students] with the excep-tion of the E. . . . After much thought, I would like to change your acronym to SOLAR, the A being for "Aim," that is, aim your head and body in the direction of your client so that when they hear your voice, be it linguistically or paralinguistically, they know that you are attending directly to what they are saying (private communication).

This counselor's comments underscore the fact that people are more sensitive to how you orient yourself to them nonverbally than you might imagine. Anything that distracts from your "being there" can harm the dialogue. The point to be stressed is that a respectful, empathic, genuine, and caring mind set might well lose its impact if the client does not see these internal attitudes reflected in your exter-nal behaviors. In the beginning, you may become overly self-conscious about the way you visibly tune in, especially if you are not used to being attentive. Still, the guidelines just presented are only that—guidelines. They should not be taken as absolute rules to be applied rigidly in all cases. Box 4-1 summarizes, in question form, the main points related to being visibly tuned in to clients. Turn to the *Exer-cises in Helping Skills* for opportunities to "practice" the skill of visibly tuning in.

## THE SHADOW SIDE OF COMMUNICATION SKILLS

Interpersonal communication competence is critical for effective everyday living. It is the principal enabling skill for just about everything we do. Yet it is, in my view, "forgotten" by society. In that respect it suffers a fate shared by a number of essential "life skills," such as problem solving, opportunity development, parenting, and managing (knowing how to make some system work), to name a few.

### Box 4-1    Questions on Visibly Tuning In

- What are my attitudes toward this client?
- How would I rate the quality of my presence to this client?
- To what degree does my nonverbal behavior indicate a willingness to work with the client?
- What attitudes am I expressing in my nonverbal behavior?
- What attitudes am I expressing in my verbal behavior?
- To what degree does the client experience me as effectively present and working with him or her?
- To what degree does my nonverbal behavior reinforce my internal attitudes?
- In what ways am I distracted from giving my full attention to this client? What am I doing to handle these distractions? How might I be more effectively present to this person?

Let me make my point. In lecturing, I have often asked audiences to answer two questions. The first question goes something like this:

> Given the importance of effective human relationships in just about every area of life, how important is it for your kids to develop a solid set of interpersonal communication skills? On a scale from 1 to 100, how high would you rate the importance?

Inevitably, the ratings are at the high end, always near 100. The second question goes something like this:

> Given the importance of these skills, where do your kids pick them up? How does society make sure that they acquire them? In what forums do they learn them?

Then the hemming and hawing begin. "Well, I guess they get some of them at home. That is, if they find good role models at home." Or, "Life itself is the best teacher of these skills. They learn them on the run." The members of the audience mill around like that for a while, until I say,

> Let me summarize what I've been hearing. And, by the way, it's no different from what I hear every place else. Although most parents rate the importance of these sets of skills very high, we live in a society that leaves their development to chance. Practically nothing is done systematically to make sure that our kids learn these skills. And, by the way, there is no assurance that they will pick them up on the run.

Children learn a bit from their parents, they might get a dash in school, perhaps a soupçon of TV helps. But, in the main, they are more often exposed to poor communicators than good ones.

Ideally, helpers-to-be would arrive at training programs already equipped with a solid set of interpersonal communication skills. Training would help them adapt these skills to the counseling process. After all, basic interpersonal communication skills are not special skills peculiar to helping. Rather, they are extensions of the kinds of skills all of us need in our everyday interpersonal interactions. However, since trainees don't ordinarily arrive so equipped, they need time to come up to speed in communication competence. This, in a strange way, creates its own problem. Some helper-training programs focus almost exclusively on interpersonal communication skills. As a result, trainees know how to communicate but not necessarily how to help. Furthermore, most adults feel that they are "pretty good" at these skills. What they actually mean is that they see themselves as good as others.

# ACTIVE LISTENING: THE FOUNDATION OF UNDERSTANDING

INADEQUATE LISTENING

    Nonlistening

    Partial listening

    Tape-recorder listening

    Rehearsing

EMPATHIC LISTENING

LISTENING TO WORDS: CLIENTS' STORIES, POINTS OF VIEW, DECISIONS, AND INTENTIONS OR PROPOSALS

  Listening to Clients' Stories

    Experiences

    Behaviors

    Affect: Feelings, emotions, and moods

  Listening to Clients' Points of View

  Listening to Clients' Decisions

  Listening to Clients' Intentions or Proposals

  "Hearing" Opportunities and Resources

LISTENING TO CLIENTS' NONVERBAL MESSAGES AND MODIFIERS

    Confirming or repeating

    Denying or confusing

    Strengthening or emphasizing

    Adding intensity

    Controlling or regulating

PROCESSING WHAT YOU HEAR: THE THOUGHTFUL SEARCH FOR MEANING

    Identify key messages and feelings

    Understand clients through context

    Hear the slant or spin: Tough-minded listening and processing

    Muse on what's missing

**LISTENING TO ONESELF: THE HELPER'S INTERNAL CONVERSATION**

**THE SHADOW SIDE OF LISTENING TO CLIENTS**

Forms of Distorted Listening
  Filtered listening
  Evaluative listening
  Stereotype-based listening
  Fact-centered rather than person-centered listening
  Sympathetic listening
  Interrupting
Myths About Nonverbal Behavior

Visibly tuning in is not, of course, an end in itself. You tune in mentally and visibly to listen to the stories, points of view, decisions, and both the intentions and proposals of your clients. Listening carefully to a client's concerns seems to be a concept so simple to grasp and so easy to do that you may wonder why it is given such explicit treatment here. Nonetheless, it is amazing how often people fail to listen to one another. Full listening means listening actively, listening accurately, and listening for meaning. Listening is not merely a skill. It is a rich metaphor for the helping relationship—indeed all relationships. An attempt to tap some of that richness will be made here.

The following case will be used to help you develop a better behavioral feel for both visibly tuning in and listening.

> Jennie, an African American college senior, was raped by a "friend" on a date. She received some immediate counseling from the university Student Development Center and some ongoing support during the subsequent investigation. But even though she was raped, it turned out that it was impossible for her to prove her case. The entire experience—both the rape and the investigation that followed—left her shaken, unsure of herself, angry, and mistrustful of institutions she had assumed would be on her side (especially the university and the legal system). When Denise, a middle-aged and middle-class African American social worker who was a counselor for a health maintenance organization, first saw her a couple of years after the incident, Jennie was plagued by a number of somatic complaints, including headaches and gastric problems. At work, she engaged in angry outbursts whenever she felt that someone was taking advantage of her. Otherwise, she had become quite passive and chronically depressed. She saw herself as a woman victimized by society and was slowly giving up on herself.

We will refer back to the interactions between Denise and Jennie to illustrate some of the main points about listening.

## INADEQUATE LISTENING

Effective listening is not a state of mind, like being happy or relaxed. It's not something that "just happens." It's an activity. In other words, effective listening requires work. Let's first take a look at the opposite of active listening. All of us have been, at one time or another, both perpetrators and victims of the following forms of inactive or inadequate listening.

**Nonlistening.** Sometimes we go through the motions of listening but are not really engaged. At times we get away with it. Sometimes we are caught. "What would you do?" Jennifer asks her colleague, Kieran, after outlining a problem the university counseling center is having with the school's administration. Embarrassed, Kieran replies, "I'm not sure." Staring him down, she says, "You haven't been listening to a word I've said." For whatever reason, he had tuned her out. Obviously, no helper sets out not to listen, but even the best can let their mind wander as they listen to the same kind of stories over and over again, forgetting that the story is unique to *this* client.

**Partial listening.** This is listening that skims the surface. The helper picks up bits and pieces but not necessarily the essential points the client is making. For instance, Janice's client, Dean, is talking to her about a date that went terribly wrong. Janice only half listens. It seems that Dean is not that interesting. Dean stops talking and looks rather dejected. Janice tries to pull together the pieces of the story she did listen to. Her attempt to express understanding has a hollow ring to it. Dean pauses and then switches to a different topic. Inadequate listening helps neither understanding nor relationships.

**Tape-recorder listening.** What clients look for from listening is not the helper's ability to repeat their words. A tape recorder could do that perfectly. People want more than physical presence in human communication; they want the other person to be present psychologically, socially, and emotionally. Sometimes a helper fails to demonstrate that visibly tuning in and listening mean being totally present. The client picks up some signals that the helper is not listening very well. How many times have you heard someone exclaim, "You're not listening to what I'm saying!"? When the person accused of not listening answers, almost predictably, "I *am too* listening; I can repeat everything you've said," the accuser is not comforted. Usually, clients are too polite or cowed or preoccupied with their own concerns to say anything when they find themselves in that situation. But it is a shame if your auditory equipment is in order but you are elsewhere. Your clients want *you*, a live counselor, not a tape recorder.

**Rehearsing.** Picture Sid, a novice counselor, sitting with Casey, a client. At one point in the conversation, when Casey talks about some "wild dreams" he is having, Sid says to himself, "I know very little about dreams; I wonder what I'm going to say?" Sid stops listening and begins rehearsing what he's going to say. Even when experienced helpers begin to mull over the perfect response to what their clients are saying, they stop listening. Effective helpers listen intently to clients and to the themes and core messages embedded in what they are saying. They are never at a loss in responding. They don't need to rehearse. And their responses are much more likely to help clients move forward in the problem-management process. When a client stops speaking, effective helpers often pause to reflect on what he or she just said and then speak. Pausing says, "I'm still mulling over what you've just said. I want to respond thoughtfully". They pause because they have listened.

# EMPATHIC LISTENING

The opposite of inactive or inadequate listening is empathic listening, listening driven by the value of empathy. Empathic listening centers on the kind of attending, observing, and listening—the kind of "being with"—needed to develop an understanding of clients and their worlds. Although it might be metaphysically impossible to actually get "inside" the world of another person and experience the world as he or she does, it is possible to approximate this.

Carl Rogers (1980) talked passionately about basic empathic listening—being with and understanding the other—even calling it "an unappreciated way of being" (p. 137). He used the word *unappreciated* because, in his view, few people in the general population developed this "deep listening" ability, and even so-called expert helpers did not give it the attention it deserved. Here is his description of empathic listening or being with:

> It means entering the private perceptual world of the other and becoming thoroughly at home in it. It involves being sensitive, moment by moment, to

the changing felt meanings which flow in this other person, to the fear or rage or tenderness or confusion or whatever that he or she is experiencing. It means temporarily living in the other's life, moving about in it delicately without making judgments. (p. 142)

Such empathic listening is selfless because helpers must put aside their own concerns to be fully with their clients. Of course, Rogers points out that this deeper understanding of clients remains sterile unless it is somehow communicated to them. Although clients can appreciate how intensely they are attended and listened to, they and their concerns still need to be understood. Empathic listening begets empathic understanding, which begets empathic responding.

Empathic participation in the world of another person obviously admits of degrees. As a helper, you must be able to enter clients' worlds deeply enough to understand their struggles with problem situations or their search for opportunities with enough depth to make your participation in problem management and opportunity development valid and substantial. If your help is based on an incorrect or invalid understanding of the client, then your helping may lead him or her astray. If your understanding is valid but superficial, then you might miss the central issues of the client's life.

Back to Jennie and Denise. Denise is a pro, so she doesn't have much of a problem with inadequate listening. She is an empathic listener par excellence. Denise knows she has to be with Jennie every step of the way.

## LISTENING TO WORDS:
## CLIENTS' STORIES, POINTS OF VIEW,
## DECISIONS, AND INTENTIONS OR PROPOSALS

Listening to what clients are saying is not a free-form activity. Helpers need to be focused. Recognizing modes of client discourse—for instance, clients tell *stories*, share *points of view*, deliver *decisions*, and state *intentions* or offer *proposals*—can help you organize your listening; that is, they can help you listen for the client's key points and relevant detail.

### Listening to Clients' Stories

Most immediately, helpers listen to clients' "stories"—that is, their accounts of their problem situations and unused opportunities. Stories tend to be mixtures of clients' experiences, behaviors, and emotions. Traditionally, human activity has been divided into three parts: thinking, feeling, and acting. A slightly different approach is taken here.

Clients talk about their *experiences*—that is, what happens to them. If a client tells you that she was fired from her job, she is talking about her problem situation as an experience. Jennie, of course, talked about being raped, belittled, and ignored. Clients talk about their *behavior*—that is, what they do or refrain from doing. If a client tells you that he smokes and drinks a lot, he is talking about his external behavior. If a different client says that she spends a great deal of time daydreaming, she is talking about her internal behavior. Jennie talked about pulling away from her family and friends after the rape investigation. Clients talk about their *affect*—that is, the feelings, emotions, and moods that arise from or are associated with their experiences and both internal and external behavior. If a client tells you how depressed she gets after fights with her fiancé, she is talking about the mood associated with her experiences and behavior. Jennie talked about her shame, her feelings of betrayal, and her anger. Since experiences, actions, and emotions are interrelated in the day-to-day lives of clients, they mix them together in telling their stories. Consider this example:

> A client says to a counselor in the personnel department of a large company, "I had one of the lousiest days of my life yesterday." At this point, the counselor knows that something went wrong and that the client feels bad about it, but she knows relatively little about the specific experiences, behaviors, and emotions that made the day such a horror for the client. However, the client continues, "Toward the end of the day, my boss yelled at me in front of some of my colleagues for not landing an order from a new customer [experience]. I lost my temper [emotion] and yelled right back at him [behavior]. He blew up and fired me on the spot [experience]. And now I feel awful [emotion] and am trying to find out if I really have been fired and, if so, if I can get my job back [behavior]."

Problem situations are much clearer when they are spelled out as specific experiences, behaviors, and feelings related to specific situations.

Since clients spend so much time telling their stories, a few words about each of these elements are in order.

**Experiences.**  Most clients spend a fair amount of time, sometimes too much time, talking about what happens *to* them.

- "My wife doesn't understand me."
- "My ulcer acts up when family members argue."
- "My boss doesn't give me feedback."
- "I get headaches a lot."

It is of paramount importance to listen to and understand clients' experiences. However, since experiences dwell on what other people do or fail to do, experience-focused stories smack a bit of passivity. At times, the implication is that others—or the world in general—are to blame for the client's problems.

- "She doesn't do anything all day. The house is always a mess when I come home. No wonder I can't concentrate at work."
- "He tells his little jokes, and I'm always the butt of them. He makes me feel bad about myself most of the time."

Some clients talk about experiences that are internal and out of their control.

- "These feelings of depression come from nowhere and seem to suffocate me."
- "I just can't stop thinking of him."

The last statement sounds like an action, but it is expressed as an experience. It is something happening to the client, at least to the client's way of thinking. One reason that some clients fail to manage the problem situations of their lives is that they are too passive or see themselves as victims, adversely affected by other people, by the immediate social settings of life such as the family, by society in its larger organizations and institutions such as government or the workplace, by cultural prescriptions, or even by internal forces. They feel that they are no longer in control of their lives or some dimension of life. Therefore, they talk extensively about these experiences.

- "Company policy discriminates against women. It's that simple."
- "The economy is booming, but the kind of jobs I want are already taken."
- "No innovative teacher gets very far around here."

Of course, some clients *are* treated unfairly; they are victimized by the behaviors of others in the social and institutional settings of their lives. Although they can be helped to cope with victimization, full management of their problem situations demands changes in the social settings themselves. For instance, a client might be helped to cope with a brutal husband, but ultimately the courts would have to intervene to keep him at bay.

For other clients, talking constantly about experiences is a way of avoiding responsibility: "It's not my fault. After all, these things are happening to me." Sykes (1992) in his book *A Nation of Victims*, was troubled by the tendency of the United States to become a "nation of whiners unwilling to take responsibility for our actions." Whether his statement is true or not, counselors must be able to distinguish "whiners" from those who are truly being victimized.

**Behaviors.**  All of us do things that get us into trouble and fail to do things that will help us get out of trouble or develop opportunities. Clients are no different.

- "When he ignores me, I think of ways of getting back at him."
- "Whenever anyone gets on my case for having a father in jail, I let him have it. I'm not taking that kind of crap from anyone."
- "Even though I feel the depression coming on, I don't take the pills the doctor gave me."
- "When I get bored, I find some friends and go get drunk."
- "I have a lot of sexual partners and have unprotected sex whenever my partner will let me."

Some clients talk freely about their experiences, what happens to them, but seem more reluctant to talk about their behaviors. One reason for this is that they can't talk about behaviors without bringing up issues of personal responsibility.

**Affect: Feelings, emotions, and moods.**  Feelings, emotions, and moods constitute a river that continually runs through us—peaceful, meandering, turbulent, or raging—often beneficial, sometimes dangerous, seldom neutral. They are certainly an important part of clients' problem situations and undeveloped opportunities (Greenberg & Paivio, 1997; Plutchik, 2001). Some have complained that psychologists—both researchers and practitioners—don't take emotions seriously enough. For instance, anger is a ubiquitous and extremely important emotion, but "it has been oddly neglected by the clinical community" (Norcross & Kobayashi, 1999, p. 275; this is the opening article of a series of articles on anger in a special section of the *Journal of Clinical Psychology*, 55, March, 1999). But there are some signs that things are changing. In 2001, a new American Psychological Association journal, *Emotion*, entered the scene because of the recognition that emotion is fundamental to so much of human life. The journal includes articles ranging from the so-called softer side of psychology through hard-nosed molecular science. Popular self-help books written by professionals have proved useful to both clients and practitioners for years (McKay & Dinkmeyer, 1994; McKay, Davis, & Fanning, 1997). These very practical books tend to take a positive-psychology approach to experiencing, regulating, and using emotions in everyday life.

Recognizing key feelings, emotions, and moods (or the lack thereof) is very important for at least three reasons. First, they pervade our lives. There is an emotional tone to just about everything we do. Feelings, emotions, and moods pervade clients' stories, points of view, decisions, and intentions or proposals. Second, they greatly affect the quality of our lives. A bout of depression can stop us in our tracks. A depressed client is an unhappy client. A client who gets out from under the burden of self-doubt breathes more freely. Third, feelings, emotions, and moods are drivers of our behavior. As Lang (1995) pointed out, they are "action dispositions" (p. 372). Clients driven by anger can do desperate things. On the other hand, enthusiastic clients can accomplish more than they ever thought they could. The good news is that we can learn how to tune ourselves in to our clients' and our own feelings, emotions, and moods (Machado, Beutler, & Greenberg, 1999) at the service of discovering how to regulate them.

Understanding the role of feelings, emotions, and moods in client's problem situations and their desire to identify and develop opportunities is central to the helping process. Emotions highlight learning opportunities.

- "I've been feeling pretty sorry for myself ever since he left me." This client learns that self-pity constricts her world and limits problem-managing action.

- "I yelled at my mother last night, and now I feel very ashamed of myself." Shame may well be a wake-up call in this client's relationship with his mother.

- "I've been anxious for the past few weeks, but I don't know why. I wake up feeling scared, and then it goes away but comes back again several times during the day." Anxiety has become a bad habit for this client. It is self-perpetuating. What can the client do to break through the vicious circle?

- "I finally finished the term paper that I've been putting off for weeks and I feel great!" Here emotion becomes a tool in this client's struggle against procrastination.

The last item in this list brings up an important point. In the psychological literature, negative emotions tend to receive more attention than positive emotions. Work is under way to study positive emotions and their beneficial effects. There are indications that we can use positive emotions to promote both physical and psychological well-being (Salovey, Rothman, Detweiler, & Steward, 2000). Studies also indicate that emotions can free up psychological resources, act as opportunities for learning, and promote health-related behaviors. In managing problems and developing opportunities, social support plays a key role. As Salovey and his colleagues note, clients are more likely to elicit social support if they manifest a positive attitude toward life. Potential supporters tend to shun clients who let their negative emotions get the best of them

Of course, clients often express feelings without talking about them. When a client says, "My boss gave me a raise and I didn't even ask for one!" you can feel the emotion in her voice. A client who is talking listlessly and staring down at the floor may not say, in so many words, "I feel depressed." A dying person may express feelings of anger and depression without talking about them. Other clients feel deeply about things but do their best to hold their feelings back. But effective helpers can usually pick up clues or hints, whether verbal or nonverbal, that indicate the feelings and emotions rumbling inside.

The point here is this: A full story is a mixture of experiences, behaviors, and feelings. And stories, often in the form of examples, are used to explain and illustrate points of view, decisions, intentions, and proposals. Your first job is to listen carefully to the mix clients use to talk about their concerns and what they would like to do about them.

## Listening to Clients' Points of View

As clients tell their stories, explore possibilities for a better future, set goals, make plans, and review obstacles to accomplishing these plans, they often share their points of view. A point of view is a client's personal estimation of something. A full point of view includes the point of view itself, the reasons for it, an illustration to bring it to life, and some indication of how open the client might be to modifying it. There is usually no direct expectation that anyone else need adopt the point of view; that would be a form of selling or persuasion. But realistically, the implication often is, "I think this way. Why don't other people think this way?" For instance, Aurora, an 80-year-old woman, is talking to a counselor about the various challenges of old age. At one point she says,

> My sister in Florida [85 years old]—Sis, we call her—is sick. She's probably dying, but she wants to stay at home. She's asked me to come down there and take care of her. I think that's asking too much. I could use some help myself these days. But she keeps calling.

Aurora's point of view is that her sister's request is not realistic because she herself needs some help to get by. But her sister persists in trying to persuade her to change her mind. Points of view reveal clients' beliefs, values, attitudes, and convictions. Clients may share their points of view about everything under the sun. The ones that are relevant to their problem situations or undeveloped opportunities need to be listened to and understood. In Jennie's case, she tells Denise,

> You just can't trust the system. They're not going to help. They take the easy way out. I don't care which system it might be. Church, government, the community, sometimes even family. They're not going to give you much help.

Denise listens carefully to Jennie's point of view and realizes how much it is influencing her behavior. In Jennie's case, it's easy to see where the point of view comes from. But Denise also knows that, at some point, she needs to challenge Jennie's point of view because it may be one of the things that is keeping her locked in her misery. Points of view have power.

## Listening to Clients' Decisions

From time to time, we all tell others about decisions we are making or that we've made. A client might say, "I've decided to stop drinking. Cold turkey." Or, "I'm tired of being alone. I'm going to join a dating service." Decisions usually have implications for the decision maker and for others. The client who has decided to go cold turkey on drink has his work cut out for him, but there are implications for his spouse. For instance, she's used to coping with a drunk, but now she may have to learn how to cope with this "new person" in the house. Commands, instructions, and even hints are, in a way, decisions about other people's behavior. A client might say to her helper, "Don't bring up my ex-husband any more. I'm finished with him."

Sharing a decision fully means spelling out the decision itself, the reasons for the decision, the implications for self and others, and some indication as to whether the decision or any part of it is open to review. For instance, Jennie, in talking with Denise about future employment, says in a rather languid tone of voice, "I'm not going to get any kind of job where I have to fight the race thing. Or the woman thing. I'm tired of fighting. I only get hurt. I know that this limits my opportunities, but I can live with that." Note that this is more than a point of view. Jenny is more or less saying, "I've made up my mind." She notes the implication for herself—a limitation of job opportunities—but the implication might be, "So that's the end of it. Don't try to convince me otherwise." Denise hears the message and the implied command. However, she believes that some of Jennie's decisions need challenging. Decisions can be tricky. Often *how* they are delivered says a great deal about the decision itself. Given the rather languid way in which Jennie delivers her decision, Denise thinks that it might not be Jennie's final decision; this is something that has to be checked out. A dialogue with Jennie about the reasons for her decision and a review of its implications are possible routes for a challenge.

## Listening to Clients' Intentions or Proposals

Finally, clients state intentions, offer proposals, or make a case for certain courses of action. Consider Lydia. She is a single parent of two young children who is a member of the "working poor." Her wages don't cover her expenses. She has no insurance. The father of her children has long since disappeared. She says to a social worker,

> I think I should quit my job. I'm making the minimum wage and, with travel expenses and all, I just can't make ends meet. I spend too much time traveling and don't see enough of my kids. Friends look after them when I'm gone, but that's hit and miss and puts a burden on them. You know if I go on welfare I could make almost as much. And then I could pick up jobs that would pay me cash. I've got friends who do this. I believe I could make ends meet. My kids and I would be better off. And I wouldn't be hassled as much.

Lydia is making a case, not announcing a decision. The case includes what she wants to do (quit her job and move into the "alternative" work economy), the reasons for doing it (the inadequacy of her current work situation, the need to make ends meet), and the implications for herself and her children (she'd be less hassled and her kids would see more of her and be better off ).

Of course, when clients talk about their concerns, they mix all these forms of discourse together. Here's an example. What follows came out through dialogue in one of Jennie's sessions with Denise. For the sake of illustration, it is presented here in summary form in Jennie's words:

> A couple of weeks ago I met a woman at work who has a story similar to mine. We talked for a while and got along so well that we decided to meet outside of work. I had dinner with her last night. She went into her story in more depth. I was amazed. At times I thought I was listening to myself! Because she had been hurt, she was narrowing her world down into a little patch so that she could control everything and not get hurt anymore. I saw right away that I'm trying to do my own version of the same thing. I know you've been telling me that, but I haven't been listening very well. Here's a woman with lots going for her, and she's hiding out. As I came back from dinner, I said to myself, "You've got to change." So, Denise, I want to revisit two areas we've talked about: my work life and my social life. I don't want to live in the hole I've dug for myself. I could see clearly some of the things she should do. So here's what I want to do. I want to engage in some little experiments in broadening my social life, starting with my family. And I want to discuss the kind of work I want without putting all the limitations on it. I want to start coming out of the hole I'm in. And I want to help my new friend do the same.

Everything is here, or elements of everything: a story about her new friend, including experiences, actions, and feelings; points of view about her new friend; decisions about where she wants her life to go; and proposals about experiments in her social life and her relationship with her friend. The point is this: Developing frameworks for listening can help you zero in on the key messages your clients are communicating and help you identify and understand the feelings, emotions, and moods that go with them.

## "Hearing" Opportunities and Resources

If you listen only for problems, you will end up talking mainly about problems. And you will shortchange your clients. Every client has something going for him or her. Your job is to spot clients' resources and help them invest these resources in managing problem situations and opportunities. If people generally use only a fraction of their potential, then there is much to be tapped. A counselor is working with a 65-year-old successful businessman who, with his wife, has raised three children. The children are well-educated and successful in their own right. The man is having difficulty coping with some health problems. The counselor learns that he was one of a group of poor inner-city boys in a longitudinal study. The boys had a mean IQ of 80 and a lot of social disadvantages (see Vaillant & Davis, 2000). As the counselor listens to the man's story, he hears a history of resilience. The man, helped to review the strategies he used to cope, finds that he can use some of them and others to cope with his health problems. At one point, the man says, "I never gave up then. Why should I start giving up now?"

## LISTENING TO CLIENTS' NONVERBAL MESSAGES AND MODIFIERS

Recall what was said about nonverbal behavior in Chapter 4. Clients send messages through their nonverbal behavior. The ability of people to read these messages can contribute to their relationship well-being (Carton, Kessler, & Pape, 1999). Helpers need to learn how to read these messages without distorting or overinterpreting them. For instance, when Denise says to Jennie, "It seems that it's hard talking about yourself," Jennie says, "No, I don't mind at all." But the real answer is probably in her nonverbal behavior, for she speaks hesitatingly while looking away and frowning. Reading such cues helps Denise understand Jennie better. Our nonverbal behavior has a way of "leaking" messages about what we really mean. The very spontaneity of nonverbal behaviors contributes to this leakage, even in the case of highly defensive clients. It is not easy for clients to fake nonverbal behavior (Wahlsten, 1991). The real messages still tend to leak out.

Besides being a channel of communication in itself, such nonverbal behavior as facial expressions, bodily motions, and voice quality often modify and punctuate verbal messages in much the same way that periods, question marks, exclamation points, and underlining punctuate written language. All the kinds of nonverbal behavior mentioned in Chapter 4 can punctuate or modify verbal communication in the following ways.

**Confirming or repeating.** Nonverbal behavior can confirm or repeat what is being said verbally. For instance, when Denise responds to Jennie with just the right degree of understanding—she hits the mark—not only does Jennie say, "That's right!" but also her eyes light up (facial expression), she leans forward a bit (bodily motion), and her voice is very animated (voice quality). Her nonverbal behavior confirms her verbal message.

**Denying or confusing.** Nonverbal behavior can deny or confuse what is being said verbally. When challenged by Denise, Jennie denies that she is upset, but her voice falters a bit (voice quality) and her upper lip quivers (facial expression). Her nonverbal behavior carries the real message.

**Strengthening or emphasizing.** Nonverbal behavior can strengthen or emphasize what is being said. When Denise suggests to Jennie that she ask her boss what he means by her "erratic behavior," Jennie says in a startled voice, "Oh, I don't think I could do that!" while slouching down and putting her face in her hands. Her nonverbal behavior underscores her verbal message.

**Adding intensity.** Nonverbal behavior often adds emotional color or intensity to verbal messages. When Jennie tells Denise that she doesn't like to be confronted without first being understood and then stares at her fixedly and silently with a frown on her face, Jennie's nonverbal behavior tells Denise something about the intensity of her feelings.

**Controlling or regulating.** Nonverbal cues are often used in conversation to regulate or control what is happening. Let's say that in a group counseling session, Nina looks at Tom and gives every indication that she is going to speak him. But he looks away. Nina hesitates and then decides not to say anything. Tom has used a nonverbal gesture to control her behavior.

In reading nonverbal behavior—*reading* is used here instead of *interpreting*—caution is a must. We listen to clients to understand them, not to dissect them. Merely reading about nonverbal behavior is not enough. Identifying relevant clues in videotaped interactions can help a great deal (Costanzo, 1992). Once you develop a working knowledge of nonverbal behavior and its possible meanings, you must learn through practice and experience to be sensitive to it and read its meaning in any given situation.

Since nonverbal behaviors can often mean a number of things, how can you tell which meaning is the real one? The key is the context in which they take place. Effective helpers listen to the entire context of the helping interview and do not become overly fixated on details of behavior. They are aware of and use the non-verbal communication system, but they are not seduced or overwhelmed by it. Sometimes novice helpers will fasten selectively on this or that bit of nonverbal behavior. For example, they will make too much of a half-smile or a frown on the face of a client. They will seize upon the smile or the frown and, in overinterpreting it, lose the person. There is no need to go overboard on listening. Remember that you are a human being listening to a human being, not a vacuum cleaner indiscriminately sweeping up every scrap. Quality, not quantity. Which brings us to the next topic.

## PROCESSING WHAT YOU HEAR: THE THOUGHTFUL SEARCH FOR MEANING

Even though we do it while we listen, we process what we hear. The trick is to become a thoughtful processor. As we shall see a bit further along, there are many less-than-thoughtful ways of processing clients' stories, points of view, and decisions. But first, what does thoughtful processing look like?

**Identify key messages and feelings.**  Denise listens to what Jennie has to say early on about her past and present experiences, actions, and emotions. She listens to Jennie's points of view and the decisions she has made or is in the process of making. She listens to Jennie's intentions and proposals. Jennie tells Denise about an intention gone awry and the emotions that went with it: "When the investigation began, I had every intention of pushing my case, because I knew that some of the men on campus were getting away with murder. But then it began to dawn on me that people were not taking me seriously because I am an African American woman. First I was angry, but then I just got numb. . . ." Later, Jennie says, "I get headaches a lot now. I don't like taking pills, so I try to tough it out. I have also become very sensitive to any kind of injustice, even in movies or on television. But I've stopped being any kind of crusader. That got me nowhere." As Denise listens to Jennie speak, questions based on the listening frameworks outlined here arise in the back of her mind:

- What are the main points here?
- What experiences and actions are most important?
- What themes are coming through?
- What is Jennie's point of view?
- What is most important to her?
- What does she want me to understand?
- What decisions are implied in what she's saying?
- What is she proposing to do?

If helpers think that everything that their clients say is key, then nothing is key. In the end, helpers make a clinical judgment as to what is key. Of course, as we shall see later, helpers have ways of checking their understanding. Denise doesn't distract herself from Jenny by asking the questions of herself directly, but they are part of her active listening and symbolize her interest in Jenny's world. What Denise has learned from both theory and experience over the years constitute the basis for the clinical judgments she makes.

**Understand clients through context.**  People are more than the sum of their verbal and nonverbal messages. Listening in its deepest sense means listening to clients themselves as influenced by the contexts in which they "live, move, and have their being." Earlier it was pointed out how important it is to interpret a client's nonverbal behavior in the context of the entire helping session. It is also essential to understand clients' stories, points of view, and decisions through the wider context of their lives. All the things that make people different—culture, personality, personal style, ethnicity, key life experiences, education, travel, economic status, and the others forms of diversity discussed in Chapter 3—provide the context for the client's problems and unused opportunities. Key elements of this context become part of the client's story whether they are mentioned directly or not. Effective helpers listen through this wider context without being overwhelmed by the details of it.

McAuliffe and Eriksen (1999) offer a where-when-how-what model for helping helpers think about their clients in context. The model is based on four questions. The first deals with background, the circumstances of the client's life: What circumstances surround the client, and how do these circumstances affect the way the client understands and deals with her or his problems and opportunities? The second question deals with developmental stage: What age-related psychosocial tasks and challenges is the client currently facing, and how does the way the client goes about these tasks affect the problem situation or opportunity? The third question deals with the client's approach to coming to know and make sense of the world about him or her: How does the client go about constructing meaning, including such things as determining what is important and what is right? The last question deals with personality: How does the client's personality style and temperament affect his or her self-understanding and his or her approach to the world? This is, of course, but one framework. You need to discover for yourself the contextual frameworks that can help you understand your clients as "people-in-systems" (Egan & Cowan, 1979). And contextual frameworks need to be updated. For instance, Arnett (2000) provides an emerging-adulthood framework for understanding the developmental tasks and challenges from the late teens through the twenties in societies that allow young people independent role exploration during those years. At the other end of the spectrum, Qualls and Abeles (2000) reframe the challenges of growing older and debunk popular misconceptions about aging.

Denise tries to understand Jennie's verbal and nonverbal messages, especially the core messages, in the context of Jennie's life. As she listens to Jennie's story, Denise says to herself right from the start something like this:

Here is an intelligent African American woman from a conservative Catholic background. She has been very loyal to the church because it proved to be a refuge in the inner city. It was a gathering place for her family and friends. It provided her with a decent primary and secondary school education and a shot at college. She did very well in her studies. Initially, college was a shock; it was her first venture into a predominantly white and secular culture. But she chose her friends carefully and carved out a niche for herself. Since studies were much more demanding, she had to come to grips with the fact that, in this larger environment, she was, academically, closer to average. The rape and investigation put a great deal of stress on what proved to be a rather fragile social network. Her life began to unravel. She pulled away from her family, her church, and the small circle of friends she had at college. At a time when she needed support the most, she cut it off. After graduation, she continued to stay "out of community." Now she is underemployed as a secretary in a small company. This does little for her sense of personal worth.

Denise listens to Jenny *through* this context without assuming that it need define Jenny. The helping context is also important. Denise needs to be sensitive about how Jennie might feel about talking to a woman who is quite different from her and also needs to understand that Jennie might well have some misgivings about the helping professions.

In sum, Denise tries to pull together the themes she sees emerging in Jennie's story and tries to see these themes in context. She listens to Jennie's discussion of her headaches (experiences), her self-imposed social isolation (behaviors), and her chronic depression (affect) against the background of her social history—the pressures of being religious in a secular society at school, the problems associated with being an upwardly mobile African American woman in a predominantly white male society. Denise sees the rape and investigation as social, not merely personal, events. She listens actively and carefully because she knows that her ability to help depends, in part, on not distorting what she hears. She does not focus narrowly on Jennie's inner world, as if Jennie could be separated from the social context of her life. Finally, although Denise listens to Jennie through the context of Jennie's life, she does not get lost in it. She uses context both to understand Jennie and to help her manage her problems and develop her opportunities more fully.

**Hear the slant or spin: Tough-minded listening and processing.** This is the kind of listening needed to help clients explore issues more deeply and to identify blind spots that need to be challenged. Skilled helpers listen not only to the client's stories, points of view, decisions, intentions, and proposals but also to any slant or spin that clients might give them. Although clients' visions of and feelings about themselves, others, and the world are real and need to be understood, their perceptions of themselves and their worlds are sometimes distorted. For instance, if a client sees herself as ugly when in reality she is beautiful, her experience of herself as ugly is real and needs to be listened to and understood. But her experience of herself does not square with the facts. This, too, must be listened to and understood. If a client sees himself as above average in his ability to communicate with others when, in reality, he is below average, his experience of himself needs to be listened to and understood, but reality cannot be ignored. Tough-minded listening includes detecting the gaps, distortions, and dissonance that are part of the client's experienced reality.

Denise realizes from the beginning that some of Jennie's understandings of herself and her world are not accurate. For instance, in reflecting on all that has happened, Jennie remarks that she probably got what she deserved. When Denise asks her what she means, she says, "My ambitions were too high. I was getting beyond my place in life." This is the slant or spin Jennie gives to her career aspirations. It is one thing to understand how Jennie might put this interpretation on what has happened; it is another to assume that such an interpretation reflects reality. To be client-centered, helpers must first be reality-centered.

Of course, helpers need not challenge clients as soon as they hear any kind of distortion. Rather, they note gaps and distortions, choose key ones, and challenge them when it is appropriate to do so (see Chapters 10 through 12).

**Muse on what's missing.** Clients often leave key elements out when talking about problems and opportunities. Having frameworks for listening can help you spot key things that are missing. For instance, they tell their stories but leave out key experiences, behaviors, or feelings. They offer points of view but say nothing about what's behind them or their implications. They deliver decisions but don't give the reasons for them or spell out the implications. They propose courses of action but don't say why they want to head in a particular direction, what the implications are for themselves or others, what resources they might need, or how flexible they are. As you listen, it's important to note what they put in and what they leave out. For instance, when it comes to stories, clients often leave out their own behavior or feelings. Jennie says, "I got a call from an old girl friend last week. I'm not sure how she tracked me down. We must have chatted away for 20 minutes. You know, catching up." Since Jennie says this in a rather matter-of-fact way, it's not clear how she felt about it at the time or feels now. Nor is there any indication of what she might want to do about it—for instance, stay in touch.

In another session, Jennie says, "I was talking with my brother the other day. He runs a small business. He asked me to come and work for him. I told him no. . . . By the way. I have to change the time of our next appointment. I forgot I've got a doctor's appointment." Denise notes the experience (being offered a job) and Jennie's behavior or reaction (a decision curtly refusing the offer). But Jennie leaves out the reasons for her refusal or the implications for herself or her brother or their relationship and moves to another topic.

Note that this is not a search for the "hidden stuff" that clients are leaving unsaid. We all leave out key details from time to time. Rather Denise, understanding what full versions of stories, points of view, and messages look like, notes what parts are missing. She then uses her clinical judgment—a large part of which is common sense—to determine whether to ask about the missing parts or not. For instance, when she asks Jennie why she refused her brother point blank, Jennie says, "Well, he's a good guy and I'd probably like the work, but this is no time to be getting mixed up with family." This is one more indication of how restricted Jennie has allowed her social life to become. It may be that she has determined that support from her family is out of bounds. Chapter 7 outlines and illustrates ways of helping clients fill out their stories with essential but missing detail related to stories, points of view, and messages.

## LISTENING TO ONESELF: THE HELPER'S INTERNAL CONVERSATION

The conversation helpers have with themselves during helping sessions is the "internal conversation." To be an effective helper, you need to listen not only to the client but also to yourself. Granted, you don't want to become self-preoccupied, but listening to yourself on a "second channel" can help you identify both what you might do to be of further help to the client and what might be standing in the way of your be-ing with and listening to the client. It is a positive form of self-consciousness.

Some years ago, this second channel did not work very well for me. A friend of mine who had been in and out of mental hospitals for a few years and whom I had not seen for over six months showed up unannounced one evening at my apart-ment. He was in a highly excited state. A torrent of ideas, some outlandish, some brilliant, flowed nonstop from him. I sincerely wanted to be with him as best I could, but I was very uncomfortable. I started by more or less naturally following the guidelines of tuning in, but I kept catching myself at the other end of the couch on which we were both sitting with my arms and legs crossed. I think that I was de-fending myself from the torrent of ideas. When I discovered myself almost literally tied up in knots, I would untwist my arms and legs, only to find them crossed again a few minutes later. It was hard work being with him. In retrospect, I realize I was concerned for my own security. I have since learned to listen to myself on the sec-ond channel a little better. When I listen to my nonverbal behavior as well as my internal dialogue, my interactions with clients are better.

Helpers can use this second channel to listen to what they are "saying" to themselves, their nonverbal behavior, and their feelings and emotions. These mes-sages can refer to the helper, the client, or the relationship.

- "I'm letting the client get under my skin. I had better do something to reset the dialogue."
- "My mind has been wandering. I'm preoccupied with what I have to do tomor-row. I had better put that out of my mind."
- "Here's a client who has had a tough time of it, but her self-pity is standing in the way of her doing anything about it. My instinct is to be sympathetic. I need to talk to her about her self-pity, but I had better go slow."
- "It's not clear that this client is interested in changing. It's time to test the waters."

The point is that this internal conversation goes on all the time. It can be a distrac-tion or it can be another tool for helping. The client, too, is having his or her inter-nal conversation. One intriguing study (Hill, Thompson, Cogar, & Denman, 1993) suggest that both client and therapist are more or less aware of the other's "covert processes." This study shows that helpers, even though they know that clients are having their own internal conversations and leave things unsaid, are not very good at determining what those things are. At times, there are verbal or nonverbal hints as to what clients' internal dialogues might be. Helping clients move key points from their internal conversations into helping dialogues is a key task and will be dis-cussed in Chapters 7, 10, 11, and 12.

## THE SHADOW SIDE OF LISTENING TO CLIENTS

Listening, as described here, is not as easy as it sounds. Obstacles and distractions abound. Some relate to listening generally. Others relate more specifically to listen-ing to and interpreting the nonverbal behavior of clients.

## Forms of Distorted Listening

The following kinds of distorted listening, as you will see from your own experience, permeate human communication. They also insinuate themselves at times into the helping dialogue. Sometimes more than one kind of distortion contaminates the helping dialogue. They are part of the shadow side because helpers never intend to engage in these kinds of listening. Rather helpers fall into them at times without even realizing that they are doing so. But they stand in the way of the kind of open-minded listening and processing needed for real dialogue.

**Filtered listening.** It is impossible to listen to other people in a completely unbiased way. Through socialization, we develop a variety of filters through which we listen to ourselves, others, and the world around us. As Hall (1977) notes: "One of the functions of culture is to provide a highly selective screen between man and the outside world. In its many forms, culture therefore designates what we pay attention to and what we ignore. This screening provides structure for the world" (p. 85). We need filters to provide structure for ourselves as we interact with the world. But personal, familial, sociological, and cultural filters introduce various forms of bias into our listening, without our being aware of it.

The stronger the cultural filters, the greater the likelihood of bias. For instance, a white middle-class helper probably tends to use white middle-class filters in listening to others. Perhaps this makes little difference if the client is also white and middle class. But if the helper is listening to an Asian client who is well-to-do and has high social status in his community, to an African American mother from an urban ghetto, or to a poor white subsistence farmer, then the helper's cultural filters might introduce bias. Prejudices, whether conscious or not, distort understanding. Like everyone else, helpers are tempted to pigeonhole clients because of gender, race, sexual orientation, nationality, social status, religious persuasion, political preferences, lifestyle, and the like. Self-knowledge on the part of helpers is essential. This includes ferreting out the biases and prejudices that distort listening.

**Evaluative listening.** Most people, even when they listen attentively, listen evaluatively. That is, as they listen, they are judging what the other person is saying as good/bad, right/wrong, acceptable/unacceptable, likable/unlikable, relevant/irrelevant, and so forth. Helpers are not exempt from this universal tendency. The following interchange takes place between Jennie and a friend of hers. Jennie recounts it to Denise as part of her story.

JENNIE: Well, the rape and the investigation are not dead, at least not in my mind. They are not as vivid as they used to be, but they are there.

FRIEND: That's the problem, isn't it? Why don't you do yourself a favor and forget about it? Get on with life, for God's sake!

Evaluative listening gives way to advice giving. It might well be sound advice, but the point here is that Jennie's friend listens and responds evaluatively. Clients should first be understood, then, if necessary, challenged or helped to challenge themselves. Evaluative listening, translated into advice giving, will just put clients off. Indeed, a judgment that a client's point of view, once understood, needs to be expanded or transcended or that a pattern of behavior, once listened to and understood, needs to be altered can be quite useful. That is, there are productive forms of evaluative listening. It is practically impossible to suspend judgment completely. Nevertheless, it is possible to set one's judgment aside for the time being in the interest of understanding clients, their worlds, their stories, their points of view, and their decisions "from the inside."

**Stereotype-based listening.** I remember my reaction to hearing a doctor refer to me as the "hernia in 304." We don't like to be stereotyped, even when the stereotype has some validity. The very labels we learn in our training—paranoid, neurotic, sexual disorder, borderline—can militate against empathic understanding. Books on personality theories provide us with stereotypes: "He's a perfectionist." We even pigeonhole ourselves: "I'm a Type A personality." In psychotherapy, diagnostic categories can take precedence over the clients being diagnosed. Helpers forget at times that their labels are interpretations rather than understandings of their clients. You can be "correct" in your diagnosis and still lose the person. In short, what you learn as you study psychology can help you to organize what you hear, but it can also distort your listening. To use terms borrowed from Gestalt psychology, make sure that your client remains "figure"—that is, in the forefront of your attention—and that models and theories about clients remain "ground"—knowledge that remains in the background and is used only in the interest of understanding and helping this unique client.

**Fact-centered rather than person-centered listening.** Some helpers ask clients many informational questions, as if clients would be cured if enough facts about them were known. It's entirely possible to collect facts but miss the person. The antidote is to listen to clients contextually, trying to focus on themes and key messages. Denise, as she listens to Jennie, picks up what is called a "pessimistic explanatory style" theme (Peterson, Seligman, & Vaillant, 1988). Concerning unfortunate events, clients with this style tend to say, directly or indirectly, such things as, "It will never go away, " "It affects everything I do," and "It is my fault." Denise knows that the research indicates that people who fall victim to this style tend to end up with poorer health than those who do not. There may be a link, she hypothesizes, between Jennie's somatic complaints (headaches, gastric problems) and this explanatory style. This is a theme worth exploring.

**Sympathetic listening.** Since most clients are experiencing some kind of misery and since some have been victimized by others or by society itself, there is a tendency on the part of helpers to feel sympathy for them. Sometimes these feelings are strong enough to distort the stories that clients are telling. Consider this case:

> Liz was counseling Ben, a man whose wife and daughter had died in a tornado. Liz had recently lost her husband to cancer. As Ben talked about his own tragedy during their first meeting, she wanted to hold him. Later that day, she took a long walk and realized how her sympathy for Ben had distorted what she heard. She heard the depth of his loss, but, reminded of her own loss, only half heard the implication that his loss now excused him from getting on with his life.

Sympathy has an unmistakable place in human relationships, but its "use," if that does not sound too inhuman, is limited in helping. In a sense, when I sympathize with someone, I become his or her accomplice. If I sympathize with my client as she tells me how awful her husband is, I take sides without knowing the complete story. Expressing sympathy can reinforce self-pity in a client. But self-pity has a way of driving out problem-managing action.

**Interrupting.** I am reluctant to add "interrupting," as some do, to this list of shadow-side obstacles to effective listening. Certainly, when helpers interrupt their clients, they, by definition, stop listening. And interrupters often say things they have been rehearsing, which means that they have been only partially listening. My reluctance, however, comes from the conviction that the helping conversation should be a dialogue. There are benign and malignant forms of interrupting. The helper who cuts the client off in mid-thought to say something important is using a malignant form. But the case is different when a helper "interrupts" a monologue with some gentle gesture and a comment such as, "You've made several points. I want to make sure that I've understood them." When interrupting promotes the kind of dialogue that serves the problem-management process, it is useful. Still, care must be taken to factor in cultural differences in storytelling.

One possible reason counselors fall prey to these kinds of shadow-side listening is the unexamined assumption that listening with an open mind is the same as approving what the client is saying. This, of course, is not the case. Rather, listening with an open mind helps you learn and understand. Whatever the reason for shadow-side listening, the outcome can be devastating because of a truth philosophers learned long ago—a small error in the beginning can lead to huge errors down the road. If the foundation of a building is out of kilter, it is hard to notice with the naked eye. But, by the time construction reaches the ninth floor, it begins to look like the Leaning Tower of Pisa. Tuning in to clients and listening both actively and with an open mind are foundation counseling skills. Ignore them and dialogue is impossible.

## Myths About Nonverbal Behavior

Myths about behavior are part of the shadow side. Richmond and McCroskey (2000) spell out the shadow side of nonverbal behavior in terms of commonly held myths (pp. 2–3):

1.  *Nonverbal communication is nonsense. All communication involves language. Therefore, all communication is verbal.* This myth is disappearing. It does not stand up under the scrutiny of common sense.

2.  *Nonverbal behavior accounts for most of the communication in human interaction.* Early studies tried to "prove" this, but they were biased. Studies were aimed at dispelling myth number 1 and overstepped their boundaries.

3.  *You can read a person like a book.* Some people, even some professionals, would like to think so. You can read nonverbal behavior, verbal behavior, and context and still be wrong.

4.  *If a person does not look you in the eye while talking to you, he or she is not telling the truth.* Tell this to liars! The same nonverbal behavior can mean many different things.

5.  *Although nonverbal behavior differs from person to person, most nonverbal behaviors are natural to all people.* Cross-cultural studies give the lie to this. But it isn't true even within the same culture.

6.  *Nonverbal behavior stimulates the same meaning in different situations.* Too often the context is the key. Yet some professionals buy the myth and base interpretive systems on it.

# SHARING EMPATHIC HIGHLIGHTS: COMMUNICATING AND CHECKING UNDERSTANDING

RESPONDING SKILLS

THE THREE DIMENSIONS OF RESPONDING SKILLS: PERCEPTIVENESS, KNOW-HOW, AND ASSERTIVENESS

Perceptiveness

Know-how

Assertiveness

SHARING EMPATHIC HIGHLIGHTS: COMMUNICATING UNDERSTANDING TO CLIENTS

THE KEY BUILDING BLOCKS OF EMPATHIC HIGHLIGHTS

The Basic Formula

Respond Accurately to Clients' Feelings, Emotions, and Moods

Use the right family of emotions and the right intensity

Distinguish between expressed and discussed feelings

Read and respond to feelings and emotions embedded in clients' nonverbal behavior

Be sensitive in naming emotions

Use different ways to share highlights about feelings and emotions

Neither overemphasize nor underemphasize feelings, emotions, and moods

Respond Accurately to the Key Experiences and Behaviors in Clients' Stories

Respond with Highlights to Clients' Points of View, Decisions, and Proposals

Communicate understanding of clients' point of view

Communicate understanding of clients' decisions

Communicate understanding of clients' intentions or proposals

PRINCIPLES FOR SHARING HIGHLIGHTS

> Use empathic highlights at every stage and step of the helping process
>
> Respond selectively to clients' core messages
>
> Respond to the context, not just the words
>
> Use highlights as a mild social-influence process
>
> Use highlights to stimulate movement in the helping process
>
> Recover from inaccurate understanding
>
> Use empathic highlights to bridge diversity gaps

TACTICS FOR COMMUNICATING HIGHLIGHTS

> Give yourself time to think
>
> Use short responses
>
> Gear your responses to the client, but remain yourself

A CAUTION: THE IMPORTANCE OF EMPATHIC RELATIONSHIPS

THE SHADOW SIDE OF SHARING EMPATHIC HIGHLIGHTS

> No response
>
> Distracting questions
>
> Clichés
>
> Interpretations
>
> Advice
>
> Parroting
>
> Sympathy and agreement
>
> Faking it

# RESPONDING SKILLS

Helpers listen to clients in order to respond to them at the service of a helping dialogue. As we have seen, the logic of listening includes visibly tuning in to clients, listening actively, processing what is heard contextually, and identifying the key ideas, messages, or points of view clients are trying to communicate—all for the sake of understanding clients. Listening, then, is a very active process that serves understanding. But helpers also respond to clients in a variety of ways. They share their understanding, they check to make sure that they've got things right, they ask questions, they probe for clarity, and they challenge clients in a variety of ways. This chapter focuses on sharing empathic highlights as a way of both communicating understanding to clients and checking to see if that understanding is accurate. When helpers communicate accurate understanding to clients, they help their clients understand themselves more fully.

## THE THREE DIMENSIONS OF RESPONDING SKILLS: PERCEPTIVENESS, KNOW-HOW, AND ASSERTIVENESS

The communication skills involved in responding to clients have three dimensions: perceptiveness, know-how, and assertiveness.

**Perceptiveness.** Your responding skills are only as good as the accuracy of the perceptions on which you base them. Consider the difference between these two examples:

> Beth is counseling Frank in a community mental-health center. Frank is scared to talk about an "ethical blunder" that he made at work. Beth senses his discomfort but thinks that he is angry rather than scared. She says, "Frank, I'm wondering what's making you so angry right now." Since Frank does not feel angry, he says nothing. In fact, he's startled by what she says and feels even more insecure. Beth takes his silence as a confirmation of his "anger." She tries to get him to talk about it.

Beth's perception is wrong and therefore disrupts the helping process. She misreads Frank's emotional state and tries to set a course based on her flawed perception. Contrast this to what happens in the following example:

> Mario, a manager, is counseling Enrique, a relatively new member of his team. During the past week, Enrique has made a significant contribution to a major project, but he has also made one rather costly mistake. Enrique's mind is on his blunder, not his success. Mario, sensing Enrique's discomfort, says, "Your ideas in the meeting last Monday helped us reconceptualize and reset the entire project. It was a great contribution. That kind of 'out of the box' thinking is very valuable here. (He pauses.) I'd also like to talk to you about Wednesday. Your conversation with Acme's purchasing agent on Wednesday made him quite angry. (He pauses briefly once more.) Something tells me that you might be more worried about Wednesday's mistake than delighted with Monday's contribution. I just wanted to let you know that I'm not." Enrique is greatly relieved. They go on to have a useful dialogue about what made Monday so good and what could be learned from Wednesday's blunder.

Mario's perceptiveness and his ability to defuse a tense situation lays the foundation for an upbeat dialogue.

The kind of perceptiveness you need to be a good helper comes from your basic intelligence, social intelligence, experience, reflecting on experience, developing wisdom, and, more immediately, tuning in to your clients, listening carefully to what they have to say, and objectively processing what they say. Perceptiveness comes with social-emotional maturity.

**Know-how.** Once you are aware of what kind of response is called for, you need to be able to deliver it. For instance, even if you are aware that a client is anxious and confused because this is his first visit to a helper, it does little good if you don't know how to translate your perceptions and your understanding into words. Let's return to Frank and Beth for a moment:

> Frank and Beth end up arguing about his "anger." Frank finally gets up and leaves. Beth, of course, takes this as a sign that she was right in the first place. The next day, Frank goes to see his minister. The minister sees quite clearly that Frank is scared and confused. His perceptions are right. He says something like this: "Frank, you seem to be very uncomfortable. It may be that whatever is on your mind might be difficult to talk about. But I'd be glad to listen to it, whatever it is. I don't want to push you into anything." Frank blurts out, "But I've done something terrible." The minister pauses and then says, "Well, let's see what kind of sense we can make of it." Frank hesitates a bit, then leans back into his chair, takes a deep breath, and launches into his story.

The minister not only is perceptive but also knows how to address Frank's anxiety and hesitation. The minister says to himself in his shadow conversation, "Here's a man who is almost exploding with the need to tell his story, but fear or shame or something like that is paralyzing him. How can I put him at ease, let him know that he won't get hurt here? I need to recognize his anxiety and gently offer an opening." He does not use these words, of course, but these are the kinds of sentiments that instinctively flit through his mind.

**Assertiveness.** Accurate perceptions and excellent know-how are meaningless if they remain locked up inside you. They need to become part of the therapeutic dialogue. For instance, if you see that self-doubt is a theme that weaves itself throughout a client's story about her frustrating search for a better relationship with her estranged brother but fail to share your hunch with her, you do not pass the assertiveness test. Consider this example:

> Nina, a young counselor in the Center for Student Development, is in the middle of the first session with Antonio, a graduate student. During the session, he mentions briefly a very helpful session he had the previous year with Carl, a middle-aged counselor on the staff. Carl has accepted an academic position at the university and is no longer involved with the center. Nina realizes that Antonio is disappointed that he couldn't see Carl and might have some misgivings about being helped by a new counselor—a younger woman at that. She has faced sensitive issues like this before and would not take it amiss if Antonio were to choose a different counselor. During a lull in the conversation, she says something like this: "Antonio, could we take a timeout here for a moment? I think you might be a bit disappointed to find out that Carl is no longer here. Or at least I probably would be if I were in your shoes. You were more or less just assigned to me and I'm not sure the fit is right. Maybe you can give that a bit of thought. Then, if you think I can be of help, you can schedule another meeting with me. But you're certainly free to review who is on staff and choose whomever you want."

In this case, perceptiveness, know-how, and assertiveness all come together. This is not to suggest that assertiveness is an overriding value in and of itself. To be assertive without perceptiveness and know-how is to court disaster.

# SHARING EMPATHIC HIGHLIGHTS: COMMUNICATING UNDERSTANDING TO CLIENTS

"Feeling empathy" for others is not helpful if the helper's perceptions are not accurate. Ickes (1993, 1997) talked about "empathic accuracy," which he defined as "the ability to accurately infer the specific content of another person's thoughts and feelings" (1993, p. 588). According to Ickes, this ability is a component of success in many walks of life.

> Empathically accurate perceivers are those who are consistently good at "reading" other people's thoughts and feelings. All else being equal, they are likely to be the most tactful advisors, the most diplomatic officials, the most effective negotiators, the most electable politicians, the most productive salespersons, the most successful teachers, and the most insightful therapists. (1997, p. 2)

The assumption is, of course, that such people not only are accurate perceivers but can weave their perceptions into their dialogues with their constituents, customers, students, and clients. Helpers do this by sharing empathic highlights with their clients.

In previous editions of *The Skilled Helper*, the same word—*empathy*—was used to denote both the value described in Chapter 3 and the communication skill described in this chapter. To avoid any confusion, the communication skill involving helpers' sharing their understanding of clients' key experiences, behaviors, and feelings has been renamed *sharing empathic highlights*—or, more simply, sharing highlights. They are *empathic* because they are driven by the helper's desire to understand the client as fully as possible and to communicate this understanding. They are *highlights* because they focus on the key points the client is making. I still avoid such terms as *paraphrasing* and *restatement*. If you are truly empathic, if you listen actively, and if you thoughtfully process what you hear, putting what the client says in its proper context, then you do more than paraphrase or restate. There is something of *you* in your response. A good response is a product of caring and hard work. Good highlights are fully human, not mechanical.

If visibly tuning in and listening are the skills that enable helpers to get in touch with the world of the client, then sharing highlights is the skill that enables them both to communicate their understanding of that world and to check the accuracy of that understanding. A secure starting point in helping others is listening to them carefully, struggling to understand their concerns, and sharing that understanding with them.

Although many people may "feel empathy" for others—that is, are motivated in many different ways by the value of empathy described in Chapter 3—the truth is that few know how to put empathic understanding into words. And so sharing empathic highlights as a way of showing understanding during conversations remains an improbable event in everyday life. Perhaps that's why it is so powerful in helping settings. When clients are asked what they find helpful in counseling sessions, being understood gets top ratings. There is such an unfulfilled need to be understood.

# THE KEY BUILDING BLOCKS OF
# EMPATHIC HIGHLIGHTS

This section is a kind of anatomy lesson; that is, we are going to take the process of sharing highlights apart and look at the pieces. Further on, we'll put them back together again.

## The Basic Formula

Basic empathic understanding can be expressed in the following stylized formula:

*You feel* . . . [here name the correct emotion expressed by the client] . . .

*because* . . . [here indicate the correct experiences and behaviors that give rise to the feelings].

For instance, Leonardo is talking with a helper about his arthritis and all its attendant ills. There is pain, of course, but more to the point, he can't get around the way he used to.

HELPER: You feel bad, not so much because of the pain but because your ability to get around—your freedom—has been curtailed.

LEONARDO: That's just it. I can take the pain. But not being able to get around is killing me! It's like being in jail.

They go on to discuss ways in which Leonardo, with the help of family and friends, can get out of "jail"—that is, become more mobile while finding ways of coping with the time he is in "jail."

The formula—"You feel . . . because . . ."—is a beginner's tool to get used to the concept of sharing highlights. It focuses on the key points of clients' stories, points of view, decisions, and proposals together with the relevant feelings, emotions, and moods associated with them. The formula is used in the following examples. For the moment, ignore how stylized it sounds. Ordinary human language will be substituted later. In the first example, a divorced mother with two young children is talking to a social worker about her ex-husband. She has been talking about the ways he has let her and their kids down.

CLIENT: I could kill him! He failed to take the kids again last weekend. This is three times out of the last six weeks.

HELPER: You feel furious because he keeps failing to hold up his part of the bargain.

CLIENT: I just have to find some way to get him to do what he promised to do. What he told the court he would do.

His not taking the kids according to their agreement (an experience for the client) infuriates her (an emotion). The helper captures both the emotion and the reason for it. And the client moves forward in terms of thinking about possible actions she could take.

In the next example, a woman who has been having a great deal of gastric and intestinal distress is going to have a colonoscopy. She is talking with a hospital counselor the night before the procedure.

PATIENT: God knows what they'll find when they go in. I keep asking questions, but they keep giving me vague answers.

HELPER: You feel troubled because you believe that you're being left in the dark.

PATIENT: In the dark about my body, my life! If they'd only tell me! Then I could prepare myself better.

They go on to discuss what she needs to do to get the kind of information she wants. The accuracy of the helper's response does not solve the woman's problems, but the patient does move a bit. She gets a chance to vent her concerns, receives a bit of understanding, and says why she wants the information. This perhaps puts her in a better position to ask for a more open relationship with her doctors.

The key elements of an empathic highlight are the same as the key elements of the client's story discussed in Chapter 5—that is, the experiences, behaviors, and feelings that make up that story. The next part of our "anatomy" lesson offers some guidelines.

## Respond Accurately to Clients' Feelings, Emotions, and Moods

The importance of feelings, emotions, and moods in our lives was discussed in Chapter 5. Helpers need to respond to clients' emotions in such a way as to move the helping process forward. This means identifying key emotions the client either expresses or discusses (helper perceptiveness) and weaving them into the dialogue (helper know-how) even when they are sensitive or part of a messy situation (helper courage or assertiveness). Remember the last time you got a problem re-solved with a good customer service representative? "I know you're angry right now because the package didn't arrive and you have every right to be. After all, we did make you a promise. Here's what we can do to make it right for you. . . ." Rather than ignoring the customer's emotions, good customer service reps face up to them as helpfully as possible.

**Use the right family of emotions and the right intensity.** In the basic highlight formula, "You feel . . ." should be followed by the correct family of emotions and the correct intensity.

*Family:* The statements "You feel hurt," "You feel relieved," and "You feel en-thusiastic" specify different families of emotion.

*Intensity:* The statements "You feel annoyed," "You feel angry," and "You're furious" specify different degrees of intensity in the same family (anger).

The words *sad, mad, bad,* and *glad* refer to four of the main families of emotion, whereas *content, quite happy,* and *overjoyed* refer to different intensities within the *glad* family.

**Distinguish between expressed and discussed feelings.** Clients both *express* emotions they are feeling during the interview and *talk about* emotions they felt at the time of some incident. For instance, consider this interchange between a client involved in a child custody proceeding and a counselor. She is talking about her husband.

CLIENT (calmly): I get furious with him [affect] when he says things, little snide things, that suggest that I don't take good care of the kids [experience].

HELPER: You feel especially angry when he intimates that you're not a good mother.

The client isn't angry right now. Rather, she is talking about the anger. The following example deals with expressed rather than discussed feelings. This woman is talking about one of her colleagues at work.

CLIENT (enthusiastically): I threw caution to the wind and confronted him about his sarcasm [action] and it actually worked. He not only apologized but behaved himself the rest of the trip [experiences for the client].

HELPER: You feel great because you took a chance and it paid off.

Clients don't always name their feelings and emotions. However, when they express emotions, it is part of the message and needs to be identified and understood.

**Read and respond to feelings and emotions embedded in clients' nonverbal behavior.** Often helpers have to read their clients' emotions—both the family and the intensity—in their nonverbal behavior. In the following example, a North American college student sits down, looks at the floor, hunches over, and haltingly begins talking with a counselor:

CLIENT: I don't even know where to start. (He falls silent.)

HELPER: It's pretty clear that you're feeling miserable. Can we talk about why?

CLIENT (after a pause): Well, let me tell you what happened. . . .

He appears depressed (affect), and his nonverbal behavior indicates that the feelings are quite intense. His nonverbal behavior reveals the broad family ("You feel bad") and the intensity ("You feel very bad"). Of course, what experiences and behaviors gave rise to these emotions are not yet known.

**Be sensitive in naming emotions.** Naming and discussing feelings threatens some clients. In such cases, it might be better to focus on experiences and behaviors and proceed only gradually to a discussion of feelings. The following client, an unmarried man in his mid-thirties who has come to talk about "certain dissatisfactions" in his life, has shown some reluctance to express or even to talk about feelings.

CLIENT (in a pleasant, relaxed voice): My mother is always trying to make a little kid out of me. And I'm 35! Last week, in front of a group of my friends, she brought out my rubber boots and an umbrella and gave me a little talk on how to dress for bad weather (laughs).

COUNSELOR A: It might be hard to admit it, but I get the feeling that down deep you were furious.

CLIENT: Well, I don't know about that. Anyway, at work. . . .

Counselor A pushes the emotion issue and is met with some resistance. The client changes the topic.

COUNSELOR B (in a somewhat lighthearted way): So she's still playing the mother role—to the hilt, it would seem.

CLIENT (with more of a bite in his voice): And the hilt includes not wanting me to grow up. But I am grown up . . . well, pretty grown up. But I don't always act grown up around her.

Counselor B, choosing to respond to the "strong mother" issue rather than the more sensitive "being kept a kid and feeling really lousy about it" issue, gives the client more room to move. This works, for the client himself moves toward the more sensitive issue: his playing the child, at least at times, when he's with his mother.

Some clients are hesitant to talk about certain emotions. One client might find it relatively easy to talk about his anger but not his hurt. The following client is talking about his disappointment at not being chosen for a special team at work:

CLIENT: I worked as hard as anyone else to get the project up and running. In fact, I was at the meeting where we came up with the idea in the first place. . . . And now they've dropped me.

COUNSELOR A: So you feel really hurt—left out of your own project.

CLIENT (hesitating): Hmm. . . . I'm really ticked off. Why shouldn't I be! . . .

Here is a client with lots of ego. He doesn't like the idea that he has been "hurt." Counselor B takes a different tact:

COUNSELOR B: So it's more than annoying to be left out of what, in many ways, is your own project.

CLIENT: How could they do that? . . . It's more than annoying. It's . . . well . . . humiliating!

Counselor B, factoring in the client's ego, sticks to the anger, allowing the client himself to name the more sensitive emotion. Contextual listening—in this case, listening to the client's emotions through the context of the pride he takes in himself and his accomplishments—is part of social intelligence. However, being sensitive to clients' sensitive emotions should not rob counseling of its robustness. Too much tiptoeing around clients' "sensitivities" does not serve them well. Remember what was said earlier: Clients are not as fragile as we sometimes make them out to be.

**Use different ways to share highlights about feelings and emotions.** Since clients express feelings in a number of different ways, helpers can communicate an understanding of feelings in a variety of ways.

- *By single words:* You feel good. You're depressed. You feel abandoned. You're delighted. You feel trapped. You're angry.

- *By different kinds of phrases:* You're sitting on top of the world. You feel down in the dumps. You feel left in the lurch. Your back's up against the wall. You're really steaming. You're really on a roll.

- *By what is implied in behavioral statements:* You feel like giving up (implied emotion: despair). You feel like hugging him (implied emotion: joy). Now that you see what he's done to you, you almost feel like throwing up (implied emotion: disgust).

- *By what is implied in experiences that are revealed:* You feel you're being dumped on (implied feeling: victimized). You feel you're being stereotyped (implied feeling: resentment). You feel you're at the top of her list (implied feeling: elation). You feel you're going to get caught (implied feeling: apprehension). Note that the implication of each can be spelled out: You feel angry because you're being dumped on. You resent the fact that you're being stereotyped. You feel great because it seems that you're at the top of her list.

Because ultimately you must discard formulas and use your own language—words that are yours rather than words from a textbook—it helps to have a variety of ways to communicate your understanding of clients' feelings and emotions. It keeps you from being wooden in your responses. Consider this example: The client tells you that she has just been given the kind of job she has been seeking for the past two years. Here are some possible responses to her emotion:

**Single word:** "You're really happy."

**A phrase:** "You're on cloud nine."

**Experiential statement:** "You feel you finally got what you deserve."

**Behavioral statement:** "You feel like going out and celebrating."

Obviously, your responses to clients should be yours, not canned responses from a textbook. With experience, you can extend your range of expression at the service of your clients. Providing variety will become second nature.

**Neither overemphasize nor underemphasize feelings, emotions, and moods.** Some counselors take an overly rational approach to helping and almost ignore clients' feelings. Others become too preoccupied with clients' emotions and moods. They pepper clients with questions about feelings and at times extort answers. To say that feelings, emotions, and moods are important is not to say that they are everything. The best defense against either extreme is to link feelings, emotions, and moods to the experiences and behaviors that give rise to them (see Anderson & Leitner, 1996).

## Respond Accurately to the Key Experiences and Behaviors in Clients' Stories

Key experiences and behaviors give rise to clients' feelings, emotions, and moods. The "because" in the empathic highlight formula is to be followed by an indication of the experiences and behaviors that underlie the client's feelings. In the following example, the client, a graduate student in law school, is venting his frustration:

CLIENT (heatedly): You know why he got an A? He took my notes and disappeared. I didn't get a chance to study them. And I never even confronted him about it.

HELPER: You feel doubly angry because not only did he steal your notes, but you let him get away with it.

The response specifies both the client's experience (the theft) and his behavior (in this case, a failure to act) that give rise to his distress. His anger is directed at not only his classmate but also himself.

In the following example, a mugging victim has been talking to a counselor to help cope with his fears of going out. Before the mugging, he had given no thought to urban problems. Now he tends to see menace everywhere.

CLIENT: This gradual approach of getting back in the swing seems to be working. Last night I went out without a companion. First time. I have to admit that I was scared. But I think I've learned how to be careful. Last night was important. I feel I can begin to move around again.

HELPER:  You feel comfortable with the one-step-at-a-time approach you've been taking. And it paid off last night when you bought back a big chunk of your freedom.

CLIENT:  That's it! I know I'm going to be free again. . . . Here's what I've been thinking of doing. . . .

The client is talking about the success of the action phase of the program. The helper's response recognizes the client's satisfaction with the success of the program and how important it is for the client to feel both safe and free. The client moves on to describe the next phase of his program.

Another client, after a few sessions spread out over six months, tells her therapist about the progress she is making in rebuilding her life after a devastating car accident. She's back at work and has been working with her husband at rebuilding their marriage.

CLIENT (talking in an animated way):  I really think that things couldn't be going better. I'm doing very well at my new job, and my husband isn't just putting up with it. He thinks it's great. He and I are getting along better than ever, even sexually, and I never expected that. We're both working at our marriage. I guess I'm just waiting for the bubble to burst.

HELPER:  You feel great because things have been going better than you ever expected—and it seems almost too good to be true.

CLIENT:  Well, a "bubble bursting" might be the wrong image. I think there's a difference between being cautious and waiting for disaster to strike. I'll always be cautious, but I'm finding out that I can make things come true instead of sitting around waiting for them to happen as I usually do. I guess I've got to keep making my own luck.

The helper's highlight captures the flavor of the client's experiences, behaviors, and feelings. By encompassing both the client's enthusiasm and her lingering fears, the response is quite useful because the client makes an important distinction between reasonable caution and expecting the worst to happen. She moves on to her need to make things happen, to become more of an agent in her life.

As a beginner's tool for understanding how to share highlights, the stylized formula of "you feel . . . because . . ." has outlived its usefulness at this point and will be dropped in most of the examples that follow. It's too wooden. Experienced trainers use it only when it sounds natural. Otherwise, they use ordinary language to share highlights.

## Respond with Highlights to Clients' Points of View, Decisions, and Proposals

By sharing highlights, you communicate to clients that you are working hard at understanding them to foster constructive change. This means not only understanding the key elements of the stories they tell but also the key elements of anything they share with you. Here are some examples that relate to clients' points of view, decisions, and proposals. It goes without saying that points of view, decisions, and proposals are, like stories, permeated to one degree or another with feelings and emotions.

**Communicate understanding of clients' points of view.**  In the following example, the client, a 45-year-old man, is a construction worker, married, with four children between the ages of 9 and 16. He has been expressing concerns about his children.

CLIENT: I don't consider myself old-fashioned, but I think kids these days suffer from overindulgence. We keep giving them things. We let them do what they want. I fall into the same trap myself. It's just not good for them. I don't think we're preparing them for what the world is really like. People assume that the economy will keep booming. Everyone keeps shouting, "Free lunch!" This isn't doing kids any good.

COUNSELOR: So you see the "do what you want" and "free lunch" messages as a lot of hogwash. It's going to backfire, and your kids could end up getting hurt.

CLIENT: Right. . . . But I'm not in control. My kids can get one set of messages from me and then get a flood of contradictory messages outside and from TV and the Internet. I don't want to be a tyrant. Or come across as a killjoy. That doesn't work anyway. At work I see problems and I take care of them. But this has got me stymied.

COUNSELOR: So the whole picture seems pretty gloomy right now. You're not exactly sure what to do about it. And it's not exactly like the problems at work. You handle those routinely. What makes these problems so different?

Once the counselor communicates understanding of the client's point of view, the client moves on to share his sense of helplessness. The counselor realizes, however, that the client needs to check out the implications of his point of view. For instance, many parents do get their children to buy a more wholesome perspective on life than the media often present. And the client is probably not as helpless as he makes himself out to be.

**Communicate understanding of clients' decisions.** When clients announce key decisions or express their resolve to do something, it's important to recognize the core of what they are saying. In the following example, a client being treated for social phobia has benefitted greatly from cognitive-behavioral therapy. For instance, in uncomfortable social situations he has learned to block self-defeating thoughts and to keep his attention focused externally—on the social situation itself and on the agenda of the people involved—instead of turning in on himself. Here is a sample of this implementation or action-arrow dialogue:

CLIENT (emphatically): I'm not going to turn back. I've had to fight to get where I am now. But I can see how easy it could be to slide back into my old habits. I bet a lot of people do. I see it all around me. People make resolutions and then they peter out.

HELPER: Even though it's possible for you to give up your hard-earned gains, you're not going to do it. You're just not.

CLIENT: But what can I do to make sure that I won't? I'm convinced I won't, but . . .

HELPER: You need some rachets. They're the things that keep roller-coaster cars from sliding back. You hear them going click, click, click on the way up.

CLIENT: Ah, right! But I need psychological ones.

HELPER: And social ones. . . . What's kept you from sliding back so far?

In a positive-psychology mode, the counselor focuses on past successes. They go on to discuss the kind of "rachets" he needs to stay on track.

**Communicate understanding of clients' intentions or proposals.** In the next example, the client, who is hearing impaired, has been discussing ways of becoming, in her words, "a full-fledged member of my extended family." The discussion between client and helper takes place through a combination of lipreading and signing.

CLIENT (enthusiastically): Let me tell you what I'm thinking of doing. . . . First of all, I'm going to stop fading to the background in family and friends' conversation groups. I'll be the best listener there. And I'll get my thoughts across even if I have to use props. That's how I really am . . . inside, you know, in my mind.

HELPER: This sounds exciting. You're thinking of getting right into the middle of things, where you belong. You might even try a bit of drama.

CLIENT: And I think that, well, socially, I'm pretty smart. So I'm not talking about being melodramatic or anything. I can do all this with finesse, not just barge in.

HELPER: So the you they'll see will be socially savvy. You'll make it all natural. Draw me a couple of pictures of what this would look like.

The client comes up with a proposal for a course of action that will help her take her "rightful place" in conversations with family and friends setting her agenda (Stage III). The helper's responses recognize her enthusiasm and sense of determination. They go on to have a dialogue about practical tactics. And providing some examples helps the client identify the implications of her proposal.

## PRINCIPLES FOR SHARING HIGHLIGHTS

Here are a number of principles that can guide you as you share highlights. Remember that these guidelines are principles, not formulas to be followed slavishly.

**Use empathic highlights at every stage and step of the helping process.** Sharing highlights is useful at every stage and every step of the helping process. Communicating and checking understanding is always helpful. Here are some examples of helpers sharing highlights at different stages and steps of the helping process.

*Stage I: Problem clarification and opportunity identification.* A teenager in his third year of high school has just found out that he is moving with his family to a different city. A school counselor responds, "You're miserable because you have to leave all your friends. But it sounds like you may even feel a bit betrayed. You didn't see this coming at all." The counselor realizes that he has to help his client pick up the pieces and move on, but sharing his understanding helps build a foundation to do so. The teen goes on to talk in positive terms about the large city they will be moving to and the opportunities it will offer. At one point, the school counselor responds, "So there's an upside to all this. Big cities are filled with things to do. You like theater and there's loads there. That's something to look forward to."

*Stage II: Evaluating goal options.* A woman has been discussing the trade-offs between marriage and career. At one point, her helper says, "There's some ambivalence here. If you marry Jim, you might not be able to have the kind of career you'd like. Or did I hear you half say that it might be possible to put both together? Sort of get the best of both worlds." The client goes on the explore the possibilities around "getting the best of both worlds." It helps her greatly in preparing for her next conversation with Jim.

*Stage III: Choosing actions to accomplish goals.* A man has been discussing his desire to control his cholesterol level without taking a medicine with side effects that worry him. He says that this drug-free approach might work. The counselor responds, "It's a relief to know that sticking to the diet and exercise might mean that you won't have to take any medicine. Hmm . . . let's explore the 'might' part. I'm not exactly sure what your doctor said." The helper recognizes an important part of the client's message but then seeks further clarification.

*The action arrow: implementation issues.* A married couple have been struggling to put into practice a few strategies to improve their communication with each other. They've both called their attempt a "disaster." The counselor replies, "OK, so you're annoyed with yourselves for not accomplishing even the simple active-listening goals you set for yourselves. . . . Let's see what we can learn from the 'disaster'" (said somewhat lightheartedly). The counselor communicates understanding of their disappointment in not implementing their plan, but, in the spirit of positive psychology, focuses on what they can learn from the failure.

Communicating understanding by sharing empathic highlights is a mode of human contact, a relationship builder, a conversational lubricant, a perception-checking intervention, and a mild form of social influence. It is always useful. Driscoll (1984), in his commonsense way, refers to highlights as "nickel-and-dime interventions which each contribute only a smidgen of therapeutic movement, but without which the course of therapeutic progress would be markedly slower" (p. 90). Since sharing highlights provides a continual trickle of understanding, it is a way of providing support for clients throughout the helping process. It is never wrong to let clients know that you are trying to understand them from their frame of reference. Of course, thoughtful listening and processing lead to highlights that are much more than "nickel-and-dime interventions." Clients who feel they are being understood participate more effectively and more fully in the helping process. Since sharing highlights helps build trust, it paves the way for stronger interventions on the part of the helper, such as challenging.

**Respond selectively to clients' core messages.** It is impossible to respond with highlights to everything a client says. Therefore, as you listen to clients, try to identify and respond to what you believe are core messages—that is, the heart of what the client is saying and expressing—especially if the client speaks at any length. Sometimes this selectivity means paying particular attention to one or two messages even though the client communicates many. For instance, a young woman, in discussing her doubts about marrying her companion, says at one time or another during a session that she is tired of his sloppy habits, is not really interested in his friends, wonders about his lack of intellectual curiosity, is dismayed at his relatively low level of career aspirations, and vehemently resents the fact that he faults her for being highly ambitious. Throughout one session, the counselor follows the client's lead, sharing a steady stream of empathic highlights to help the client herself identify what is core. His summary highlight at the end allows her to question the direction in which she and her friend are headed:

COUNSELOR: The whole picture doesn't look very promising, but the mismatch in career expectations is especially troubling.

CLIENT: You know, I'm beginning to think that Jim and I would be pretty good friends, even *because* we're so different. But partners? Maybe that's pushing it.

Of course, since clients are not always so obliging, helpers must continually ask themselves as they listen, "What is key? What is most important here?" and then find ways of checking it out with clients. This helps clients sort out things that are not clear in their own minds.

Responding selectively sometimes means focusing on experiences or actions or feelings rather than all three. Consider the following example of a client who is experiencing stress because of his wife's poor health and concerns at work:

CLIENT: This week I tried to get my wife to see the doctor, but she refused, even though she fainted a couple of times. The kids had no school, so they were underfoot almost constantly. I haven't been able to finish a report my boss expects from me next Monday.

HELPER: It's been a lousy, overwhelming week all the way around.

CLIENT: As bad as they come. When things are lousy both at home and at work, there's no place for me to relax. I just want to get the hell out of the house and find some place forget it all. . . . Almost run away. . . . But I can't. . . . I mean I won't.

The counselor chooses to emphasize the feelings of the client, because she believes that his feelings of frustration and irritation are uppermost in his consciousness right now. This helps him move deeper into the problem situation—and then find a bit of resolve at the bottom of the pit.

At another time or with another client, the emphasis might be quite different. In the next example, a young woman is talking about her problems with her father:

CLIENT: My dad yelled at me all the time last year about how I dress. But just last week I heard him telling someone how nice I looked. He yells at my sister about the same things he ignores when my younger brother does them. Sometimes he's really nice with my mother and other times, too much of the time, he's just awful—demanding, grouchy, sarcastic.

HELPER: The inconsistency is killing you.

CLIENT: Absolutely! It's hard for all of us to know where we stand. I hate coming home when I'm not sure which "dad" will be there. Sometimes I come late to avoid all this. But that makes him even madder.

In this response, the counselor emphasizes the client's experience of her father's inconsistency. It hits the mark and she explores the problem situation further.

**Respond to the context, not just the words.** A good empathic response is not based solely on the client's immediate words and nonverbal behavior. It also takes into account the context of what is said, everything that "surrounds" and permeates a client's statement. This client may be in crisis. That client may be doing a more leisurely "taking stock" of life. You are listening to clients in the context of their lives. The context modifies everything the client says.

Consider this case. Jeff, a white teenager, is accused of beating a black youth whose car stalled in a white neighborhood. The beaten youth is still in a coma. When Jeff talks to a court-appointed counselor, the counselor listens to what Jeff says in light of Jeff's upbringing and environment. The context includes the racist attitudes of many people in his blue-collar neighborhood, the sporadic violence there, Jeff's gang membership, the fact that his father died when Jeff was in primary school, a somewhat indulgent mother with a history of alcoholism, and easy access to soft drugs. The following interchange takes place:

JEFF: I don't know why I did it. I just did it, me and these other guys. We'd been drinking a bit and smoking up a bit—but not too much. It was just the whole thing.

HELPER: Looking back, it's almost like it's something that happened rather than something you did, and yet you know, somewhat bitterly, that you actually did it.

JEFF: More than bitter! I've screwed up the rest of my life. It's not like I got up that morning saying that I was going to bash someone that day.

The counselor's response is in no way an attempt to excuse Jeff's behavior, but it does factor in some of the environmental realities. Later on he will challenge Jeff to decide whether his environment—prejudices, gang membership, family history—is to own him or whether, to the degree that this is possible, he is to own his environment.

**Use highlights as a mild social-influence process.** Because helpers cannot respond with highlights to everything their clients say, they are always searching for core messages. They are forced into a selection process that influences the course of the therapeutic dialogue. So even sharing highlights can be part of the social-influence process outlined in Chapter 3. Helpers believe that the messages they select for attention are core primarily because they are core for the client. But helpers also believe, at some level, that certain messages should be important for the client. In the following example, an incest victim turned incest perpetrator is in jail awaiting trial. In a session with a counselor, he is trying to exonerate himself by blaming what happened to him in the past. He has been talking so quickly that the helper finds it difficult to interrupt. Finally, the helper, who has a pretty good working relationship with the client, breaks in.

HELPER: You've used some strong language to describe yourself. Let me see if I have some of them right. You said something about being "structurally deformed." I believe you also used the term "automatic reactions." You describe yourself as "haunted" and "driven."

CLIENT: Well . . . I guess it's strong language. . . . Makes me sound like a psychological freak. Which I'm not.

The helper wants the client to listen to himself. So his "let me get this straight" response is a kind of empathic highlight form of challenge. It hits the mark because the client pulls himself up short. Of course, helpers need to be careful not to put words in clients' mouths.

**Use highlights to stimulate movement in the helping process.** Sharing highlights is an excellent tool for building the helping relationship. But it also needs to serve the goals of the helping process. Therefore, sharing highlights is useful to the degree that it helps the client move forward. What does "move forward" mean? That depends on the stage or step in focus. For instance, sharing highlights helps clients move forward in Stage I if it helps them explore a problem situation or an undeveloped opportunity more realistically. It helps clients move forward in Stage II to the degree that it helps them identify and explore possibilities for a better future, craft a change agenda, or discuss commitment to that agenda. Moving forward in Stage

III means clarifying action strategies, choosing specific things to do, and setting up a plan. In the action phase, moving forward means identifying obstacles to action, overcoming them, and accomplishing goals.

In the following example, a somewhat stressed trainee in a counseling program is talking to his supervisor:

TRAINEE: I don't think I'm going to make a good counselor. The other people in the program seem brighter than I am. Others seem to be picking up the knack of sharing highlights faster than I am. I'm still afraid of responding directly to others. I think I should reevaluate my participation in the program.

TRAINER: So you catch yourself saying to yourself things like, "She's better than I am at this" and "He's picking this stuff up faster than I am." And this adds up to, "Maybe I shouldn't be here at all." Is it something like that?

TRAINEE: Well, yes, but . . . I know that my tendency to get down on myself and give up is part of the problem, part of my style. I'm not the brightest, but I'm certainly not dumb either. My communication skills are a lot better than when I first arrived. And I'm using them more in everyday life. . . ."

When the trainer "hits the mark," it jolts the trainee into focusing on his strengths rather than his weaknesses.

In the next example, a young woman visits the student services center at her college to discuss an unwanted pregnancy:

CLIENT: And so here I am, two months pregnant. I don't want to be pregnant. I'm not married, and I don't even love the father. To tell the truth, I don't even think I like him. Oh, Lord, this is something that happens to other people, not me! I wake up thinking this whole thing is unreal. Now people are trying to push me toward abortion.

HELPER: You're still so amazed that it's almost impossible to accept that it's true. To make things worse, people are telling you what to do.

CLIENT: Amazed? I'm stupefied! Mainly, at my own stupidity for getting myself into this. I've never had such an expensive lesson in my life. But I've decided one thing. No one, no one is going to tell me what to do now. I'll make my own decisions.

After the helper's highlight, self-recrimination over her lack of self-responsibility helps the client make a stand. She says she wants to capitalize on a very expensive mistake. It often happens that sharing highlights that hit the mark puts pressure on clients to move forward. So sharing highlights, even though it is a communication of understanding, is also part of the social-influence process.

**Recover from inaccurate understanding.** Although helpers should strive to be accurate in the understanding they communicate, all helpers can be inaccurate at times. You may think you understand the client and what he or she has said only to find out, when you share your understanding, that you were off the mark. Therefore, sharing highlights is a perception-checking tool. If the helper's response is accurate, the client often tends to confirm its accuracy in two ways. The first is some kind of verbal or nonverbal indication that the helper is right. That is, the client nods or gives some other nonverbal cue, or uses some assenting word or phrase such as "that's right" or "exactly." This happens in the following example, in which a client who has been arrested for selling drugs is talking to his probation officer.

HELPER: So your neighborhood makes it easy to do things that can get you into trouble.

CLIENT: You bet it does! For instance, everyone's selling drugs. You not only end up using them, but you begin to think about pushing them. It's just too easy.

The second and more substantive way in which clients acknowledge the accuracy of the helper's response is by moving forward in the helping process—for instance, by clarifying the problem situation or preferred-scenario possibilities more fully. In the preceding example, the client not only acknowledges the accuracy of the helper's empathy verbally—"You bet it does"—but, more importantly, also outlines the problem situation in greater detail. By again responding with a shared highlight, the helper leads the client to the next cycle: further clarification of the problem or opportunity, or moving on to goal setting or some kind of problem-managing action.

On the other hand, when a response is inaccurate, the client often lets the counselor know in different ways. He or she may stop dead, fumble around, go off on a different tangent, tell the counselor, "That's not exactly what I meant," or even try to get the helper back on track. Helpers need to be sensitive to all these cues. In the following example, Ben, a man who lost his wife and daughter in a train crash, has been talking about the changes that have taken place since the accident:

HELPER: So you don't want to do a lot of the things you used to do before the accident. For instance, you don't want to socialize much anymore.

BEN (pausing a long time): Well, I'm not sure that it's a question of wanting to or not. I mean that it takes much more energy to do a lot of things. It takes so much energy for me just to phone others to get together. It takes so much energy sometimes being with others that I just don't try.

HELPER: It's like a movie of a man in slow motion—it's so hard to do almost anything.

BEN: I'm in low gear, grinding away. And I don't know how to get out of it.

Ben says that it is not a question of motivation but of energy. The difference is important to him. By picking up on it, the helper gets the interview back on track. Ben wants to regain his old energy, but he doesn't know how. His "lack of energy" is most likely some form of depression, and there are a number of ways to help clients deal with depression. This provides an opening for moving the helping process forward.

If you are intent on understanding your clients, they will not be put off by occasional inaccuracies on your part. Figure 6-1 indicates two different paths: one when helpers hit the mark in sharing highlights the first time, the other when they are inaccurate and then recover.

**Use empathic highlights to bridge diversity gaps.** This principle is a corollary of the preceding two. Highlights based on effective tuning in and listening constitute one of the most important tools you have in interacting with clients who differ from you in significant ways. Sharing highlights is one way of telling clients that you are a learner, especially if the client differs from you in significant ways. Scott and Borodovsky (1990) refer to empathic listening as "cultural role taking." They could say "diversity role taking." In the following example, a young white male

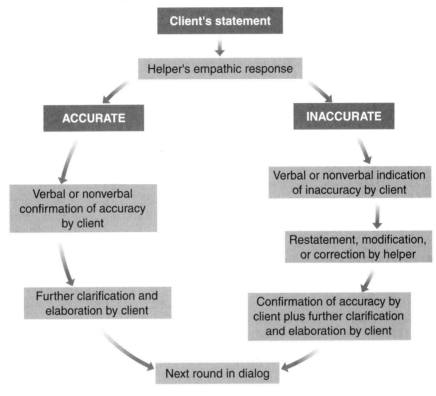

**FIGURE 6-1**
**The Movement Caused by Accurate and Inaccurate Highlights**

counselor is talking with an elderly African American woman who has recently lost her husband. She is in the hospital with a broken leg.

CLIENT: I hear they try to get you out of these places as quick as possible. But I seem to be lying around here doing nothing. Jimmy [her late husband] wouldn't even recognize me.

HELPER: It's pretty depressing to have this happen so close to losing your husband.

CLIENT: Oh, I'm not depressed. I just want to get out of here and get back to doing things at home. Jimmy's gone, but there's plenty of people around there to help me take care of myself.

HELPER: Getting back into the swing of things is the best medicine for you.

CLIENT: Now you got it right. What I need right now is to know when I can go home and what I need to do for my leg once I get there. I've got to get things in order. That's what I do best.

The helper makes assumptions that might be true for him and his culture, but they miss the mark with the client. She's taking her problems in stride and counting on her social system and a return to everyday household life to keep her going. The helper's second response hits the mark and she, in Stage II fashion, outlines some of things she wants.

## TACTICS FOR COMMUNICATING HIGHLIGHTS

The principles just outlined provide strategies for sharing empathic highlights. Here are a few hints—tactics, if you will—to help you improve the quality of your responses:

**Give yourself time to think.** Beginners sometimes jump in too quickly with an empathic response when the client pauses. "Too quickly" means that they do not give themselves enough time to reflect on what the client has just said to identify the core message being communicated. Watch some experts on tape. They often pause and allow themselves to assimilate what the client is saying.

**Use short responses.** I find that the helping process goes best when I engage the client in a dialogue rather than give speeches or allow the client to ramble. In a dialogue, the helper's responses can be relatively frequent but should be lean and trim. In trying to be accurate, the beginner is often long-winded, especially if he or she waits too long to respond. Again, the question "What is the core of what this person is saying to me?" can help you make your responses short, concrete, and accurate.

**Gear your responses to the client, but remain yourself.** If a client speaks animatedly, telling you how he finally got his partner to listen to his point of view about a new venture, and you reply accurately but in a flat, dull voice, your response is not fully empathic. This does not mean that you should mimic your clients, go overboard, or not be yourself. It means that part of being with the client is sharing in a reasonable way in his or her emotional tone. Consider this example:

TWELVE-YEAR-OLD CLIENT: My teacher started picking on me from the first day of class. I don't fool around more than anyone else in class, but she gets me anytime I do. I think she's picking on me because she doesn't like me. She doesn't yell at Bill Smith, and he acts funnier than I do.

COUNSELOR A: This is a bit perplexing. You wonder why she singles you out for so much discipline.

Counselor A's language is stilted, not in tune with the way a 12-year-old speaks. Here's a different approach.

COUNSELOR B: You're mad because the way she picks on you seems unfair.

On the other hand, helpers should not adopt a language that is not their own just to be on the client's wavelength. An older counselor using "hip" language or slang with a young client sounds ludicrous.

## A CAUTION: THE IMPORTANCE OF EMPATHIC RELATIONSHIPS

In day-to-day conversations, sharing empathic highlights is a tool of civility. Making an effort to get in touch with your conversational partner's frame of reference sends a message of respect. Therefore, sharing highlights plays an important part in building relationships. However, the communication skills as practiced in helping settings don't automatically transfer to the ordinary social settings of everyday life.

In everyday life, understanding does not necessarily have to be put into words. Given enough time, people establish empathic relationships with one another in which understanding is communicated in a variety of rich and subtle ways without necessarily being put into words. A simple glance across a room as one spouse sees the other trapped in a conversation with a person he or she does not want to be with can communicate worlds of understanding. The glance says, "I know you feel caught. I know you don't want to hurt the other person's feelings. I can feel the struggles going on inside you. But I also know that you'd like me to rescue you, as soon as I can do so tactfully."

People with empathic relationships often express empathy in actions. An arm around the shoulders of someone who has just suffered a defeat expresses both empathy and support. I was in the home of a poor family when the father came bursting through the front door shouting, "I got the job!" His wife, without saying a word, went to the refrigerator, got a bottle of beer with a makeshift label on which "Champagne" had been written and offered it to her husband. Beer never tasted so good.

On the other hand, some people enter caringly into the world of their relatives, friends, and colleagues and are certainly "with" them but don't know how to communicate understanding through words. When a wife complains, "I don't know whether he really understands," she is not necessarily saying that her relationship with her husband is not mutually empathic. She is more likely saying that she would appreciate it if he were to put his understanding into words more often. In general, it is highly desirable to use empathic highlights more frequently in everyday life, especially when relationships are not going as well as they might. Sharing highlights plays an important role in developing empathic relationships. Box 6-1 summarizes the main points about the use of empathy as a communication skill.

## THE SHADOW SIDE OF SHARING EMPATHIC HIGHLIGHTS

Some helpers are poor communicators without even realizing it. Many responses that novice or inept helpers make are really poor substitutes for sharing accurate empathic highlights. Here is an example that illustrates a range of such responses. Robin is a young woman who has just started a career in law. This is her second visit to a counselor in private practice. In the first session, she said she wanted to "talk through" some issues relating to the "transition" from school to business life. She appeared quite self-confident. In this session, after talking about a number of transition issues, she begins speaking in a rather strained voice and avoids eye contact with the counselor.

ROBIN: Something else is bothering me a bit. . . . Maybe it shouldn't. After all, I've got the kind of career that a lot of women would die for. Well—I'm glad that none of my feminist colleagues is around—I don't like the way I look. I'm neither fat nor thin, but I don't really like the shape of my body. And I'm uncomfortable with some of my facial features. Maybe this is a strange time of life to start thinking about this. In two years, I'll be 30. . . . I bet I seem like an affluent, self-centered yuppie.

Robin pauses and looks at a piece of art on the wall. What would you do or say? Here are some possibilities that are better avoided.

### Box 6-1   Suggestions for Sharing Empathic Highlights

1. Remember that empathy is a value, a way of being, that should permeate all communication skills.
2. Tune in carefully, both physically and psychologically, and listen actively to the client's point of view.
3. Make every effort to set your judgments and biases aside for the moment and walk in the client's shoes.
4. As the client speaks, listen especially for core messages.
5. Listen to both verbal and nonverbal messages and their context.
6. Respond with highlights fairly frequently, but briefly, to the client's core messages.
7. Be flexible and tentative enough that the client does not feel pinned down.
8. Use highlights to keep the client focused on important issues.
9. Move gradually toward the exploration of sensitive topics and feelings.
10. After sharing a highlight, attend carefully to cues that either confirm or deny the accuracy of your response.
11. Determine whether your highlights are helping the client remain focused and are stimulating the clarification of key issues.
12. Note signs of client stress or resistance; try to judge whether these arise because you are inaccurate or because you are too accurate in your responses.
13. Keep in mind that the communication skill of sharing empathic highlights, however important, is just one tool to help clients see themselves and their problem situations more clearly with a view to managing them more effectively.

**No response.** It can be a mistake to say nothing, though cultures differ widely in how they deal with silence (Sue, 1990). In North American culture, generally speaking, if the client says something significant, respond to it, however briefly. Otherwise, the client might think that what he or she has just said doesn't merit a response. Don't leave Robin sitting there stewing in her own juices. A skilled helper would realize that a person's nonacceptance of his or her body could generalize to other aspects of life (Dworkin & Kerr, 1987; Worsley, 1981) and therefore should not be treated as just a "vanity" problem.

**Distracting questions.** Some helpers, like many people in everyday life, cannot stop themselves from asking questions. Instead of responding with an empathic highlight, a counselor might ask something like, "Is this something new now that you've started working?" This response ignores what Robin has said and the feelings she has expressed and focuses rather on the helper's mistaken agenda to get more information. More about this in Chapter 7.

**Clichés.** A counselor might say, "The workplace is competitive. It's not uncommon for issues like this to come up." This is cliché talk. It turns the helper into an insensitive instructor and probably sounds dismissive to the client. Clichés are hollow. The helper is saying, in effect, "You don't really have a problem at all, at least not a serious one." Clichés are a very poor substitute for understanding.

**Interpretations.** For some helpers, interpretive responses based on their theories of helping seem more important than expressing understanding. Such a counselor might say something like this: "Robin, my bet is that your body-image concerns are probably just a symptom. I've got a hunch that you're not really accepting yourself. That's the real problem." The counselor fails to respond to the client's feelings and also distorts the content of the client's communication. The response implies that what is really important is hidden from the client.

**Advice.** In everyday life, giving unsolicited advice is extremely common. It happens in counseling, too. For instance, a counselor might say to Robin, "Hey, don't let this worry you. You'll be so involved with work issues that these concerns will disappear." Advice giving at this stage is out of order and, to make things worse, the advice given has a cliché flavor to it. Furthermore, giving advice robs clients of self-responsibility.

**Parroting.** Sharing a highlight does not mean merely repeating what the client has said. Such parroting is a parody of sharing empathic highlights. Review what Robin said about herself at the beginning of this section. Then evaluate the following "empathic" response.

COUNSELOR: So Robin, even though you have a great job, one that many people would envy, it's your feelings about your body that bother you. The feminist in you recoils a bit from this news. But there are things you don't like—your body shape, some facial features. You're wondering why this is hitting you now. You also seem to be ashamed of these thoughts. "Maybe I'm just self-centered" is what you're saying to yourself.

Most of this is accurate, but it sounds awful. Mere repetition, or restatement, or paraphrasing carries no sense of real understanding of, no sense of being with, the client. Real understanding, because it passes through you, should convey some part of you. Parroting doesn't. To avoid parroting, tap into the processing you've been doing as you listened, come at what the client has said from a slightly different angle, use your own words, change the order, refer to an expressed but unnamed emotion—in a word, do whatever you can to let the client know that you are working at understanding.

**Sympathy and agreement.** Being empathic is not the same as agreeing with the client or being sympathetic. An expression of sympathy has much more in common with pity, compassion, commiseration, and condolence than with empathic understanding. Although these are fully human traits, they are not particularly useful in counseling. Sympathy denotes agreement, whereas empathy denotes understanding and acceptance of the person of the client. At its worst, sympathy is a form of collusion with the client. Note the difference between Counselor A's response to Robin and Counselor B's response.

COUNSELOR A: This is not an easy thing to struggle with. It's even harder to talk about. It's even worse for someone who is as self-confident as you usually are.

ROBIN: I guess so.

Note that Robin does not respond very enthusiastically to collusion talk. She is interested in managing her problem. The helping process does not move forward. Let's see a different approach.

COUNSELOR B: You've got some misgivings about how you look, yet you wonder whether you're even justified talking about it.

ROBIN: I know. It's like I'm ashamed of my being ashamed. What's worse, I get so preoccupied with my body that I stop thinking of myself as a person. It blinds me to the fact that I more or less like the person I am.

Counselor B's response gives Robin the opportunity to deal with her immediate anxiety and then to explore her problem situation more fully.

**Faking it.** Clients are sometimes confused, distracted, and in a highly emotional state. All these conditions affect the clarity of what they are saying about themselves. Helpers may fail to pick up what clients are saying because of the clients' confusion or because clients are not stating their messages clearly. Or the helpers themselves may have become distracted in one way or another. In any case, it's a mistake to feign understanding. Genuine helpers admit that they are lost and then work to get back on track again. A statement like "I think I've lost you. Could we go over that once more?" indicates that you think it's important to stay with the client. It is a sign of respect. Admitting that you're lost is infinitely preferable to such clichés as "uh-huh," "um," and "I understand." On the other hand, if you often catch yourself saying that you don't understand, you'd better find out what's going on. Faking it is never a substitute for competence.

# THE ART OF PROBING AND SUMMARIZING

NONVERBAL AND VERBAL PROMPTS
> Nonverbal prompts
> Vocal and verbal prompts

DIFFERENT FORMS OF PROBES
> Statements
> Requests
> Questions
> Words or phrases that are, in effect, questions or requests

USING QUESTIONS EFFECTIVELY
> Ask a limited number of questions
> Ask open-ended questions

PRINCIPLES IN THE USE OF PROBES
> Use probes to help clients engage as fully as possible in the therapeutic dialogue
> Use probes to help clients achieve concreteness and clarity
> Use probes to help clients complete the picture
> Use probes to help clients get a balanced view of problem situations and opportunities
> Use probes to help clients move into more beneficial stages and steps of the helping process
> Use probes to help clients move forward within some step of the helping process
> Use probes to explore and clarify clients' points of view, decisions, and proposals
> Use probes to challenge clients and help them challenge themselves

THE RELATIONSHIP BETWEEN SHARING HIGHLIGHTS AND USING PROBES

THE ART OF SUMMARIZING: PROVIDING FOCUS AND DIRECTION

At the beginning of a new session

During a session that is going nowhere

When a client gets stuck

When a client needs a new perspective

HOW TO BECOME PROFICIENT IN USING COMMUNICATION SKILLS

SHADOW-SIDE REALITIES OF COMMUNICATION SKILLS

In most of the examples used in the discussion of sharing empathic highlights, the clients demonstrated a willingness to explore themselves and their behavior relatively freely. Obviously, this is not always the case. Although it is essential that helpers respond with highlights when their clients do reveal themselves, it is also necessary at times to encourage, prompt, and help clients explore their concerns when they fail to do so spontaneously. Therefore, the ability to use prompts and probes well is another important communication skill. If sharing highlights is the lubricant of dialogue, then probes provide often-needed nudges.

Prompts and probes are verbal and sometimes nonverbal tactics for helping clients talk more freely and concretely about any issue at any stage or step of the helping process. Counselors using probes help clients identify and explore opportunities they have been overlooking, clear up blind spots, translate dreams into realistic goals, come up with strategies for accomplishing goals, and work through obstacles to action. Probes, judiciously used, provide focus and direction for the entire helping process. They make it more efficient. But let's begin by looking at prompts.

## NONVERBAL AND VERBAL PROMPTS

Prompts are brief verbal or nonverbal interventions designed to let clients know that you are with them and to encourage them to talk further.

**Nonverbal prompts.** You can use various behaviors—bodily movements, gestures, nods, eye movement, and the like—as nonverbal prompts. For example, a client who has been talking about how difficult it is to make a peace overture to a neighbor she is at odds with says, "I just can't do it!" The helper says nothing but simply leans forward attentively and waits. The client pauses and then says, "Well, you know what I mean. It would be very hard for me to take the first step. It would be like giving in. You know, weakness." They go on to explore how such an overture, properly done, could be a sign of strength rather than weakness.

**Vocal and verbal prompts.** You can use responses "um," "uh-huh," "sure," "yes," "I see," "ah," "okay," and "oh" as prompts, provided you use them intentionally and they are not simply a sign that your attention is flagging, you don't know what else to do, or you are on automatic pilot. In the following example, the client, a 33-year-old married woman, is struggling with perfectionism both at work and at home:

CLIENT (hesitatingly): I don't know whether I can "kick the habit" . . . you know, just let some trivial things go at work and at home. I know I've made a contract with myself. I'm not sure that I can keep it.

HELPER: Huh. [The helper utters this briefly and then remains silent.]

CLIENT (laughs): Here I am deep into perfectionism and I hear myself saying that I can't do something! How ironic. Of course, I can. I mean it's not going to be easy . . . at least at first.

The helper's "huh" prompts the client to reconsider what she has just said. Prompts should never be the main course. They are part of the therapeutic dialogue only as a condiment.

## DIFFERENT FORMS OF PROBES

Probes, used judiciously, help clients name, take notice of, explore, clarify, or further define any issue at any stage or step of the helping process. Designed to provide clarity and move things forward, probes can take different forms:

**Statements.** One form of probe is a statement indicating the need for further clarity. For instance, a helper, talking to a client who is having problems with his 25-year-old daughter who is still living at home, says: "It's still not clear to me whether you want to challenge her to leave the nest or not." The client replies, "Well, I want to, but I don't know how to do it without alienating her. I don't want it to sound like I don't care about her and that I'm just trying to get rid of her." Probes in the form of statements often take the form of the helper's confessing that he or she is in the dark in some way: "I'm not sure I understand how you intend . . ." or "I guess I'm still confused about. . . ." This kind of request, putting the burden on the helper, has the advantage of not accusing clients of failing to cough up the truth.

**Requests.** Probes can take the form of direct requests for further information or more clarity. A counselor, talking to a woman living with her husband and her mother-in-law, says, "Tell me what you mean when you say that three's a crowd at home." She answers, "I get along fine with my husband, I get along fine with my mother-in-law. But the chemistry among the *three* of us is very unsettling." Obviously, requests should not sound like commands. "Come on, just tell me what you are thinking." Tone of voice and other paralinguistic and nonverbal cues help to soften requests.

**Questions.** Direct questions are perhaps the most common type of probe: "How do you react when he flies off the handle?" "What keeps you from making a decision?" "Now that the indirect approach to letting him know your needs is not working, what might Plan B look like?" Consider this case: A client has come for help in controlling her anger. With the help of a counselor she comes up with a solid program. In the next session, the client gives signs of backtracking. The counselor says, "You seemed enthusiastic about the program last week. But now, unless I'm mistaken, I hear a bit of hesitancy in your voice. What obstacles do you see standing in your way?" The client responds, "Well, after taking a second look at the program, I'm afraid it will make me look like a wimp. My fellow workers could get the wrong idea and begin pushing me around." The counselor says, "So there's something about yourself you don't want to lose. What might that be?" The client hesitates for a moment and then says, "Spunk!" The counselor replies, "Well, let's see how you can keep your spunk and still get rid of the outbursts that get you in trouble." They go on to discuss the difference between assertiveness and aggression.

**Words or phrases that are, in effect, questions or requests.** Sometimes single words or simple phrases are, in effect, probes. A client talking about a difficult relationship with her sister at one juncture says, "I really hate her." The helper responds simply and unemotionally, "Hate." The client responds, "Well, I know that *hate* is too strong a term. What I mean is that things are getting worse and worse." Another client, troubled with irrational fears, says, "I've had it. I just can't go on like this. No matter what, I'm going to move forward." The counselor replies, "Move forward to . . . ?" The client says, "Well . . . to not indulging myself with my fears. That's what they are, a form of self-indulgence. From our talks, I've learned that it's a bad habit. A very bad habit." They go on to discuss ways of controlling such thoughts.

Whatever form probes take, they are often, directly or indirectly, questions of some sort. Therefore, a word about the use of questions is in order.

## USING QUESTIONS EFFECTIVELY

Helpers, especially novices and inept counselors, tend to ask too many questions. When in doubt about what to say or do, they ask questions that add no value. It is as if gathering information were the goal of the helping interview. Social intelligence calls for restraint. When judiciously used, however, questions can be an important part of your interactions with clients. Here are two guidelines.

**Ask a limited number of questions.** When clients are asked too many questions, they feel grilled, and that does little for the helping relationship. Furthermore, many clients instinctively know when questions are just filler, used because the helper does not have anything better to say. I have caught myself asking questions the answers to which I didn't even want to know. Let's assume that the helper working with Robin, the young woman exploring her concerns about her looks and body image discussed in Chapter 6, asks her a whole series of questions:

"When did you first feel like this?"

"Have you discussed this with anyone?"

"What do you do to improve your looks?"

"What is it about your looks that you think others don't like?"

Robin would have every right to say, "Good-bye, no thanks" in response to these intrusions. Such questions constitute a random search for information that is of little value. Helping sessions were never meant to be question-and-answer sessions that go nowhere.

**Ask open-ended questions.** As a general rule, ask open-ended questions—that is, questions that require more than a simple yes or no or similar one-word answer. Not, "Now that you've decided to take early retirement, do you have any plans?" but, "Now that you've decided to take early retirement, what are your plans?" Counselors who ask closed questions find themselves asking more and more questions. One closed question begets another. Of course, when a specific piece of information is needed, then a closed question may be called for. A career counselor might ask, "How many jobs have you had in the past two years?" The information is relevant to helping the client draw up a resumé and a job-search strategy. And occasionally, a sharp closed question can have the right impact. For instance, when a client finishes outlining what he is going to do to get back at his "ungrateful" son, the counselor asks, "Is that what you really want?" In moderation, open-ended questions at every stage and step of the helping process help clients fill in what is missing.

## PRINCIPLES IN THE USE OF PROBES

Here, then, are some principles that can guide you in the use of all probes, whatever form they may take.

**Use probes to help clients engage as fully as possible in the therapeutic dialogue.** As noted earlier, many clients do not have the all the communication skills needed to engage in the problem-managing and opportunity-developing dialogue. However, if you have these skills, you can use them to help your clients "play"—that is, engage in the kind of turn taking, connecting, mutual influence, and co-creation of outcomes that characterize dialogue. Probes are the principal tools needed to help all clients engage in the give-and-take of the helping process. The following exchange is between a counselor at a church parish center and a parishioner who has been struggling to tell her story about her attempts to get her insurance company to respond to the claim she filed after a car accident:

CLIENT: They just won't do anything. I call and get the cold shoulder. They ignore me, and I don't like it!

HELPER: You're angry with the way you're being treated. And you want to get to the bottom of it. . . . Give me a brief overview of what you've done so far.

CLIENT: Well, they sent me forms that I didn't understand very well. I did the best I could. I think they were trying to show that it was my fault. I even kept copies. I've got them with me.

HELPER: You're not sure you can trust them. . . . I'd like to take a look at your copies of the forms.

The forms turn out to be standard claim forms. Given that this is the client's first encounter with an insurance company and that she has poor communication skills, the counselor gains some insight into what the phone conversations between her and the insurance company must be like. By sharing highlights and using probes, the counselor gets her to see that her experience might well be normal. The outcome is that the client gets help from someone in the parish who has gone through a similar experience.

Helping clients engage in dialogue is not some form of manipulation. You can encourage dialogue without in any way being patronizing or condescending. This is a robust use of probes, often very useful in interacting with nonassertive and reluctant clients. You can't force your clients to do anything, but your invitations can be strong. Social intelligence will tell you how far you can go.

**Use probes to help clients achieve concreteness and clarity.** Probes can help clients turn what is abstract and vague into something concrete and clear—something you can get your hands on and work with. In the next example, a man is talking about an intimate relationship that has turned sour:

CLIENT: She treats me badly, and I don't like it!

HELPER: Tell me what she actually does.

CLIENT: She talks about me behind my back. I know she does. Others tell me what she says. She also cancels dates when something more interesting comes up.

HELPER: That's pretty demeaning. . . . How have you been reacting to all this?

CLIENT: Well, I think she knows that I know. But we haven't talked about it much.

In this example, the helper's probe leads to a clearer statement of the client's experience and behavior. By sharing highlights and using probes, the helper discovers that the client puts up with a great deal because he is afraid of losing her. He goes on to help the client deal with the psychological "economics" of such a one-sided relationship.

In the next example, a man who is dissatisfied with living a somewhat impoverished social life is telling his story. A simple probe leads to a significant revelation.

CLIENT: I do funny things that make me feel good.

HELPER: What kinds of things?

CLIENT: Well, I daydream about being a hero, a kind of tragic hero. In my daydreams, I save the lives of people I like but who don't seem to know I exist. And then they come running to me, but I turn my back on them. I choose to be alone! I come up with all sorts of variations of this theme.

HELPER: So in your daydreams, you play a character who wants to be liked or loved but who gets some kind of satisfaction from rejecting those who haven't loved him back. . . . I'm not sure I've got that right.

CLIENT: Well . . . yeah . . . I sort of contradict myself . . . I do want to be loved, but I don't do very much to get a real social life. It's all in my head.

The helper's probe leads to a clearer statement of the client's internal behaviors. Helping the client to explore his fantasy life could be a first step toward finding out what he really wants from relationships.

The next client has become the breadwinner since her husband suffered a stroke. Someone takes care of her husband during the day.

CLIENT: Since my husband had his stroke, coming home at night is rather difficult for me. I just . . . Well, I don't know.

HELPER: It really gets you down. . . . What's it like?

CLIENT: When I see him sitting immobile in the chair, I'm filled with pity for him, And the next thing I know, it's pity for myself, and it's mixed with anger or even rage. But I don't know what or whom to be angry at. I don't know how to focus my anger. Good God, he's only 42, and I'm only 40!

In this case, the helper's probe leads to a fuller description of the intensity of the client's feelings and emotions.

In each of these cases, the client's story gets more specific. Of course, the goal is not to get more and more detail. Rather, it is to get the kind of detail that makes the problem or unused opportunity clear enough to see what can be done about it.

**Use probes to help clients complete the picture.**    Probes further the therapeutic dialogue by helping clients identify missing pieces of the puzzle—experiences, behaviors, and feelings that would help both clients and helpers get a better fix on problem situations, possibilities for a better future, or drawing up plans of action. The client in the following example is at odds with his wife over his mother-in-law's upcoming visit:

HELPER: I realize now that you often get angry when your mother-in-law stays for more than a day. But I'm still not sure what she does that makes you angry.

CLIENT: First of all, she throws our household schedule out and puts in her own. Then she provides a steady stream of advice on how to raise the kids. My wife sees this as an "inconvenience." For me it's a total family disruption. When she leaves, there's a lot of emotional cleaning up to be done.

Just what the client's mother-in-law does to get him going has been missing. Once the behavior has been spelled out in some detail, it is easier to help him come up with some remedies. Still missing, however, is what *he* does as a result of his mother-in-law's behavior. The helper continues:

HELPER: So when she takes over, everything gets turned upside down. . . . How do you react in the face of all this turmoil?

CLIENT: Well . . . well . . . I guess I go silent. Or I just get out of there, go somewhere, and fume. After she's gone, I take it out on my wife, who still doesn't see what all the fuss is about.

So now it's clear that the client does little to change things. It is also obvious that he is a little taken aback by being asked how he handles the situation.

In the next example, a divorced woman is talking about the turmoil that takes place when her ex-husband visits the children. It has some similarities with the case we've just seen.

HELPER: The Sundays your ex-husband exercises his visiting rights with the children end in his taking verbal potshots at you, and you get these headaches. I've got a fairly clear picture of what he does when he comes over, but it might help if you could describe what you do.

CLIENT: Well, I do nothing.

HELPER: So last Sunday he just began letting you have it for no particular reason. Or just to make you feel bad.

CLIENT: Well . . . not exactly. I asked him about increasing the amount of the child support payments. And I asked him why he's dragging his feet about getting a better job. He's so stupid. He can't even take a bit of sound advice.

Through probes, the counselor helps the client fill in a missing part of the picture—her own behavior. She keeps describing herself as total victim and her ex-husband as total aggressor. But that doesn't seem to be the full story.

Next we have Iolanda, a mother of four kids, two of whom are in their early teens, talking to a friend, Vivian, who does a lot of volunteer work at a community social center. Her complaint is that her husband and kids don't provide much help in getting household chores done. Vivian has a lot of social savvy.

IOLANDA (in a matter-of-fact voice): At the end of the day, what with the kids and dinner and cleaning up, I'm not at my best.

VIVIAN: When you ask for help, how do they respond?

IOLANDA: I shouldn't have to ask! They can see what needs to be done.

VIVIAN: It sounds like some bad habits are in place. . . . If yours were the ideal household, what would be happening? What would the division of labor look like?

IOLANDA: Hmm. . . . Well, first of all . . . .

Her friend realizes that the "family culture" is probably filled with bad habits. Her probe indicates that the mother resents having to ask for help. So she uses another probe to put the conversation on a different tack—looking at possibilities for a better future rather than digging deeper into the problem situation itself. Her strategy is to help Iolanda determine what she wants and then help her see what needs to be done to get it.

**Use probes to help clients get a balanced view of problem situations and opportunities.** Clients, in their eagerness to discuss issues or make points, often describe one side of a picture or one viewpoint. Probes can be used to help them fill out the picture. In the following example, the client—a manager who has to work with a bright, highly ambitious, aggressive young woman who plays politics to further her own interests—has been agonizing over his plight:

COUNSELOR: I've been wondering whether you see any upside to this, any hidden opportunities.

CLIENT: I'm not sure what you mean. It's just a disaster.

COUNSELOR: Well, you strike me as a pretty bright guy. I'm wondering if there are any lessons for you hidden in all this.

CLIENT (pausing): Oh, well, you know I tend to ignore politics around here, but now it's in my face. Where there are people, there are politics. I think she's being political to serve her own career. But I don't want to play her game. There must be some other kind of game or something that would let me keep my integrity. The days of avoiding all of this are probably over.

The problem situation has a flip side. It is an opportunity for rethinking and learning. As such, problems are incentives for constructive change. The client can learn something from all this. It's an opportunity to come to grips with the male-female dynamics of the workplace and a chance to explore "positive" political skills.

**Use probes to help clients move into more beneficial stages and steps of the helping process.** Probes can be used to open up new areas for discussion. They can be used help clients engage in dialogue about any part of the helping process—telling their stories more fully, attacking blind spots, setting goals, formulating action strategies, discussing obstacles to action, and reviewing actions taken. Many clients do not easily move into whatever stage or step of the helping process might be most useful for them. Probes can help them do so. In the following example, the counselor uses a probe to help a middle-aged couple—Sean and Fiona, who have been complaining about each other—move on to the Stages II and III. Besides complaining, they have talked vaguely about "reinventing" their marriage. Part of this reinvention might focus on doing more things in common.

COUNSELOR: What kinds of things do you like doing together? What are some possibilities?

FIONA: I can think of something, though it might sound stupid to you [she says in an aside to her husband]. We both like doing things for others, you know, caring about other people. Before we were married we talked about spending some time in the Peace Corps together, though it never happened.

SEAN: I wish we had. . . . But those days are past.

COUNSELOR: Are they? The Peace Corps may not be an option, but there must be other possibilities. [Neither Fiona nor Sean says anything.] I tell you what. Here are a couple of pieces of paper. Jot down three ways of helping others. Do your own list. Forget what your partner might think.

The counselor uses probes to get Sean and Fiona to brainstorm possibilities for some kind of service to others. He turns it into a written exercise. This gets them away from what was proving to be a tortuous process of problem exploration and moves them toward opportunity development.

The next client has been talking endlessly about the affair her husband is having. He knows that she knows.

COUNSELOR: You've said you're not going to do anything about it because it might hurt your son. But doing nothing is not the only possible option. Let's just name some others. Who knows? We might find a gem.

CLIENT: Hmm. . . . I'm not sure I know.

COUNSELOR: Well, you know people in the same predicament. You've read novels, seen movies. What are some of the standard things people do? I'm not saying do them. Let's just review them.

CLIENT: Hmm. . . . Well, I knew someone who did an outrageous thing. She knew her daughter knew. So one night at dinner she just said, "Let's all talk about the affair you're having and how to handle it. It's certainly not news to any of us."

COUNSELOR: All right, that's one way. Let's hear some more.

The client's current way of handling the affair is just one way. In their discussion, she says that she's pretty sure her son knows. So they also explore possibilities based on the assumption that he does know.

In the following example, Jill, the helper, and Justin, the client, have been discussing how Justin is letting his impairment—he has lost a leg in a car accident—get in the way of his picking up his life again. The session has bogged down a bit.

JILL: Let's try a bit of drama. I'm going to be Justin for a while. You're going to be Jill. As my counselor, ask me some questions that you think might make a difference for me . . . that is, Justin.

JUSTIN (pausing a long time): I'm not much of an actor, but here goes. . . . Why are you taking the coward's way out? Why are you on the verge of giving up? [His eyes tear up.]

Jill gets Justin to formulate the probes. It's her way of asking Justin to "move forward" and take responsibility for his part of the session. Justin's "probes" turn out to be challenges, almost accusations, certainly much stronger than anything Jill, at this stage, would have tried. However painful this is for Justin, it's a breakthrough.

In the following example, George, a single man in his sixties, is seeing a counselor to deal with the anxiety he is experiencing after an unexpected ten-day stay in the hospital for an intestinal disorder. George is a hardworking, very independent man. The counselor would like him to use his illness as time-out to reflect on his current lifestyle. The illness and the subsequent anxiety could be used as a springboard for creating a different way of living that would include a less haphazard social life. Developing a better social system could help him deal with the natural and inevitable ups and downs of growing older.

HELPER: George, describe to me the current balance between your work life and your leisure and social life.

GEORGE (pausing): I never stop to think about it. . . . Leisure and social life are things that get shoe-horned into my work schedule. On paper it all looks terribly unbalanced, but that's not how I feel. . . . Except. . . .

HELPER: Except?

CLIENT: A lot of times, when I finish a project and am looking for something to do with my friends, I realize that I do everything by appointment. . . . A social life by appointment. That doesn't sound very balanced . . . even to me.

The helper's probe around work–social-life balance has the impact of a "What's going on?" probe. George says, in effect, "Maybe I should take a more serious look at this."

**Use probes to help clients move forward within some step of the helping process.** Probes can be used not only to help clients move to a different stage or step but also to move within a step. Contrast the two following approaches to probing. The client, a woman in the middle of an acrimonious divorce, has recently learned that her breast cancer has reappeared. She is seeing the counselor after a long interlude.

CLIENT (toward the end of the session): Well, now we're up-to-date. You know the full miserable story.

COUNSELOR A: I haven't seen you for a while. When did you find out about the reappearance of the cancer?

CLIENT: Let's see. . . . Oh, who knows and who cares! . . . Well, I have to go.

The probe is a useless one, mere filler. It does nothing but annoy the client. Let's replay the scene with another counselor.

COUNSELOR B: I haven't seen you for a while. I've been wondering whether the reappearance of the cancer has altered your thinking about the divorce in any way.

CLIENT: Only in one way. If I die, I don't want to die married to him. I just don't. It would be dishonest. Our relationship, as you know, died a long time ago.

COUNSELOR B: So you're sticking to your guns. In fact, finding out about the cancer has increased your resolve. . . . But the divorce proceedings up to now have been pretty bitter for you. I'm wondering how going through divorce proceedings fits in with your resolve to be kinder to yourself.

CLIENT: Hmm. . . . The divorce *is* driving me into the ground. . . . I tell you what. Caving in is off limits. Trade-offs . . . well, that's a different story. Let's talk about trade-offs the next time we meet. I have to think about it.

The helper probes to see whether her decision to pursue the divorce (a Stage II activity) is irrevocable. She's not going to "cave in." But there might be some trade-offs on the way she pursues the divorce that might spare her much needless pain.

**Use probes to explore and clarify clients' points of view, decisions, and proposals.** Clients often fail to clarify their points of view, decisions, and intentions or proposals. For instance, the decision itself might be unclear, and the reason behind it and the implications for the client and others are not spelled out. In the following case, the client, driving under the influence, has had a bad automobile accident. Luckily, he was the only one hurt. He is recovering physically, but his psychological recovery has been slow. The accident opened up a Pandora's box of psychological problems—not the least of which is a lack of self-responsibility—that had been waiting to pop out. A counselor is helping him work through some of the key problems. Here is part of an early session.

CLIENT: I don't think that the laws around driving under the influence should be as tough as they are. I'm scared to death of what might happen to me if I ever had an accident again.

COUNSELOR: So you feel you're in jeopardy. . . . What makes you think that the laws are too tough?

CLIENT: Well, they bully us. One little mistake and bingo! Your freedom goes out the window. Laws should make people free.

COUNSELOR: Let's say all laws on driving under the influence were dropped. What then?

The counselor knows that the client is running away from taking responsibility for his actions. Using probes to get him to spell out the implications of his point of view on DUI laws and its implications is the beginning of an attempt to help the client face up to himself.

In a later session, the client talks about the legal ramifications of the accident. He has to go to court.

CLIENT: I've been thinking about this. I'm going to get me a really good lawyer and fight this thing. I talked with a friend, and he thinks he knows someone who can get me off. I need a break. It might cost me a bundle. After all, I messed up someone's property a bit, but I didn't hurt anyone.

COUNSELOR: What's the best thing that could happen in court?

CLIENT: I'd get off scot-free. Well, maybe a slap on the wrist of some kind. A warning.

COUNSELOR: And what's the worst thing that could happen?

CLIENT: I haven't given that a lot of thought. I don't really know much about the laws or the courts or how tough they might be. That sort of stuff. But with the right lawyer . . . .

COUNSELOR: Hmm. . . . I'm trying to put myself in your shoes. . . . I think I'd try to find out what the most likely deal in court might be before I made up my mind about lawyers and things. What do you think?

The counselor is using probes to help the client explore the implications of a decision he's making.

The state has very tough DUI laws. In the end the client, because of the alcohol level in his blood, has his license suspended for six months, is fined heavily, and has to spend a month in jail. All of this is very sobering. The counselor visits him in jail, and they talk about the future.

CLIENT: I feel like I've been hit by a train.

COUNSELOR: You had no idea that it would be this bad.

CLIENT: Right. No idea. . . . I know you tried to warn me in your own way, but I wasn't ready to listen. . . . Now I have to begin to put my life back together. Though I don't feel like it.

COUNSELOR: But now you've had the wake-up call, a horrible wake-up call. So what does the future hold . . . even if you don't feel up to looking at it?

CLIENT: I have been thinking. One thing I want to do is to make some sort of apology to my family. They're hurting as bad as I am. I feel so awkward. I know how to act in cocky mode. Humble mode I'm not used to. Do I write a long letter? Do I wait and just apologize through my actions? Do I take each one of them aside? I don't know, but I've just got to do it.

COUNSELOR: Somehow you have to make things right with them. Just how, well that's another matter. Maybe we could start by finding out what you want to accomplish through an apology, however it's done.

Here we find a much more sober and cooperative client. He proposes, roughly, a course of action. The counselor supports his need to move beyond past stupidities and present misery. It's about the future, not the past. The counselor's last statement is a probe aimed at giving substance and order to the client's proposal: "What do you want to accomplish?"

**Use probes to challenge clients and help them challenge themselves.** In the last chapter, we saw that sharing highlights can act as a mild form of social influence or challenge. We also saw that effective highlights often act as probes. That is, they can be indirect requests for further information or ways of steering a client toward a different stage or step of the helping process. And, as you have probably noticed in the examples used in this chapter, probes can have an edge of challenge in them. Many probes are not just requests for relevant information. They often place some kind of demand on the client to respond, reflect, review, or reevaluate. Such probes are challenges of one kind or another. Or at least they serve as a bridge between communicating understanding to clients and helping them challenge themselves. The following client, having committed himself to standing up to some of his mother's possessive ways, now shows signs of weakening in his resolve:

HELPER: The other day you talked of "having it out with her"—though that might be too strong a term. A little while ago you mentioned something about "being reasonable with her." Tell me how these two differ.

CLIENT (pausing): Well, I think you might be witnessing a case of cold feet. . . . She's a very strong woman.

The counselor helps the client revisit his decision to "get tough" with his mother and, if this is what he really wants, what he can do to strengthen his resolve. Using probes as mild forms of challenge is perfectly legitimate, provided you know what you are doing.

Probes can also be used to help clients remain focused on relevant and important issues. Some clients meander because that's their communication style. Probes help them stay focused. Other clients wander because the topic at hand is getting too uncomfortable. Probes are then gentle nudges to keep them focused on real issues. On the other hand, probes should not be ways of extorting from clients things they don't want to give. Statements that have the flavor of "Oh, come on, tell me! It's really not going to hurt" move into dangerous territory. High-quality probes increase rather than decrease the client's sense of self-responsibility. Challenge and the wisdom that should permeate it are discussed in Chapters 10, 11, and 12.

## THE RELATIONSHIP BETWEEN SHARING HIGHLIGHTS AND USING PROBES

The trouble with dealing with skills one at a time is that each skill is taken out of context. In the give-and-take of any helping session, however, a helper must intermingle the skills in a natural way. In actual sessions, skilled helpers continually tune in, listen actively, and use a mix of probes and empathy to help clients clarify and come to grips with their concerns, deal with blind spots, set goals, make plans, and get things done. There is no formula for the right mix; it depends on the client, client needs, the problem situation, possible opportunities, the stage, and the step.

A word about the relationship between sharing highlights and using probes. Here is a basic guideline: After using a probe to which a client responds, share a highlight that expresses and checks your understanding. Be hesitant to follow one probe with another. The logic of this is straightforward. First, if a probe is effective, it will yield information that needs to be listened to and understood. Second, a shared highlight, if accurate, tends to place a demand on the client to explore further. It puts the ball back in the client's court. Years ago during a seminar Bob Carkhuff suggested, with his usual edge, that if helpers find themselves asking two questions in a row, they may just have asked two stupid questions.

In the following example, the client is a young Chinese American woman whose father died in China and whose mother is now dying in the United States. She has been talking about the traditional obedience of Chinese women and her fears of slipping into a form of passivity in her American life. She talks about her sister, who gives everything to her husband without looking for anything in return. The first counselor sticks to probes.

COUNSELOR: To what degree is this self-effacing role rooted in your culture?

CLIENT: Well, being somewhat self-effacing is certainly in my cultural genes. And yet I look around and see many of my North American counterparts adopt a different style . . . a style that frankly appeals to me. But last year, when I took a trip back to China with my mother to meet my half sisters, the moment I landed I wasn't American. I was totally Chinese again.

COUNSELOR A: What did you learn there?

CLIENT: That I am Chinese!

The client says something significant about herself, but instead of responding with understanding, the helper uses another probe. This elicits only a repetition, with some annoyance, of what she had just said. Now a different approach:

COUNSELOR B: You learned just how deep your cultural roots go.

CLIENT: And if these roots are so deep, what does that mean for me here? I love my Chinese culture. I want to be Chinese and American at the same time. How to do that . . . well, I haven't figured that out yet. I thought I had, but I haven't.

In this case, an empathic highlight works much more effectively than another probe. Counselor B helps the client move forward.

In the next example, a single middle-aged woman working in a company that has reinvented itself for the so-called new economy still has a job, but the pay is much less and she is doing work she does not enjoy. She does not have the computer and Internet-related skills for the better jobs. She feels both stressed and depressed.

CLIENT: Well, I suppose that I should be grateful for even having a job. But now I work longer hours for less pay. And I'm doing stuff I don't even like. My life is no longer mine!

HELPER: So the extra pressure and stress make you wonder just how "grateful" you should feel.

CLIENT: Precisely. And the future looks pretty bleak.

HELPER: What could you change in the short term to make things more bearable?

CLIENT: Hmm. . . . Well, I know one way. We all keep complaining to one another at work. And this seems to make things even worse. I can get out of that loop. It's a simple way of making life a bit less miserable.

HELPER: So one way is to stop contributing to your own misery by staying away from the complaining chorus. . . . What might you start doing?

CLIENT: Well, there's no use sitting around hoping that what has happened is going to be reversed. I've been really jolted out of my complacency. I assumed with the economy humming and unemployment so low I'd just motor on as usual. But I'm still young enough to acquire some more skills. And I do have some skills that I haven't needed to use before. I'm a good communicator and I've got a lot of common sense. I work well with people. There are probably lots of jobs around here that require those skills.

HELPER: So, given the wake-up call, you think it might be possible to take unused skills and reposition yourself at work. And, of course, you could develop some technology skills.

CLIENT: Repositioning. Hmm, I like that word. It makes a lot of pictures dance through my mind. . . . Yes, I need to reposition myself. For instance . . .

This combination of highlights and probing gets things moving. Instead of focusing on the misery of the present situation, the client names a few possibilities for a better future—a Stage II activity. Box 7-1 summarizes the main points about the use of probes.

On the other hand, you should be careful not to become either an empathic highlight "machine," grinding out one highlight after another, or an "interrogator," peppering your clients continually with needless probes. All responses to clients, including probes and challenges, are empathic if they are based on solid understanding of clients' core messages and points of view. All responses that build on and add to clients' remarks are implicitly empathic. Since these responses are empathic in effect, they cut down on the need for a steady stream of highlights.

# THE ART OF SUMMARIZING: PROVIDING FOCUS AND DIRECTION

The communication skills of visibly tuning in, listening, sharing highlights, and probing need to be orchestrated in such a way that they help clients focus their attention on issues that make a difference. The ability to summarize and help clients summarize the main points of a helping interchange or session is a skill that can be used to provide both focus and challenge.

Brammer (1973) lists a number of goals that can be achieved by judicious use of summarizing: "warming up" the client, focusing scattered thoughts and feelings, bringing the discussion of a particular theme to a close, and prompting the client to explore a theme more thoroughly. There are certain times when summaries prove particularly useful: at the beginning of a new session, when a session seems to be going nowhere, and when a client gets stuck.

**At the beginning of a new session.**  When a summary is used at the beginning of a new session, especially when a client seems uncertain about how to begin, it prevents the client from merely repeating what has already been said before. It puts the client under pressure to move on. Consider this example: Liz, a social worker, begins a session with a rather overly talkative man with a summary of the main points

## Box 7-1    Suggestions for the Use of Probes

1. Keep in mind the goals of probing:

- To help clients engage as fully as possible in the therapeutic dialogue.
- To help nonassertive or reluctant clients tell their stories and engage in other behaviors related to managing their problems and developing their opportunities.
- To help clients identify experiences, behaviors, and feelings that give focus to their stories.
- To open up new areas for discussion.
- To help clients explore and clarify points of view, decisions, and proposals.
- To help clients be as concrete and specific as possible.
- To help clients remain focused on relevant and important issues.
- To help clients move on to further stages or steps in the helping process.
- To mildly challenge clients to examine the way they think, behave, and act both within helping sessions and in their daily lives as they try to manage problems and develop opportunities.

2. Make sure that probing is done in the spirit of empathy.
3. Use a mix of statements—open-ended questions, prompts, and requests— not questions alone.
4. Do not engage clients in question-and-answer sessions.
5. If a probe helps a client reveal relevant information, follow it up with an empathic highlight rather than another probe.
6. Use whatever judicious mixture of highlights and probing is needed to help clients clarify problems, identify blind spots, develop new scenarios, search for action strategies, formulate plans, and review outcomes of actions.

of the previous session. This serves several purposes: First, it shows the client that she had listened carefully to what he had said in the last session and that she had reflected on it after the session. Second, the summary gives the client a jumping-off point for the new session. It gives him an opportunity to add to or modify what was said. Finally, it places the responsibility for moving forward on the client. The implied sentiment of the summary is "Now where do you want to go with this?" Summaries put the ball in the clients' court and give them an opportunity to exercise initiative.

**During a session that is going nowhere.** A summary can give focus to a session that seems to be going nowhere. One of the main reasons sessions go nowhere is that helpers allow clients to keep "going 'round the mulberry bush"—that is, saying the same things over and over again—instead of helping them either go more deeply into their stories, focus on possibilities and goals, or discuss strategies that will help clients get what they need and want. For instance, a counselor provides assistance to the staff of a shelter for the homeless. One of the staff members is showing signs of burnout. In a second meeting with the counselor, she keeps going over the same ground, talking endlessly about stressful incidents that have taken place over the last few months. At one point the counselor provides a summary.

COUNSELOR: Let's see if I can pull together what you've been saying. The work here, by its very nature, is stressful. You've mentioned a whole string of incidents such as being hit by someone you were trying to help and heated arguments with some of your coworkers. But I believe you've intimated that these are the kinds of things that happen in these places. Shelters are prone to them. They are part of the furniture. They're not going to stop. But they can be very punishing. At times, you wish you weren't here. But if they are not going to stop, maybe the next question is, "How do I cope with them? How do I do my work and get some ongoing satisfaction from it?"

The purpose of the summary is to help the client move beyond "poor me" and find ways of coping with this kind of work. The challenge in places like shelters is creating a supportive work environment, developing a sense of organizational and personal purpose, promoting the kind of teamwork that fits the institution's mission, and fostering a culture of coping strategies (see Brown & O'Brien, 1998).

**When a client gets stuck.** A summary can be used when a client doesn't seem to know where to go next, either in the helping session itself or in an action program out there in the real world. In such a case, the helper can, of course, use probes to help the client move on. A summary, however, has a way of keeping the ball in the client's court. Moreover, the helper does not always have to provide the summary. Often it is better to ask the client to pull together the major points. This helps the client own the helping process, pull together the salient points, and move on. Since this is not meant to be a way of testing the client, the counselor should offer the client help to stitch the summary together. Consider, for example, a client who has lost her job and her boyfriend because she has outbreaks of anger when she drinks. She has been talking about "not being able to stick to the program." The counselor asks her to summarize what she's been doing and the obstacles she has been running into. With the help of the counselor, she stumbles through a summary. At the end of it she says, "I guess it's clear to both of us that I haven't been doing a very good job sticking to the program. On paper, the 12 steps look like a snap. But it seems that I don't live on paper." The counselor now sees that it's time to go back to the drawing board on the action steps, what it takes to stick to the program, and how to cope with obstacles.

**When a client needs a new perspective.** Often when scattered elements are brought together, the client sees the "bigger picture" more clearly. In the following example, a man who has been reluctant to go to a counselor with his wife has, in a solo session with the counselor, agreed to a couple of sessions "to please her." In the session, he talks a great deal about his behavior at home, but in a rather disjointed way. Much of it is caring.

COUNSELOR:  I'd like to pull a few things together. You've encouraged your wife in her career, especially when things are difficult for her at work. You also encourage her to spend time with her friends as a way of enjoying herself and letting off steam. You also make sure that you spend time with the kids. In fact, time with them is important for you.

CLIENT:  Yeah. That's right.

COUNSELOR:  Also, if I have heard you correctly, you currently take care of the household finances. You are usually the one who accepts or rejects social invitations, because your schedule is tighter than hers. And now you're about to ask her to move because you can get a better job in Boston.

CLIENT:  When you put it all together like that, it sounds as if I'm running her life. . . . She never tells me I'm running her life.

COUNSELOR:  Maybe we could talk a little about this when the three of us get together.

IGNATIUS:  Hmm. . . . well, I'd . . . hmm . . . [laughs]. I'd better think about all of this before the next session.

The summary provides the client with a mild jolt. He realizes that he needs to face up to the "I am making all the big decisions for her" theme implied in the summary.

## HOW TO BECOME PROFICIENT IN USING COMMUNICATION SKILLS

Understanding communication skills and how they fit into the helping process is one thing. Becoming proficient in their use is another. Some trainees think that they can learn these "soft" skills easily and fail to put in the kind of hard work and practice needed to become "fluent" in them (Binder, 1990; Georges, 1988). Doing the exercises on communication skills in the manual that accompanies this book and practicing these skills in training groups can help, but that isn't enough to make the skills second nature. If you trot out your skills of tuning in, listening, processing, sharing highlights, and probing only for helping encounters your sessions are likely to have a hollow ring to them. You need to exercise these skills in all your relationships, making them part of your everyday communication style.

After providing some initial training in communication skills, I tell students, "Now, go out into your real lives and get good at these skills. I can't do that for you." In the beginning, it may be difficult to practice all these skills in everyday life, not because they are very difficult, but because they are relatively rare in conversations. Take sharing highlights. Listen to the conversations around you. If you were to use an unobtrusive counter, pressing the plunger every time you heard someone share an empathic highlight, you might go days without pressing the plunger. But you can make sharing highlights a reality in your everyday life. And those who interact with you will notice the difference. They probably will not call it empathy or sharing highlights. Rather, they will says such things as "She really listens to me" or "He takes me seriously."

On the other hand, you will hear many probes in everyday conversations. People are much more comfortable asking questions than providing understanding. However, many of these probes are aimless. Worse, many will be disguised criticisms: "Why on earth did you do *that*?" Learning how to integrate purposeful probes with highlights demands practice in everyday life. Life is your lab. Every conversation is an opportunity.

I once ran a training program on these skills for a CPA firm. Although the director of training believed in their value in the business world, many of the account executives did not. They resisted the whole process. I got a call one day from one of the more notable resisters. "I owe you this call," he said. "Really?" I replied with an edge of doubt in my voice. "Really," he said. He went on to tell me how he had recently called on a potential client. This client, dissatisfied with its current audit firm, was interviewing for a new one. During the interview, he said to himself, "Since we don't have the slightest chance of getting this account, why don't I amuse myself by trying these communication skills?" In his phone call to me, he went on to say, "This morning I got a call from that client. He gave us the account, but in doing so he said, 'You're not getting the account because you were the low bidder. You were not. You're getting the account because we thought that you were the only one that really understood our needs.' So, almost literally, I owe you this call." I forgot to ask him for a share of the fee.

## SHADOW-SIDE REALITIES OF COMMUNICATION SKILLS

Some helpers tend to overidentify the helping process with the communication skills—that is, with the tools that serve it. This is true not only of tuning in, listening, processing, sharing highlights, and probing but also of the skills of challenging that are the focus of Chapters 10, 11, and 12. Being good at communication skills is not the same as being good at helping. Moreover, an overemphasis on communication skills can turn helping into a great deal of talk with very little action, and few outcomes.

Communication skills are essential, of course, but they still must serve both the process and the outcomes of helping. These skills certainly help you establish a good relationship with clients. And a good relationship is the basis for the kind of social-emotional reeducation that has been outlined earlier. But you can be good at communication, good at relationship building, even good at social-emotional reeducation and still shortchange your clients, because they need more. Some helpers who overestimate the value of communication skills tend to see a skill such as sharing highlights as some kind of "magic bullet." Others overestimate the value of information gathering. This is not a broad indictment of the profession. Rather, it is a caution for beginners.

On the other hand, some practitioners underestimate the need for solid communication skills. There is a subtle assumption that the "technology" of their approach, such as manualized treatments, is sufficient. They listen and respond through their theories and constructs rather than through their humanity. They become technologists instead of helpers. They are like some medical doctors who become more and more proficient in the use of medical technology and less and less in touch with the humanity of their patients. Some years ago, I spent ten days in a hospital (an eternity in these days of managed care). The staff were magnificent in addressing my medical needs. But no one ever addressed the psychological needs that sprang from my anxiety about my illness. Unfortunately, my anxieties were often expressed through physical symptoms. Then those symptoms were treated medically. I asked, "When you have conferences during which patients are discussed, do you say, 'Well, we've thoroughly reviewed his medical status and needs. Now let's turn our attention to what he's going through. What can we do to help him through this experience?'" One resident said, "No, we don't have time." Don't get me wrong. These were dedicated, generous people who had my interests at heart. But they ignored many of my needs. We have a long way to go.

# Stage I of the Helping Model and Advanced Communication Skills

The basic communication skills reviewed in Part Two are critical tools. With them, you can help clients engage in all the stages and steps of the helping model. But those communication skills are not the helping process itself. Part Three is a detailed exposition and illustration of Stage I of the helping model, together with its three steps. Step I-A, helping clients tell their stories, is discussed in Chapter 8. Chapter 9 deals with reluctant and resistant clients. Step I-B focuses on advanced communication skills—those related to helping clients challenge themselves. These skills are discussed and illustrated in Chapters 10, 11, and 12. Step I-C introduces the concept of leverage—helping clients choose the right issues to work on—which is the focus of Chapter 13.

# Step I-A:
# "What Are My Concerns?"
# Helping Clients Tell
# Their Stories

**An Introduction to Stage I: Identifying and Exploring Problems and Opportunities**

**Step I-A: "What's Going On?"**

**Helping Clients Explore Problem Situations and Unexploited Opportunities**

Learn to Work with All Styles of Storytelling

Start Where Your Clients Start

Help Clients Clarify Key Issues

Assess the Severity of Clients' Problems

Help Clients Talk Productively About the Past

Help clients talk about the past to make sense of the present

Help clients talk about the past to be liberated from it

Help clients talk about the past to prepare for action in the future

As Clients Tell Their Stories, Search for Resources, Especially Unused Resources

Help Clients Spot and Develop Unused Opportunities

See Every Problem as an Opportunity

**Step I-A and Action**

The Importance of Proactivity

Using the Time Between Sessions Productively

Appreciating the Self-Healing Nature of Clients

**Is Step 1-A Enough?**

A declaration of intent and the mobilization of resources

Coming out from under self-defeating emotions

**The Shadow Side of Step I-A**

Clients as Storytellers

The Nature of Discretionary Change

**Evaluation Questions for Step I-A**

## AN INTRODUCTION TO STAGE I:
## IDENTIFYING AND EXPLORING PROBLEMS
## AND OPPORTUNITIES

Clients come to helpers because they need help in managing their lives more effectively. Stage I has three ways of helping clients understand themselves, their problem situations, and their unused opportunities with a view to managing them more effectively. The three steps can be stated in three principles:

- *Step I-A: Stories.* Help clients tell their stories in terms of problem situations and unused opportunities.

- *Step I-B: Blind spots.* Help clients identify and move beyond blind spots to new perspectives on their problem situations and opportunities.

- *Step I-C: Leverage.* Help clients choose issues that will make a difference in their lives.

These principles are not restricted to Stage I for three reasons: First, clients don't tell all of their stories at the beginning of the helping process. Often the full story "leaks out" over time. Second, blind spots can appear at any stage or step of the helping process. Blind spots affect choosing goals, setting strategies, and implementing programs. Third, leverage deals with the "economics" of helping. Choosing the right problem or opportunity to work on is just one way of delivering effectiveness and efficiency. Figure 8-1 highlights the three steps of Stage I.

Stage I of the helping process can be seen as the assessment stage—finding out what's going wrong, what opportunities lie fallow, what resources are not being used. Client-centered assessment means helping clients understand themselves, find out "what's going on" with their lives, see what they have been ignoring, and make sense out of the messiness of their lives. Assessment, then, is not something helpers do to clients: "Now that I have my secret information about you, I can fix you." Rather, it is a kind of learning in which, ideally, both client and helper participate through their ongoing dialogue.

In medicine, assessment is often a separate phase. In helping, there is an interplay between assessment and intervention. Helpers who continually listen to clients in context are engaging in ongoing assessment. In this sense, assessment is part and parcel of all stages and steps of the helping model. As in medicine, though, some initial assessment of the seriousness of the client's concerns is called for.

Members of the helping profession use many different procedures to assess clients; however, the "clinical interview," the dialogue between client and helper, is the most common assessment procedure. Other forms of assessment, such as psychological testing and applying psychiatric diagnostic categories (American Psychiatric Association, 1994), are beyond the scope of this book.

## STEP I-A: "WHAT'S GOING ON?"

The importance of helping clients tell their stories well should not be underestimated. As Pennebaker (1995b) notes, "An important . . . feature of therapy is that it allows individuals to translate their experiences into words. The disclosure

**The Skilled-Helper Model**

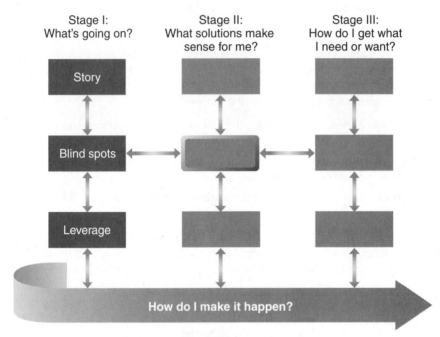

FIGURE 8-1
The Helping Model—Stage 1

process itself, then, may be as important as any feedback the client receives from the therapist" (p. 3). Self-disclosure provides the grist for the mill of problem solving and opportunity development  Here, then, are four goals for Step I-A:

- *Initial stress reduction.* Help clients "get things out on the table." This can and often does have a cathartic effect that leads to stress reduction. Some clients carry their secrets around for years. Helping them unburden themselves is part of the social-emotional reeducation process alluded to earlier.
- *Clarity.* Help clients spell out their problem situations and unexploited opportunities with the kind of concrete detail—specific experiences, behaviors, and emotions—that enables them to do something about them. Clarity opens the door to more creative options in living. Vague stories lead to vague options and actions.
- *Relationship building.* Help clients tell their stories in such a way that the helping relationship develops and strengthens. The communication skills outlined in earlier chapters—suffused, of course, with the values of respect, genuineness, empathy, and empowerment—are basic tools for both clarity and relationship building.
- *Action.* Right from the beginning, help clients act on what they are learning. Clients do not need "grand plans" before they can act on their own behalf.

Later in this chapter, more will be said about the "bias toward client action" needed in the helping process.

Clients differ radically in their abilities to talk about themselves and their problem situations. Reluctance to disclose themselves within counseling sessions is often a window into clients' inability to share themselves with others and to be reasonably assertive in the social settings of everyday life. For such clients, one of the goals of the entire counseling process is to help them develop the skills, confidence, and courage they need to share themselves appropriately.

## HELPING CLIENTS EXPLORE PROBLEM SITUATIONS AND UNEXPLOITED OPPORTUNITIES

As with values and communication skills, there are a number of principles that can guide you as you help clients tell their stories.

### Learn to Work with All Styles of Storytelling

There are both individual and cultural differences (Wellenkamp, 1995) in clients' willingness to talk about themselves. Both affect storytelling. Some clients are highly verbal and quite willing to reveal almost everything about themselves at the first sitting. Take the case of Martina:

> Martina, 27, asks a counselor in private practice for an appointment to discuss "a number of issues." Martina is both verbal and willing to talk, and her story comes tumbling out in rich detail. Although the helper uses the skills of attending, listening, sharing highlights, and probing, she does so sparingly. Martina is too eager to tell her story.
>
> Although trained as a nurse, Martina is currently working in her uncle's business because of an offer she "could not turn down." She is doing very well financially, but she feels guilty because service to others has always been a value for her. And although she likes her current job, she also feels hemmed in by it. A year of study in Europe during college whetted her appetite for "adventure." She feels that she is nowhere near fulfilling the great expectations she has for herself.
>
> She also talks about her problems with her family. Her father is dead. She has never gotten along well with her mother, and now that she has moved out of the house, she feels that she has abandoned her younger brother, who is twelve years younger than she is and whom she loves very much. She is afraid that her mother will "smother" her brother with too much maternal care.
>
> Martina is also concerned about her relationship with a man who is two years younger than she. They have been involved with each other for about three years. He wants to get married, but she feels that she is not ready. She still has "too many things to do" and would like to keep the arrangement they have.
>
> This whole complex story—or at least a synopsis of it—comes tumbling out in a torrent of words. Martina feels free to skip from one topic to another. The way Martina tells her story is part of her enthusiastic style. At one point she stops, smiles, and says, "My, that's quite a bit, isn't it!"

As the helper listens to Martina, he learns a number of things about her. She is young, bright, and verbal and has many resources; she is eager and impatient; some of her problems are probably of her own making; she has some blind spots that could stand in the way of her grappling more creatively with her problems; she has many unexplored options, many unexploited opportunities. That said, the counselor surmises that Martina would make her way in life, however erratically, with no counseling at all.

Contrast Martina's story with the following one of a man who comes to a local mental-health center because he feels he can no longer handle his 9-year-old boy:

Nick is referred to the center by a doctor in a local clinic because of the trouble he is having with his son. He has been divorced for about two years and is living in a housing project on public assistance. After introductions and preliminary formalities have been taken care of, he just sits there and says nothing; he does not even look up. Since Nick offers almost nothing spontaneously, the counselor uses a relatively large number of probes to help him tell his story. Even when the counselor responds with empathic highlights, Nick volunteers very little. Every once in a while, he tears up a bit. When asked about the divorce, he says he does not want to talk about it. "Good riddance" is all he can say about his former wife. Gradually, almost torturously, the story gets pieced together. Nick talks mostly about the "trouble" his son is getting into and how uncontrollable he seems to be getting and how helpless this makes Nick feel.

Martina's story is full of possibilities, whereas Nick's is mainly about limitations. In both content and communication style, they are at opposite ends of the scale.

Each client is different and approaches the telling of the story in a different way. Some clients come voluntarily; others are sent. Some of the stories you help clients tell are long and detailed, others short and stark. Some are filled with emotion; others are told coldly, even though they are stories of terror. Some stories are, at least at first blush, single-issue stories—"I want to get rid of these headaches"— whereas others, like Martina's, are multiple-issue stories. Some stories deal with the inner world of the client almost exclusively—"I hate myself," "I'm depressed," "I feel lonely"—whereas others deal with the outer world—problems with finances, work, or relationships. Still others deal with a combination of inner and outer concerns.

Some clients tell their stories readily, letting them tumble out. They need little help from you. Other clients tell their stories grudgingly and need all the help you can give them. Some clients tell the core story immediately, while others tell a secondary story first to test your reactions. Some clients make it clear that they trust you just because you are a helper, but you can read mistrust in the eyes of others, sometimes just because you are a helper.

In all these cases, your job is to establish a working relationship with your clients and help them tell their stories as a prelude to helping them manage the problems and take advantage of the opportunities buried in those stories. A story that is brought out into the open is the starting point for possible constructive change. Often the very airing of the story is a solid first step toward a better life.

When a client like Martina pours out a story, you may let him or her go on, or you may insist on some kind of dialogue. When a client tells the "whole" story in a more or less nonstop fashion, it is impossible for you to share highlights relating to every core issue the client brings up. But you can help the client review the most salient points in some orderly way. Using expressions like the following can help the client review the core parts of the story:

"You've said quite a bit. Let's see if I've understood some of the main points you've made. First of all . . ."

At this point, the highlights you share will let the client know that you have been listening intently and that you are concerned about him or her. With a client like Nick, however, it's a different story. A client who lacks the skills needed to tell the

### Box 8-1   Problem Finding

Here are some questions counselors can help clients ask themselves to "find" and specify problem situations:

- What are my concerns?
- What's problematic in my life?
- What issues do I need to face?
- What's troubling me?
- What would those who know me best tell me?
- What's keeping me back from being what I want to be? from doing what I want to do?
- What do I need to resolve?

story well or who is reluctant to do so presents a different kind of challenge. Engaging in dialogue with such a client can be tough work.

Snippets of stories have been used in the last few chapters to illustrate basic communication skills. So you have some idea of the variety of stories you will run into. Box 8-1 provides questions clients can ask themselves to identify problem situations.

## Start Where Your Clients Start

Clients have different starting points when they launch into their stories. They can start with any stage or step of the helping process. Therefore, in Step I-A, *story* is used in its widest sense. It does not mean, narrowly, "This is what happened to me, here's how I reacted, and now this is how I feel." Your job is to stay with your clients no matter where they are. Consider the following:

- Martha starts by saying, "I thought I knew how to handle my son when he reached his teenage years. 'He's going to want to try all sorts of crazy things, so keep the reins tight,' I said to myself. Well, now things are awful. It's not working. He's out of control." Her starting point is a *failed solution* that has spawned a new problem. Her version of problem management went wrong somewhere.

- Thad says, "I don't know whether I want to be a doctor or a politician, or at least a political scientist. I love both, but I can't do both. I mean I have to make my mind up this coming year and choose my college courses. I hate being stuck with a decision." Thad's starting point is *choosing a goal*. He has an approach-approach conflict. He wants both goals.

- Kimberley, a human resources executive for a large company, says, "I've found out that our chief executive has been involved in some unethical and, I think, immoral behavior. He's due to retire within the next six months. I don't know whether it's best to bring all this to light or just monitor him till he goes. If I move on him, this could blow up into something big and hurt the reputation of the company itself. If I just monitor him till he goes, he gets away with it. I want to do what's best for the company." Her starting point is a dilemma about which *strategy* to use.

• Owen is having problems sticking to his resolve to restrain himself when one of his neighbors on his block "does something stupid." He says, "I know when I speak up [his euphemism for flying off the handle] things tend to get worse. I know I should leave it to others who are more tactful than I am. But they don't move quickly enough. Or forcefully enough." His starting point is difficulty in *implementing* a course of action to which he has committed himself.

Helping the client explore the context and background of each of these issues may well bring you to other steps of the helping process.

## Help Clients Clarify Key Issues

To clarify means to discuss problem situations and unused opportunities—including possibilities for the future, goals, strategies for accomplishing goals, plans, implementation issues, and feelings about all of these—as concretely as possible. Vagueness and ambiguity lead nowhere.

Consider this case. Janice's husband has been suffering from severe depression for more than a year and is a seeing a therapist regularly. All the attention is on him. One day, after suffering a fainting spell, Janice talks with a counselor herself. At first, feeling guilty about her husband, Janice is hesitant to discuss her own concerns. She says only that her social life is "a bit restricted by my husband's illness." With the help of empathic highlights and probing on the part of the helper, her story emerges. "A bit restricted" turns, bit by bit, into the following broader story. This is a summary; Janice did not say this all at once.

> John has some sort of "general fatigue" illness that no one has been able to figure out. It's like nothing I've ever seen before. I move from guilt to anger to indifference to hope to despair. I have no social life. Friends avoid us because it is so difficult being with John. I feel I shouldn't leave him, so I stay at home. He's always tired, so we have little interaction. I feel like a prisoner in my own home. Then I think of the burden he's carrying, and the roller-coaster emotions start all over again. Sometimes I can't sleep; then I'm as tired as he. He is always saying how hopeless things are and, even though I'm not experiencing what he is, some kind of hopelessness creeps into my bones. I feel that a stronger woman, a more selfless woman, a smarter woman would find ways to deal with all of this. But I end up feeling that I'm not dealing with it at all. From day to day I think I cover most of this up, so that neither John nor the few people who come around see what I'm going through. I'm as alone as he is.

Here, then, is the fuller story spelled out in terms of specific experiences, behaviors, and feelings. The actions Janice takes—staying at home, covering up her feelings—are part of the problem, not the solution. But now that the story is out in the open, there is some possibility of doing something about it.

In another case, a bulimic woman, now under psychiatric care, says that she acted "a little erratically at times" with some of her classmates in law school. The counselor, sharing highlights and using probes, helps her tell her story in much greater detail. Like Janice, she does not say all of this at once, but this is the fuller picture of "a little erratic."

I usually think about myself as plain looking, even though when I take care of myself, some say that I don't look that bad. Ever since I was a teenager, I've preferred to go it alone, because it was safer. No fuss, no muss, and especially no rejection. In law school, right from the beginning, I entertained romantic fantasies about some of my classmates who I didn't think would give me a second look. I pretended to have meals with those who attracted me and then I'd have fantasies of having sex with them. Then I'd purge, getting rid of the fat I got from eating and getting rid of the guilt. But all of this didn't just stay in my head. I'd go out of my way to run into my latest imagined partner in school. And then I'd be rude to him to "get back at him" for what he did to me. That was my way of getting rid of him.

She was not really delusional, but gradually her external behavior, with a kind of twisted logic, began to reinforce her internal fantasies. However, once her story became "public"—that is, once she began talking about it openly with her helper—she began to take back control of her life.

Sometimes helping clients explore the background or context of the concern they bring up helps clarify things. Consider the following case:

In a management development seminar, YK tells his counselor that he is a manager in a consulting firm. The firm is global, and he works in one of its offices in Southeast Asia. He says that he is already overworked, but now his new boss wants him to serve on a number of committees that will take away even more of his precious time. He is also having trouble with one of his subordinates, himself a manager who, YK says, is undermining YK's authority in the wider team.

So far we have a garden-variety story, one that could be repeated thousands of times throughout the world.

The counselor, however, suspects there is more to this. Since he is Canadian and the client is Asian, he wants to make sure that he is a learner. He knows that there is an overlay of Western culture in these consulting firms, but he wants to deal with his client as a full person. In sharing highlights and by using a few probes, he learns enough about the background of YK's story to cast a new light on the problem situation. Here is the more complete story that emerges:

Not only is YK a manager in the firm, but he also has just been made a partner. The structure in these firms is relatively flat, but the culture is quite hierarchical. As a newly minted partner, YK gets the clients that are, in large part, considered the "dogs" of the region. His boss is an American who has been in this post for only four months. YK knows that his boss is going to stay for only one more year; nearing retirement, this is his "fling" in Asia. Though a decent man, he is quite distant and offers YK scant help. So YK believes that his real boss is his boss's boss. But he can't approach him because of company and cultural protocol. The subordinate who is giving him trouble is also a partner. In fact, he has been a partner for several years but has not delivered the goods. This man thought that he should have been made the manager of the unit YK is running. A bit of sabotage is brewing behind YK's back.

On the upside, a few probes reveal that YK's boss thinks that his young manager has a bright future with the company. Putting him on the various committees is his way of giving YK wide exposure. A probe in the form of a managerial personality survey dealing with career derailers reveals that YK scores relatively high on the "dramatic" scale. Since this does not seem to fit with this mild-mannered man, the counselor, using a couple of further probes, discovers that he is not as mild-mannered as he seems. At work, he does get angry, but it is controlled; he is seen as colorful and is considered to have spunk.

A search for some background quickly takes the client's story out of the "routine" category. Of course, you should not be looking for background just for the sake of looking. The right kind and amount of background provides both richness and context. As the counselor coaches YK, he takes a completely different approach from the one he would have taken had YK's preliminary story been the only one. Here are some questions (adapted from a checklist devised by John Scherer, reported in *Training*, January 1993, p. 14) that you might turn into judicious probes to get at background issues relating to the problem or opportunity being discussed:

> What is it about the problem situation that moved the client to seek help in the first place?
>
> Who is being affected, besides the client?
>
> What is the problem or the failure to develop an opportunity costing either the client or others?
>
> To what degree is a larger problem lurking behind the symptoms the client is talking about?

## Assess the Severity of Clients' Problems

Clients come to helpers with problems of every degree of severity. Objectively, problems run from the inconsequential to the life threatening. Subjectively, however, a client can experience even a relatively inconsequential problem as severe. If a client *thinks* that a problem is critical, even though by objective standards the problem does not seem to be that bad, then for him or her it *is* critical. In such a case, the client's tendency to "catastrophize"—to judge a problem situation to be more severe than it actually is—becomes an important part of the problem situation. One of the tasks of the counselor in this case will be to help the client put the problem in perspective or to teach him or her how to distinguish between degrees of problem severity. Howard (1991) puts it well:

> In the course of telling the story of his or her problem, the client provides the therapist with a rough idea of his or her orientation toward life, his or her plans, goals, ambitions, and some idea of the events and pressures surrounding the particular presenting problem. Over time, the therapist must decide whether this problem represents a minor deviation from an otherwise healthy life story. Is this a normal, developmentally appropriate adjustment issue? Or does the therapist detect signs of more thorough-going problems in the client's life story? Will therapy play a minor, supportive role to an individual experiencing a low point in his or her life course? If so, the orientation and major themes of the life will be largely unchanged in the therapy experience. But if the trajectory of the life story is problematic in some fundamental way, then more serious, long-term story repair might be indicated. So, from this perspective, part of the work between client and therapist can be seen as life-story elaboration, adjustment, or repair. (p. 194)

A savvy therapist not only gains an understanding of the severity of a client's problem or the extent of the client's unused resources, but also understands the limits of helping. What Howard calls life-story adjustment or repair is not the same as attempting to redo the client's personality.

Years ago, Mehrabian and Reed (1969) suggested the following formula as a way of determining the severity of any given problem situation. It is still useful today.

$$\text{Severity} = \text{Distress} \times \text{Uncontrollability} \times \text{Frequency}$$

The multiplication signs in the formula indicate that these factors are not just additive. Even low-level anxiety, if it is uncontrollable or persistent, can constitute a severe problem; that is, it can severely interfere with the quality of a client's life.

One way to view helping is to see it as a process in which clients are helped to control the severity of their problems in living. The severity of any given problem situation will be reduced if the stress can be reduced, if the frequency of the problem situation can be lessened, or if the client's control over the problem situation can be increased. Consider the following example:

> Indira is greatly distressed because she experiences migraine-like headaches, sometimes two or three times a week, and seems to be able to do little about them. No painkillers seem to work. She has even been tempted to try strong narcotics to control the pain, but she fears she might become an addict. She feels trapped. For her, the problem is quite severe because stress, uncontrollability, and frequency are all high.
>
> Eventually, Indira is referred to a doctor who specializes in treating headaches. He is an expert in both medicine and human behavior. He first helps her to see that the headaches are getting worse because of her tendency to "catastrophize" whenever she experiences one. That is, the self-talk she engages in ("How intolerable this is!") and the way she fights the headache actually add to its severity. He helps her control her stress-inducing self-talk and teaches her relaxation techniques. Neither of these gets rid of the headaches, but they help reduce the degree of stress she feels.
>
> Second, he helps her identify the situations in which the headaches seem to develop. They tend to come at times when she is not managing other forms of stress well, as when she lets herself become overloaded and gets behind at work. He helps her see that her headaches are part of a larger problem situation. Once she begins to control and reduce other forms of stress in her life, the frequency of the headaches diminishes.
>
> Third, the doctor also helps Indira spot cues—early warning signals—that a headache is beginning to develop. Though she can do little to control the headache once it is in full swing, there are things she can do during the beginning stage. For instance, she learns that medicine that has no effect when the headache is in full force does help if it is taken as soon as the symptoms appear. Relaxation techniques are also helpful if they are used soon enough. Indira's headaches do not disappear completely, but they are no longer the central reality of her life.

The doctor helps the client manage a severe problem situation much better than she has been doing by helping her reduce stress and frequency while increasing control. In other words, he helps the client learn. Of course, if using Mehrabian and Reed's formula leads you to believe that a client's problem is too severe for you to handle, you need to refer the client to another helper.

## Help Clients Talk Productively About the Past

Some schools of psychology suggest that problem situations are not clear and fully comprehended until they are understood in the context of their historic roots. Therefore, helpers in these schools spend a great deal of time helping clients uncover the past. Others disagree with that point of view. Glasser (2000) puts it this way:

> Although many of us have been traumatized in the past, we are not the victims of our past unless we presently choose to be. The solution to our problem is rarely found in explorations of the past unless the focus is on past successes. (p. 23)

Fish (1995) suggests that attempts to discover the hidden root causes of current problem behavior may be unnecessary, misguided, or even counterproductive. Constructive change does not depend on causal connections in the past. There is evidence to support Fish's contention. Long ago, Deutsch (1954) noted that it is often almost impossible, even in carefully controlled laboratory situations, to determine whether event B, which follows event A in time, is actually caused by event A. Therefore, asking a client to come up with causal connections between current unproductive behavior and past events can be an exercise in futility. Among the reasons this process can be frustrating is that causal connections cannot be proved; they remain hypothetical. Also, there is little evidence suggesting that understanding past causes changes present behavior. Moreover, talking about the past often focuses mostly on what happened to the client (experience) rather than on what he or she did about what happened (behavior) and therefore interferes with the bias toward action the client needs to manage the current problem.

This is not to say that a person's past does not influence current behavior. Nor does it imply that a client's past should not be discussed. But the fact that past experiences may well *influence* current behavior does not mean that they necessarily *determine* present behavior. Kagan (1996) has challenged what may be called the "scarred for life" assumption: "If orphans who spent their first years in a Nazi concentration camp can become productive adults and if young children made homeless by war can learn adaptive strategies after being adopted by nurturing families" (p. 901), that means there is hope for us all. As you can imagine, this is one of those issues that members of the helping professions argue about endlessly. Therefore, this is not a debate that is to be settled with a few words here. Instead, there are a couple of principles based on hope that might help.

**Help clients talk about the past to make sense of the present.** Many clients come expecting or wanting to talk about the past. There are ways of talking about the past that help clients make sense of the present. But making sense of the present needs to remain center stage. Thus, *how* the past is discussed is more important than whether it is discussed. In the following example, a man has been discussing how his interpersonal style gets him into trouble. His father, now dead, played a key role in the development of his son's style.

HELPER: So your father's unproductive interpersonal style is, in some ways, alive and well in you.

CLIENT: Until we began talking, I had no idea about how alive and well it is. For instance, even though I hated his cruelty, it lives on in me in much smaller ways. He beat my brother. Now I just cut my brother down to size verbally. He told my mother what she could do and couldn't do. I try to get my mother to adopt my "reasonable" proposals "for her own good"—without, of course, listening very carefully to her point of view. There's a whole pattern that I haven't noticed. I've inherited more than his genes.

HELPER: That's quite an inheritance. . . . But now what?

CLIENT: Well, now that I see what's happened, I'd like to change things. A lot of this is ingrained in me, but I don't think it's genetic in any scientific sense. I've learned a lot of bad habits.

It really does not make any difference whether the client's behavior has been "caused" by his father or not. In fact, if, by hooking the present into the past, he feels in some way that his current nasty style is not his fault, then he has a new problem. Helping is about the future. Now that the problem has been named and is out in the open, it is possible to do something about it. It is about "bad habits," not sociobiological determinism.

In sum, if the past can add clarity to current experiences, behaviors, and emotions, let it be discussed. If it can provide hints as to how self-defeating thinking and behaving can be changed now, let it be discussed. The past, however, should never become the principal focus of the client's self-exploration. When it does, helping tends to slow down needlessly.

**Help clients talk about the past to be liberated from it.** A potentially dangerous logic can underlie discussions of the past. It goes something like this: "I am what I am today because of my past. But I cannot change my past. So how can I be expected to change today?"

CLIENT: I was all right until I was about 13. I began to dislike myself as a teenager. I hated all the changes—the awkwardness, the different emotions, having to be as "cool" as my friends. I was so impressionable. I began to think that life actually must get worse and worse instead of better and better. I just got locked into that way of thinking. That's the same mess I'm in today.

That is not liberation talk. The past is still casting its spell. Helpers need to understand that clients may see themselves as prisoners of their past. But in the spirit of Kagan's earlier comments, helper's need to assist clients in moving beyond such self-defeating beliefs.

The following case provides a different perspective. It is about the father of a boy who has been sexually abused by a minister of their church. He finds that he can't deal with his son's ordeal without revealing his own abuse by his father. In a tearful session, he tells the whole story. In a second interview he has this to say:

CLIENT: Someone said that good things can come from evil things. What happened to my son was evil. But we'll give him all the support he needs to get through this. Though I had the same thing happen to me, I kept it all in until now. It was all locked up inside. I was so ashamed, and my shame became part of me. When I let it all out last week, it was like throwing off a dirty cloak that I'd been wearing for years. Getting it out was so painful, but now I feel so different, so good. I wonder why I had to hold it in for so long.

This is liberation talk. When counselors help or encourage clients to talk about the past, they should have a clear idea of what their objective is. Is it to learn from the past? Is it to be liberated from it? To assume that there is some "silver bullet" in the past that will solve today's problem is to search for magic.

**Help clients talk about the past to prepare for action in the future.** The well-known historian A. J. Toynbee had this to say about history: "History not used is nothing, for all intellectual life is action, like practical life, and if you don't use the stuff—well, it might as well be dead." As we will soon see, any discussion of problems or opportunities should lead to constructive action, starting with Step I-A and going all the way through to implementation. The insights you help clients get from the past should in some way stir them to action. When one client, Christopher, realized how much his father and one of his high school teachers had done to make him feel inadequate, he made this resolve: "I'm not going to do anything to demean anyone around me. You know, up to now I think I have, but I called it something else—wit." Help clients invest the past proactively in the future.

## As Clients Tell Their Stories, Search for Resources, Especially Unused Resources

Incompetent helpers concentrate on clients' deficits. Skilled helpers, as they listen to and observe clients, do not blind themselves to deficits, but they are quick to spot clients' resources, whether used, unused, or even abused. These resources can become the building blocks for the future. Consider this example: Terry is a young woman in her late teens who has been arrested several times for street prostitution. She is an involuntary, or "mandated," client as a result of her latest arrest for possession of drugs. She ran away from home when she was 16 and is now living in a large city on the East Coast. Like many other runaways, she was seduced into prostitution "by a halfway decent pimp." Now she is very street-smart. She is also cynical about herself, her occupation, and the world. She is forced to see a counselor as part of the probation process. As might be expected, Terry is quite hostile during the interview. She has seen other counselors and sees the interview as a game. The counselor already knows a great deal of the story because of the court record. The dialogue is not easy. Some of it goes like this:

TERRY: If you think I'm going to talk with you, you've got another think coming. What I do is my business.

COUNSELOR: You don't have to talk about what you do outside. We could talk about what we're doing here in this meeting.

TERRY: I'm here because I have to be. You're here either because you're dumb and like to do stupid things or because you couldn't get a better job. You people are no better than the people on the street. You're just more "respectable." That's a laugh.

COUNSELOR: So nobody could really be interested in you.

TERRY: I'm not even interested in me!

COUNSELOR: So, if you're not interested in yourself, then no one else could be, including me.

Terry has obvious deficits. She is engaged in a dangerous and self-defeating lifestyle. But as the counselor listens to Terry, he spots many different resources. Terry is a tough, street-smart woman. The very virulence of her cynicism and self-hate, the very strength of her resistance to help, and her almost unchallengeable determination to go it alone are all signs of resources. Many of her resources are currently being used in self-defeating ways. They are misused resources, but they are resources nevertheless.

Helpers need a resource-oriented mind-set in all their interactions with clients. This is part of positive psychology. Let's contrast two different approaches.

CLIENT: I practically never stand up for my rights. If I disagree with what anyone is saying, especially in a group, I keep my mouth shut.

HELPER: So clamming up is the best policy. . . . What happens when you do speak up?

CLIENT (pausing): I suppose that on the rare occasions when I do speak up, the world doesn't fall in on me. Sometimes others do actually listen to me. But I still don't seem to have much impact on anyone.

COUNSELOR A: So speaking up, even when it's safe, doesn't get you much.

CLIENT: No, it doesn't.

Counselor A, sticking to sharing highlights, misses the resource mentioned by the client. Although it is true that the client habitually fails to speak up, he has some impact when he does speak. Others do listen, at least sometimes, and this is a resource. Counselor A emphasizes the deficit. Let's try another counselor.

COUNSELOR B: So when you do speak up, you don't get blasted. You even get a hearing. Tell me what makes you think you don't exercise much influence when you speak.

CLIENT (pauses): Well, maybe influence isn't the issue. Usually, I don't want to get involved. Speaking up gets you involved.

Note that both counselors share a highlight, but they focus on different parts of the client's message. Counselor A emphasizes the deficit; Counselor B notes the asset and follows up with a probe. This produces a significant clarification of the client's problem situation. Not wanting to get involved is an issue to be explored.

The search for resources is especially important when the story being told is bleak. I once listened to a man's story that included a number of bone-jarring life defeats—a bitter divorce, false accusations that led to his dismissal from a job, months of unemployment, serious health concerns, months of homelessness, and more. The only emotion the man exhibited during his story was depression. Toward the end of the session, we had this interchange:

HELPER: Just one blow after another, grinding you down.

CLIENT: Grinding me down, and almost doing me in.

HELPER: Tell me a little more about the "almost" part of it.

CLIENT: Well, I'm still alive, still sitting here talking to you.

HELPER: Despite all these blows, you haven't fallen apart. That seems to say something about the fiber in you.

At the word *fiber*, the client looked up, and there seemed to be a glimmer of something besides depression in his face. I put a line down the center of a newsprint pad. On the left side I listed the life blows the man had experienced. On the right I put "Fiber." Then I said, "Let's see if we can add to the list on the right side." We came up with a list of the man's resources. His fiber included his musical talent, his honesty, his concern for and ability to talk to others, and so forth. After about a half hour, he smiled weakly, but he did smile. He said that it was the first time he could remember smiling in months.

## Help Clients Spot and Develop Unused Opportunities

Early in the history of modern psychology, William James remarked that few people bring to bear more than about 10% of their human potential on the problems and challenges of living. Others since James, though changing the percentages somewhat, have said substantially the same thing, and few have challenged their statements (Maslow, 1968). It is probably not an exaggeration to say that unused human potential constitutes a more serious social problem than emotional disorders, since it is more widespread. Maslow suggests that what is usually called "normal" in psychology "is really a psychopathology of the average, so undramatic and so widely spread that we don't even notice it ordinarily" (p. 16). Many clients you will see, besides having more or less serious problems in living, will also probably be chronic victims of a self-inflicted psychopathology of the average.

Clients are much more likely to talk about problem situations than about un-used opportunities. That's a pity because clients can manage many problems better by developing unused resources instead of dealing directly with their problems. Here are some examples:

- Lech was lonely. And he was somewhat socially inept. The counselor could have helped him take a look at the roots of his loneliness, determine what kind of social life he wanted, learn some social skills, and set up a program to get himself "into community." This would mean spending a lot of time living with "the prob-lem." Instead, the counselor asked him what he liked to do most. Lech said, "Read popular books on architecture. I love buildings of all sorts." A quick Internet search showed that there were a half dozen clubs or groups or programs for architecture af-ficionados. Once Lech was with people who shared his passion, he lost much of his social awkwardness. He went to lectures and discussions. He went on tours. He had some meals with his cohorts. He did not cultivate many closer friendships, but he was no longer lonely. He had a ready-made community. This opportunity-development approach worked for him more efficiently than the problem-solving route.

- Jasmin, a single woman in her mid-twenties, was arrested when she broke into a government computer. She was fined and spent some time in jail, a defeated and demoralized hacker. She became friends with Marion, who one day convinced her to go to a religious service. "Just this one time," Jasmin said. She discovered that the minister was a man with lots of street smarts and "not too pious." She liked the service more than she wanted to admit. After a few weeks, she saw this man "for a bit of advice." Once he found out that she was a hacker, he rolled his eyes and said, "Boy, could we use you." She saw him on and off and discussed what she'd do "on the outside." Once out, she helped him set up a training program in com-puter skills for disadvantaged kids. The fact that she had been a hacker appealed to the kids. They listened to her. She supported herself by working as a consultant in the computer security business. This kept her in touch with the latest in hacking techniques. This, of course, was still her passion. Fortunately, the minister saw tal-ent, however misdirected, rather than depravity.

- A therapist in a state mental hospital also led a marriage therapy group for members of the local community. When one couple "graduated" or dropped out, another couple took their place. One new couple almost undid the group. They fought like cats and dogs, disrupted group meetings, and made no progress. If any-thing, they got worse. The therapist, at his wits' end, had a special meeting with them. As they entered the room, a light went on in his head. He said something like this: "I've never seen anyone as bad as you two. But it just struck me that you're so bad that you'd be good for something. I'd like to film you in the raw, as it were. Just as you are. Fight. Scream. Refuse to do anything constructive. I'd like to say things like this to the audience: 'If you let your marital problems fester, this is what could happen to you' and 'If this couple could get better, anyone could.' How about it?" The couple looked at each other and said, "Why not?" The therapist added, "No national rollout. I'll use the film in these kinds of groups. With your per-mission, of course." However, when the couple tried to be bad on the films, they couldn't. When they watched the films, they discovered how sadly comic they were. This proved to be the turning point for them in the group.

Of course, the opportunities you help clients spot and develop need not have this kind of dramatic flair. If you are to help others spot opportunities and unused resources in their lives, you must become good at spotting them in your own.

### See Every Problem as an Opportunity

Clients don't come with just problems *or* opportunities. They come with a mixture of both. Although there is no justification for romanticizing pain, the flip side of human problems is human opportunities. Here are a few examples:

- Kevin used his diagnosis of AIDS as a starting point for reintegrating himself into his extended family and challenging the members of his family to come to grips with some of their own problems, problems they had been denying for a long time.

- Beatrice used her divorce as an opportunity to develop a new approach to men based on mutuality. Because she was on her own and had to make her own way, she discovered that she had entrepreneurial skills. She started an arts and crafts company.

- Jerome, after an accident, used a long convalescence period to review and reset some of his values and life goals. He began to visit other patients in the rehabilitation center. This gave him deep satisfaction. He began to explore opportunities in the helping professions.

- Sheila used her incarceration for shoplifting—she called it "time out"—as an opportunity to finish her high school degree and get a head start on college.

- An actor suffering from a traumatic disability found new life by becoming a public advocate for those suffering a similar fate.

- A couple mourning the death of their only child started a day-care center in conjunction with other members of their church.

William Miller (1986) talked about one of the worst days of his life. Everything was going wrong at work. Projects were not working out, people were not responding, the work overload was bad and getting worse—nothing but failure all around. Later that day, over a cup of coffee, he took some paper, put the title "Lessons Learned and Relearned" at the top, and wrote down as many entries as he could. Some hours and seven pages later, he had listed 27 lessons. The day turned out to be one of the best of his life. So he began to keep a daily "Lessons Learned" journal. It helped him avoid getting caught up in self-blame and defeatism. Subsequently, on days when things were not working out, he would say to himself, "Ah, this will be a day filled with learnings!"

Sometimes helping a client spot a small opportunity, be it the flip side of a problem or a standalone, provides enough positive-psychology leverage to put him or her on a more constructive tack. Box 8-2 provides some questions clients can ask themselves to identify unused resources and opportunities.

## STEP I-A AND ACTION

Shakespeare, in the person of Hamlet, talks about important enterprises—what he calls "enterprises of great pith and moment"—losing "the name of action." Helping, an "enterprise of great pith and moment," can lose "the name of action." One of the principal reasons clients do not manage the problem situations of their lives

### Box 8-2   Opportunity Finding

Here are some questions counselors can help clients ask themselves to identify unused opportunities:

- What are my unused skills/resources?
- What are my natural talents?
- How could I use some of these?
- What opportunities do I let go by?
- What ambitions remain unfulfilled?
- What could I accomplish if I put my mind to it?
- What could I become good at if I tried?
- Which opportunities should I be developing?
- Which role models could I be emulating?

effectively is their failure to act intelligently, forcefully, and prudently in their own best interests. Covey (1989), in his immensely popular *The Seven Habits of Highly Effective People*, named "proactivity" as the first habit: "It means more than merely taking initiative. It means that as human beings, we are responsible for our own lives. Our behavior is a function of our decisions, not our conditions. We can subordinate feelings to values. We have the . . . responsibility to make things happen" (p. 71).

## The Importance of Proactivity

Inactivity can be bad for body, mind, and spirit. Consider the following workplace example:

> A counselor at a large manufacturing concern realized that inactivity did not benefit injured workers. If they stayed at home, they tended to sit around, gain weight, lose muscle tone, and suffer from a range of psychological symptoms such as psychosomatic complaints unrelated to their injuries. Taking a people-in-systems approach (Egan & Cowan, 1979), she worked with management, the unions, and doctors to design temporary, physically light jobs for injured workers. In some cases, nurses or physical therapists visited these workers on the job. Counseling sessions helped to get the right worker into the right job. The workers, active again, felt better about themselves, and the company benefitted.

Counselors add value by helping their clients become proactive. Helping too often entails too much talking and too little action.

Here is a brief overview of a case that illustrates the almost-magic of action. Ricardo's "helpers" are his sister-in-law and a friend.

> Life was ganging up on Ricardo. Only recently "retired" from his job as office manager of a brokerage firm because "they" did not think that he would fit into the new e-commerce strategy of the business, he discovered that he had a form of cancer the course of which was unpredictable. He had started treatment and was relatively pain free, but he was always tired. He was

also beginning to have trouble walking, but the medical specialists did not know why. It may or may not have been related to the cancer. Healthwise, his future was uncertain. After losing his job, he took a marketing position with a software company. But soon, health problems forced him to give it up. His daughter had been expelled from high school for using drugs. She was sullen and uncommunicative and seemed indifferent to her father's plight. His wife and daughter did not get along well at all, and the home atmosphere was tense whenever they were together. Discussions with his wife about their daughter went nowhere. To top things off, one of his two married sons announced that he was getting a divorce from his wife and was going to fight for custody of their 3-year-old son.

Given all these concerns, Ricardo, though very independent and self-reliant, more or less said to himself, "I've got to take charge of my life or it could fall apart. And I could use some help." He found help in two people: Sarah, an intelligent, savvy, no-nonsense sister-in-law who was a lay minister of their church and, Sam, a friend of many years who happened to be a counselor.

Sarah helped Ricardo stay in touch with the religious principles that had guided his life up to that point. Her nonjudgmental style helped him turn these principals into practical guides for everyday living.

Sam helped Ricardo maintain his self-reliance by supporting his finding out as much as he could about his illness through the Internet. Without becoming preoccupied with his search, Ricardo learned enough about his illness and possible treatments to become a partner with his doctors in choosing the way forward. He also drew up a living will and gave Sam power of attorney to avoid getting caught in the trap of using every possible remedy in a futile attempt to prolong his life at the end. Sam also helped him pursue some intellectual interests—art, literature, theater—that Ricardo hadn't had time for because of business.

Here's a man who, "with a little help from his friends," uses adversity as an opportunity to get more fully involved with life. He redefines life, in part, as good conversations with family and friends. Refusing to become a victim, he stays active. Ricardo died two years later, but he managed to live until he died.

## Using the Time Between Sessions Productively

If the future of helping is brief, or at least briefer, therapy sessions, then counselors must help clients use the time between sessions as productively as possible. "How can I leverage what I do within the session to have an impact on what the client does the rest of the week?" Or month, as the case may be. This does not deny that good things are happening within the session. On the contrary, it capitalizes on whatever learning or change takes place there.

Many helpers give their clients homework from time to time as a way of helping them act on what they're learning in the sessions (Kazantzis, 2000; Kazantzis & Deane, 1999). Mahrer and his associates (1994) reviewed the methods helpers used to do this. They came up with 16 methods, including the following:

- Mention some homework task, and ask the client to carry it out and report back because he or she is now a "new person."

- Wait until the client comes up with a postsession task, and then help the client clarify and focus it.

- Highlight the client's readiness or seeming willingness to carry out some task, but leave the final decision to the client.

- Use some contractual agreement to move the client to some appropriate activity.

Counselors provide help in defining the activity and custom fitting it to the client's situation. Some helpers exhort clients to carry out some action. Some even mandate some kind of action as a condition for further sessions: "I'll see you after you've attended your first Alcoholics Anonymous meeting."

Broder (2000) has further suggestions on how to incorporate homework into your helping sessions. He not only offers many techniques but discusses why some clients resist the idea of homework and suggest ways of addressing such resistance. Broder, in conjunction with Albert Ellis, the originator of Rational Emotive Behavior Therapy, has developed a series of audiotapes dealing with many of the problems clients encounter. Clients use these tapes to further their understanding of what they learned in the therapy session and to engage in activities that will help them resolve their conflicts (see www.therapistassistant.com).

There is no need to call what the client does between sessions homework. The term *homework* puts some clients off. It is also too "teacherish" for some helpers. It sounds like an add-on rather than something that flows organically from what takes place within the helping sessions. But *flows* is a wide term. For instance, if you are using some kind of manualized treatment (see Chapter 1) for treating anxiety, then work between sessions, an essential dimension of the treatment, is part of the flow. The principle behind homework is more important than the name, and the principle is clear. Use every stage and every step of the helping process as a stimulus for problem-managing and opportunity-developing action. Use the term *homework* if it works for you; "assign" it if that works for you and your client. But have a clear picture of why you are assigning any particular task. Don't routinely assign homework for its own sake.

Homework often has a predetermined cast to it. But encouraging clients to act in their own behalf can be a much more spontaneous exercise. Take the case of Mildred.

> Mildred, 70, single, was a retired teacher. In fact, she had managed to find ways of teaching even after the mandatory retirement. So she was newly retired. Given her savings, her pension, and social security, she was ready for retirement financially. But she had not prepared socially or emotionally. She was soon depressed and described life as "aimless." She wandered around for a few months and finally, at the urging of a friend, saw a counselor, someone about her age. The counselor knew that she had many internal resources, but they had all seemed to go dormant. In the second session, he said to her, "Mildred, you're allowing yourself to become a wreck. You'd never stand for this kind of behavior from your students. You need to get off your butt and seize life again. Tell me what you're going to do. I mean right after you get out of this session with me." This was a bit of shock treatment for Mildred, except that no electricity was involved. After a brief pause, a revived Mildred looked at him and asked, "Well, what do *you* do?" They went on to discuss what life can offer after 70. Her first task, she decided, was to get into community in some way. The school had been her community, but it was gone.

Whether you use homework assignments or find other, perhaps more organic, ways of helping clients move as quickly as possible to action is not the issue. Your role demands that you be a catalyst for client action.

## Appreciating the Self-Healing Nature of Clients

Helpers should not underestimate the "agency" of their clients. Bohart and Tallman (1999), in a book on "how clients make therapy work," lay out the principles of client self-healing, principles which, they say, must be respected if therapy is to be a collaborative enterprise. These principles are part of a positive-psychology approach to helping. I have taken their principles and have both reworded and reordered them while making every attempt to retain their spirit (see pp. 227–235 in their book for a full description of each of their principles):

- Respect clients' ability to act in their own best interest.
- Be aware that clients have widely differing world views.
- Support clients' capability for thinking for themselves.
- Respect clients' wisdom and actively elicit their ideas and intuitions.
- Give clients time to think things through and avoid premature closure.
- Support clients' efforts to generate new ways of thinking about themselves and their problems and to act on what they discover.
- Believe that clients are capable of learning and help them amplify small signs of learning and change.
- Be convinced that clients prefer positive, proactive rather than immature, defensive solutions.
- Help clients in their efforts to understand and deal creatively with the constraints that keep them stuck in old ways.
- Remember that clients are agents in their own lives and do not always need our guidance.

While the ideal might be "the client as active self-healer, co-director, and collaborator in therapy" (Bohart & Tallman, p. 141), the authors realize that neither all clients nor all therapists live up to the ideal. But one message is clear—don't sell clients short.

# IS STEP I-A ENOUGH?

Some clients seem to need only Step I-A. That is, they spend a relatively limited amount of time with a helper, they tell their story in greater or lesser detail, and then they go off and manage quite well on their own. Mildred, once blasted out of her torpor, was one of those. Here are two ways in which clients may need only the first step of the helping process.

**A declaration of intent and the mobilization of resources.**  For some clients, the very fact that they approach someone for help is sufficient to help them begin to pull together the resources needed to manage their problem situations more effectively. For these clients, going to a helper is a declaration not of helplessness but of intent: "I'm going to do something about this problem situation."

> Declan was a young man from Ireland living illegally in New York. Even though he had a community of fellow countrymen to provide him support, he felt, as he put it, "hunted." He lived with the fear that something terrible was going to happen. He talked to an older friend about his concerns. The friend listened carefully but had no advice to give him. But the talk provided the stimulus Declan needed. He acted. He returned to Ireland, became skilled in computers, moved to England, and got a job with the British arm of a German computer firm. Just as important, he felt "whole" again.

Merely seeking help can trigger a resource-mobilization process in some clients. Once they begin to mobilize their resources, they begin to manage their lives quite well on their own.

**Coming out from under self-defeating emotions.** Some clients come to helpers because they are incapacitated, to a greater or lesser degree, by negative feelings and emotions. Often when helpers show such clients respect, listen carefully to them, and understand them in a nonjudgmental way, those self-defeating feelings and emotions subside. That is, the clients benefit from the counseling relationship as a process of social-emotional reeducation and repair. Once that happens, they are able to call on their own inner and environmental resources and begin to manage the problem situations that precipitated the incapacitating feelings and emotions. In short, they move to action. These clients, too, seem to be "cured" merely by telling their stories.

> Katrina was very depressed after undergoing a hysterectomy. She had only two relatively brief sessions with a counselor. Katrina, helped to sort through the emotions she was feeling, discovered that the predominant emotion was shame. She felt wounded and exposed to herself and guilty about her incompleteness. Once she saw what was going on, she was able to pick up her life once more.

Of course, not all clients fall into these two categories. Many need the kind of help provided by one or more of the other stages and steps of the helping model.

## THE SHADOW SIDE OF STEP I-A

There are a number of shadow-side dimensions to Step I-A. Two are discussed here. The first relates to the ways clients tell their stories. The second deals with the nature of discretionary change.

### Clients as Storytellers

When clients present themselves to helpers, there is no instant or easy way of reading what is in their hearts. This is revealed over the course of helping sessions. And some helpers are much better than others in discovering "what is really going on." As we have seen, clients approach storytelling in quite different ways. Let's consider a couple of shadow-side versions.

Some clients who tell stories that are general, partial, and ambiguous may or may not have ulterior motives. For instance, one subtext in the shadows is, "If I tell my story too clearly and reveal myself warts and all, I will be expected to do something about it." The accountability issue lurks in the background. Another issue is the accuracy of the story. At one end are clients who tell their stories as honestly as possible. At the other end are clients who, for whatever reason, lie—or fudge. Anyone who has done any marital counseling doesn't have to prove this. They just watch it.

Fudging seems to have something to do with self-image. Some clients are not especially concerned about what their helpers think of them. They have no particular need to be seen in a favorable light. Other clients ask themselves, at least subconsciously, "What will the helper think of me?" Some are extremely concerned about what their helpers think of them and will skew their stories to present themselves in the best light. Kelly (2000a; see also Kelly, Kahn, & Coulter, 1996) sees therapy, at least in part, as a self-presentational process. She suggests that clients benefit by perceiving that their therapists have favorable views of them. Therefore, if therapy contributes to clients' positive identity development, we should expect some fudging. Hiding some of the less desirable aspects of themselves, intentionally or otherwise (see Kelly, 2000b), becomes a means to an end. Hill, Gelso, and Mohr (2000) object to this hypothesis, suggesting that research shows that clients don't hide much from their therapists. Arkin and Hermann (2000) suggest that all of this is really more complicated than the others realize. So, here we have an example of how you must use your own common sense in coming to grips with client self-presentation.

Take an example. Relatively few male victims of childhood sexual abuse discuss the residue of that in adulthood (Holmes, Offen, & Waller, 1997). This used to be explained by the now-discounted myths that few males are sexually abused and that abuse has little impact on males. Rather it seems that it is just harder for males, at least in North American society, to admit both the abuse and its effects. We know that some clients fudge and some clients lie. We can only hypothesize why. But we do know that the vagaries of client self-presentation muddy the waters.

Throw all these factors together with their various combinations and permutations, add in all other variables that might affect client storytelling, and you have an "infinite" number of self-presentation styles. In the end, each client has his or her own style. Add the enormous diversity found in clients and in story content, and it is clear to savvy helpers that each client represents an N = 1 (sole subject) research project. The competent, caring, and knowledgeable helper tackles each project without naiveté and without cynicism.

## The Nature of Discretionary Change

Step I-A is the first step—that is, the first logical step—in a process of constructive change. And so the distinction between discretionary and nondiscretionary change is critical for helpers. Nondiscretionary change is mandated change. If the courts say to a divorced man negotiating visiting rights with his children, "You can't have visiting rights unless you stop drinking," the change is nondiscretionary. There will be no visiting rights without the change.

In contrast, a husband and wife having difficulties with their marriage are not under the gun to change the current pattern. Change here is discretionary. "*If you want more productive relationships, then* you must change in the following ways." The shadow-side principle here is quite challenging.

> In both individual and organizational affairs, the track
> record for discretionary change is quite poor.

It's the Okavango/Kalahari phenomenon. The what? The Okavango River, which rises in the highlands of Angola, gives way to a beautiful inland delta, a blue-green wilderness of fresh water teeming with life. Where does it go? It never finds the sea.

Its water disappears, somehow, into the Kalahari desert. It's a metaphor for much of both personal and organizational change. "Whatever happened to that [lush] management-development program we started two years ago?" The answer: "It's in the Kalahari." Much of counseling is planning. Don't let clients take the (blue-green) planning they do with you and let it disappear into the Kalahari.

The fact of discretionary change is central to the psychopathology of the average. If we don't *have* to change, very often we don't. We need merely review the track record of our New Year's resolutions. Unfortunately, in helping situations, clients probably see most change as discretionary. They may talk about it as if it were nondiscretionary, but deep down, a great deal of "I don't really have to change" pervades the helping process. "Other people should change; the world should change. But I don't have to." This is not cynical. It's the way things are. The sad state of discretionary change is not meant to discourage you but to make you more realistic about the challenges you face as a helper and about the challenges you help your clients face.

A pragmatic bias toward client action on your part—rather than merely talking about action—is a cardinal value. Effective helpers tend to be active with clients and see no particular value in mere listening and nodding. They engage clients in a dialogue. During that dialogue, they constantly ask themselves, "What can I do to raise the probability that this client will act on her own behalf intelligently and prudently?" I know a man who years ago went "into therapy" (as "into another world") because, among other things, he was indecisive. Over the years, he became engaged several times to different women and each time broke it off. So much for decisiveness. Again, savvy helpers know that in some ways, or at least in some cases, the deck is stacked against them from the start. But if we can believe the outcome studies reviewed in Chapter 1, the best helpers consistently win against the odds.

## Evaluation Questions for Step I-A

How effectively am I doing the following?

### Establishing a Working Alliance

- Developing a collaborative working relationship with the client.
- Using the relationship as a vehicle for social-emotional reeducation.
- Not doing for clients what they can do for themselves.

### Helping Clients Tell Their Stories

- Using a mix of tuning in, listening, empathy, probing, and summarizing to help clients tell their stories, share their points, discuss their decisions, and talk through their proposals as concretely as possible.
- Using probes when clients get stuck, wander about, or lack clarity.
- Understanding blocks to client self-disclosure and providing support for clients who have difficulty talking about themselves.
- Helping clients talk productively about the past.

### Building Ongoing Client Assessment into the Helping Process

- Getting an initial feel for the severity of a client's problems and his or her ability to handle them.
- Noting and working with client resources, especially unused resources.
- Understanding clients' problems and opportunities in the larger context of their lives.

### Helping Clients Move to Action

- Helping clients develop an action orientation.
- Helping clients spot early opportunities for changing self-defeating behavior or engaging in opportunity-development behavior.

### Integrating Evaluation into the Helping Process

- Keeping an evaluative eye on the entire process with the goal of adding value through each interaction and making each session better.
- Finding ways of getting clients to participate in and own the evaluation process.

# RELUCTANT AND RESISTANT CLIENTS

RELUCTANCE: MISGIVINGS ABOUT CHANGE
> Fear of intensity
> Lack of trust
> Fear of disorganization
> Shame
> Fear of change

RESISTANCE: REACTING TO COERCION

PRINCIPLES FOR MANAGING RELUCTANCE AND RESISTANCE
> Avoid Unhelpful Responses to Reluctance and Resistance
> Develop Productive Approaches to Dealing with Reluctance and Resistance
>> Explore your own reluctance and resistance
>> See some reluctance and resistance as normal
>> Accept and work with the client's reluctance and resistance
>> See reluctance as avoidance
>> Examine the quality of your interventions
>> Be realistic and flexible
>> Establish a "just society" with your client
>> Help the client search for incentives for moving beyond resistance
>> Employ the client as a helper

PSYCHOLOGICAL DEFENSES: THE SHADOW
SIDE OF RELUCTANCE AND RESISTANCE

It is impossible to be in the business of helping people for long without encountering both reluctance and resistance (Clark, 1991; Ellis, 1985; Fremont & Anderson, 1986; Friedlander & Schwartz, 1985; Harris, 1995; Kottler, 1992; Otani, 1989). Mahalik (1994) developed a Client Resistance Scale to measure clients' opposition to dealing with painful emotions, disclosing intimate or painful material, developing a working alliance with the helper, dealing with blind spots, developing new perspectives, and embracing constructive change. In this book, a distinction is made between reluctance and resistance.

• *Reluctance* refers to clients' hesitancy to engage in the work demanded by the stages and steps of the helping process. Problem management and opportunity development involve a great deal of work. Therefore, there are sources of reluctance in all clients—indeed, in all human beings. For instance, part of problem management is trying new behaviors. Many clients are reluctant to do so. Unused opportunities also provide challenges. Developing unused opportunities means venturing into unknown waters. Although this is a charming idea for some, it strikes something akin to terror into others. Many clients are reluctant to talk about themselves, especially about themselves as flawed. The results are the shadow-side behaviors discussed in Chapter 8.

• *Resistance* refers to the push-back by clients when they feel they are being coerced. Clients who think that they are being mistreated by their helpers in some way tend to resist. Clients who believe that their cultural beliefs, values, and norms— whether group or personal—are being violated by the helper can be expected to resist. For instance, since individual and cultural norms regarding self-disclosure differ widely (see Wellenkamp, 1995), clients who believe that self-disclosure is being extorted from them might well resist.

Although the behaviors through which reluctance and resistance are expressed might either seem or be the same, the distinction is still a useful one. The seeds of reluctance are in the client, whereas the stimulus for resistance is in the helper (Bischoff & Tracey, 1995) or the social setting surrounding the helping process. In practice, of course, a mixture of reluctance and resistance is often found in the same client. If therapy is to become more efficient, counselors need to find ways of helping their clients deal with reluctance and resistance as expeditiously as possible.

## RELUCTANCE: MISGIVINGS ABOUT CHANGE

Reluctance refers to the ambiguity clients feel when they know that managing their lives better is going to exact a price. Clients are not sure whether they want to pay that price. Incentives for not changing often drive out or stand in the way of incentives for changing. This accounts for the sad record for discretionary change mentioned in the last chapter.

Clients exercise reluctance in many, often covert, ways. They talk about only safe or low-priority issues, seem unsure of what they want, benignly sabotage the helping process by being overly cooperative, set unrealistic goals and then use them as an excuse for not moving forward, don't work very hard at changing their behavior, and are slow to take responsibility for themselves. They tend to blame others or the social settings and systems of their lives for their troubles and play games with helpers. Reluctance admits of degrees; clients come "armored" against change to a greater or lesser degree.

The reasons for reluctance are many. They are built into the human condition. Here is a sampling.

**Fear of intensity.** If the counselor uses high levels of tuning in, listening, sharing empathic highlights, and probing, and if the client cooperates by exploring the feelings, experiences, behaviors, points of view, and intentions related to his or her problems in living, the helping process can be an intense one. This intensity can cause both helper and client to back off. Skilled helpers know that counseling is potentially intense. They are prepared for it and know how to support a client who is not used to such intensity.

**Lack of trust.** Some clients find it very difficult to trust anyone, even a most trustworthy helper. They have irrational fears of being betrayed. Even when confidentiality is an explicit part of the client-helper contract, clients can be very slow to reveal themselves. A combination of patience, encouragement, and challenge is demanded of the helper.

**Fear of disorganization.** Some clients fear self-disclosure because they feel that they cannot face what they might find out about themselves. They feel that the façades they have constructed, no matter how much energy must be expended to keep them propped up, are still less burdensome than exploring the unknown. Such clients often begin well but retreat once they begin to be overwhelmed by the data produced in the problem-exploration process. Digging into one's inadequacies always leads to a certain amount of disequilibrium, disorganization, and crisis. But growth takes place at crisis points. A high degree of disorganization immobilizes some clients, whereas a very low degree of disorganization is often indicative of a failure to get at clients' core concerns. A helping model provides clients with "channels" that act as safety devices that help them contain their fears. When helpers challenge clients to take "baby steps" that don't end in disaster, they help clients build confidence.

**Shame.** Shame is a much overlooked variable in human living (Egan, 1970; Kaufman, 1989; Lynd, 1958). It is an important part of disorganization and crisis. The root meaning of the verb *to shame* is "to uncover, to expose, to wound"—a meaning that suggests a process of painful self-exploration. Shame is not just being painfully exposed to another; it is primarily an exposure of self to oneself. In shame experiences, particularly sensitive and vulnerable aspects of the self are exposed, especially to one's own eyes. Shame is often sudden; in a flash, one sees previously unrecognized inadequacies without being ready for such a revelation. Shame is sometimes touched off by external incidents, such as a casual remark someone makes, but it could not be touched off by such insignificant incidents unless, deep down, one was already ashamed. A shame experience might be defined as an acute emotional awareness of a failure to be in some way. Once more, empathy and support help clients deal with whatever shame they might experience.

**Fear of change.** Some people are afraid of taking stock of themselves because they know, however subconsciously, that if they do, they will have to change—that is, surrender comfortable but unproductive patterns of living, work more diligently, suffer the pain of loss, acquire skills needed to live more effectively, and so on. For instance, a husband and wife may realize, at some level of their being, that if they see a counselor, they will have to reveal themselves and that once the cards are on the table, they will have to go through the agony of changing their style of relating to each other.

In a counseling group, I once dealt with a man in his sixties who complained about constant anxiety. He told the story of how he had been treated brutally by his father until he finally ran away from home. He kept saying implicitly: "No one who grows up with scars like mine can be expected to take charge of his life and live responsibly." He used his mistreatment as a youth as an excuse to act irresponsibly at work (he had an extremely poor job record), in his relationship with himself (he drank excessively), and in his marriage (he had been uncooperative and unfaithful and yet expected his wife to support him). The idea that he could change, that he could take responsibility for himself even at his age, frightened him, and he wanted to run away from the group. But since his anxiety was so painful, he stayed. And he changed.

The pace and price of change cause problems for some clients. Some clients think that change is impossible, so why try? Others are dissatisfied with the pace of change. For some, counseling is too fast, for others too slow. Some clients are looking for short-term relief. Some come with the idea that counseling is magic and are put off when change proves to be hard work. Future rewards are not compelling.

Each one of us needs only to look at his or her own struggles with growth, development, and maturity to add to this list of the roots of reluctance.

## RESISTANCE: REACTING TO COERCION

Resistance refers to the reaction of clients who in some way feel coerced. Although reluctance is usually passive, resistance can be both active and passive. It is the client's way of fighting back (see Dimond, Havens, & Jones, 1978; Driscoll, 1984). For instance, spouses who feel forced to come to marriage counseling sessions are often resistant. They resist because they resent what they see as a power play. Some clients see coercion where it does not exist. But because people act on their perceptions, the result is still some form of active or passive fighting back.

Resistant clients, feeling abused, let everyone know that they have no need for help, show little willingness to establish a working relationship with helpers, and frequently try to con counselors. They are often resentful, make active attempts to sabotage the helping process, or terminate the process prematurely. They can be testy or actually abusive and belligerent. Resistance to helping is, of course, a matter of degree, and not all these behaviors in their most virulent forms are seen in all resistant clients.

Involuntary clients—the term *w* clients is sometimes used—are often resisters. For instance, a high school student gets into trouble with a teacher and sees being sent to a counselor as a form of punishment. Or a disgruntled employee is told that some reward—for instance, a promotion—will be conferred if he or she sees a coach or counselor. Clients like these are found in schools, especially schools below college level; in correctional settings; in marriage counseling, especially if it is court-mandated; in employment agencies; in welfare agencies; in court-related settings; and in other social agencies. But any client who feels that he or she is being coerced or treated unfairly can become a resister.

There are all sorts of reasons for resisting, which is to say that clients can experience coercion in a wide variety of ways. The following kinds of clients are likely to be resistant:

- Clients who see no reason for going to helpers in the first place.
- Clients who resent third-party referrers (parents, teachers, correctional facilities, social service agencies) and whose resentment carries over to helpers.
- Clients in medical settings who are asked to participate in counseling.
- Clients who feel awkward in participating, who do not know how to be "good" clients.
- Clients who have a history of being rebels.
- Clients who see the goals of helpers or helping systems as different from their own. For instance, the goal of counseling in a welfare setting may be to help clients become financially independent, whereas some clients may be satisfied with financial dependency.
- Clients who have developed negative attitudes about helping and helping agencies and who harbor suspicions about helping and helpers. Sometimes these clients refer to helpers in derogatory and inexact terms ("shrinks").
- Clients who believe that going to a helper is the same as admitting weakness, failure, and inadequacy. They feel that they will lose face by going. By resisting the process, they preserve their self-esteem.
- Clients who feel that counseling is something that is being done to them. They feel that their rights are not being respected.
- Clients who feel that they have not been invited to participate in the decisions that will affect their lives. This includes decisions about the helping process itself and decisions about their future.
- Clients who feel a need for personal power and find it through resisting powerful figures or agencies: "I may be relatively powerless, but I still have the power to resist" is the subtext.
- Clients who dislike their helpers but do not discuss their dislike with them.
- Clients who differ from their helpers about the degree of change needed.
- Clients who differ greatly from their helpers—for instance, a poor kid with an older middle-class helper.

Of course, resistance can be a healthy sign. Clients are standing up for their rights and fighting back.

Many sociocultural variables—gender, prejudice, race, religion, social class, upbringing, cultural and subcultural blueprints, and the like—can play a part in resistance. For instance, a man might instinctively resist being helped by a woman and vice versa. An African American might instinctively resist being helped by a white and vice versa. A person with no religious affiliation might instinctively think that help coming from a minister will be "pious" or will automatically include some form of proselytizing.

## PRINCIPLES FOR MANAGING RELUCTANCE AND RESISTANCE

Because both reluctance and resistance are such pervasive phenomena, helping clients deal with them is part and parcel of all our interactions with clients (Kottler, 1992). Here are some things counselors can do to create a climate which allows clients to face up to both reluctance and resistance.

### Avoid Unhelpful Responses to Reluctance and Resistance

Helpers, especially beginning helpers who are unaware of the pervasiveness of reluctance and resistance, are often disconcerted when they encounter uncooperative clients. Such helpers are prey to a variety of emotions—confusion, panic, irritation, hostility, guilt, hurt, rejection, depression. Distracted by these unexpected feelings, they react in any of several unhelpful ways.

- They accept their guilt and try to placate clients.
- They become impatient and hostile and manifest these feelings either verbally or nonverbally.
- They do nothing in the hope that the reluctance or the resistance will disappear.
- They lower their expectations of themselves and proceed with the helping process, but in a halfhearted way.
- They try to become warmer and more accepting, hoping to win clients over by love.
- They blame clients and end up in a power struggle with them.
- They allow themselves to be abused by clients, playing the role of a scapegoat.
- They lower their expectations of what can be achieved by counseling.
- They hand the direction of the helping process over to clients.
- They give up.

In short, when helpers engage "difficult" clients, they experience stress, and some give in to self-defeating "fight or flight" approaches to handling it.

The source of this stress is not just the behavior of clients; it also comes from the helper's own self-defeating attitudes and assumptions about the helping process. Here are some of them:

- All clients should be self-referred and adequately committed to change before appearing at my door.
- Every client must like me and trust me.

- I am a consultant and not a social influencer; it should not be necessary to place demands on clients or even help them place demands on themselves.
- Every unwilling client can be helped.
- No unwilling client can be helped.
- I alone am responsible for what happens to this client.
- I have to succeed completely with every client.

Effective helpers neither court reluctance and resistance nor are surprised by them.

## Develop Productive Approaches to Dealing with Reluctance and Resistance

In a book like this, it is impossible to identify every possible form of reluctance and resistance, much less provide a set of strategies for managing each. Here are some principles and a general approach to managing reluctance and resistance in whatever forms they take.

**Explore your own reluctance and resistance.**  Examine reluctance and resistance in your own life. How do you react when you feel coerced? What do you do when you feel you are being treated unfairly? How do you run away from personal growth and development? If you are in touch with the various forms of reluctance and resistance in yourself and are finding ways of overcoming them, you are more likely to help clients deal with theirs.

**See some reluctance and resistance as normal.**  Help the client see that reluctance or resistance is not "bad" or odd. After all, yours isn't. Beyond that, help the client see the positive side of reluctance or resistance. It may well be an indication of fiber, a sign of affirmation of self.

**Accept and work with the client's reluctance and resistance.**  This is a central principle. Start with the frame of reference of the client. Accept both the client and his or her reluctance or resistance. Do not ignore it or be intimidated by what you find. Let the client know how you experience it, and then explore it together. Model openness to challenge. Be willing to explore your own negative feelings. The skill of direct, mutual talk (called immediacy, to be discussed later) is extremely important here. Help clients work through the emotions associated with reluctance or resistance. Avoid moralizing. Befriend the reluctance or the resistance instead of reacting to it with hostility or defensiveness.

**See reluctance as avoidance.**  Reluctance is a form of avoidance that is not necessarily tied to client ill will. Therefore, you need to understand the principles and mechanisms underlying avoidance behavior, which is often discussed in texts dealing with the principles of behavior (see Watson & Tharp, 1997). Some clients avoid counseling or give themselves only halfheartedly to it because they see it as lacking in suitable rewards or even as punishing. If that is the case, the counselor has to help the client search for suitable incentives. Constructive change is usually more rewarding than a miserable status quo, but that might not be the perception of the client, especially in the beginning. Find ways of presenting the helping process as rewarding. Talk about life-enhancing outcomes.

**Examine the quality of your interventions.**  Without setting off on a guilt trip, examine your helping behavior. What are you doing that might seem unfair to the client? In what ways does the client feel coerced? For example, you may have become too directive without realizing it. Furthermore, take stock of the emotions that are welling up in you because the client lashes back or becomes bogged down. How are these emotions "leaking out"? No use denying such feelings. Rather, own them and find ways of coming to terms with them. Do not overpersonalize what the client says and does. If you are allowing a hostile client to get under your skin, you are probably reducing your effectiveness. Of course, the client might be resistant, not because of you but because he or she is under pressure from others to deal with the problems. But you take the brunt of it. Find out, if you can.

**Be realistic and flexible.**  Remember that there are limits to what a helper can do. Know your own personal and professional limits. If your expectations for growth, development, and change exceed the client's, you can end up in an adversarial relationship. Rigid expectations of the client and of yourself become self-defeating.

**Establish a "just society" with your client.**  Deal with the client's feelings of coercion. Provide what Smaby and Tamminen (1979) called a "two-person just society" (p. 509). A just society is based on mutual respect and shared planning. Therefore, establish as much mutuality as is consonant with helping goals. Invite participation. Help the client participate in every step of the helping process and in all the decision making. Share expectations. Discuss and get reactions to helping procedures. Explore the helping contract with your client, and encourage the client to contribute to it.

**Help the client search for incentives for moving beyond resistance.**  Help the client find incentives for participating in the helping process. Use client self-interest and brainstorming to discover possible incentives. For instance, the realization that he or she is going to remain in charge of his or her own life may be an important incentive for a client. Do not see yourself as the only helper in the life of your client. Engage significant others, such as peers and family members, in helping the client face reluctance or resistance. For instance, lawyers who belong to Alcoholics Anonymous may be able to deal with a fellow lawyer's reluctance to join a treatment program more effectively than you can.

**Employ the client as a helper.**  If possible, find ways to get a reluctant or resistant client into situations to help others. The change of perspective can help the client come to terms with his or her own unwillingness to work. One tactic is to take the role of the client in the interview and manifest the same kind of reluctance or resistance he or she does. Have the client take the counselor role and help you overcome your unwillingness to work or cooperate. One person who did a great deal of work for Alcoholics Anonymous had a resistant alcoholic go with him on his city rounds, which included visiting hospitals, nursing homes for alcoholics, jails, flophouses, and down-and-out people on the streets. The alcoholic saw through all the lame excuses other alcoholics offered for their plight. After a week, he joined AA himself. Group counseling, too, is a forum in which clients become helpers.

Hanna, Hanna, and Keys (1999) have drawn up a list of 50 strategies—some original, many from the literature—for counseling defiant, aggressive adolescents (also see Sommers-Flanagan & Sommers-Flanagan, 1995). Many of the strategies have wider application to reluctant and resistant clients of all ages. They divide the strategies into three categories: reaching clients, accepting them, and relating to them. Here is the selection I have made (see the article for descriptions of each strategy). I have reworded some of them.

- **Reaching clients:**
  - Avoid desks.
  - Be genuine and unpretentious.
  - Show deep respect.
  - Keep and use your sense of humor.
  - Be able to laugh at yourself.
  - Educate clients about counseling.
  - Avoid being a symbol of authority.
  - Avoid taking an expert stance until relationships are fairly stable.
  - Avoid asserting your credentials.
  - Avoid thinking in clinical labels.
  - Convey a brief-therapy attitude.
  - Let clients "circle in" on more sensitive issues.
  - Balance insight and action.
  - Admire terrific defensive behavior.
  - Address the hurt behind the anger.
  - Encourage resistance. "Fight back!"
  - "What percentage of you is made up of that part that is worried about you?"

- **Accepting clients:**
  - Be clear about the boundaries of acceptable behavior in counseling sessions.
  - Avoid power struggles.
  - Deal nondefensively with verbal disrespect: "I wonder whom you are really mad at."
  - Validate clients' perceptions when they are accurate.
  - Deal with issues in counseling relationships.
  - Treat shocking statements with equanimity. Reframe messages embedded in such statements.

- **Relating to clients:**
  - Admit when you are confused or uninformed.
  - Expect crises to happen in clients' lives.
  - Tell stories of other clients in similar situations who made changes in their lives.

- Let clients know how much you are learning from your sessions with them.
- Stay in touch with similar problems you have had.
- If you think another counselor will be a better fit with a client, think of switching.
- Use sound bites rather than paragraphs when communicating points.
- Share only things about yourself that you have worked through and would not mind being repeated.
- Do not allow the depth of caring to interfere with clearly understanding your clients and their problem situations.
- Encourage clients to establish therapeutic peer cultures.
- Identify victimization, whatever its source.
- When clients seek attention, give it: "All right, now that you have my full attention, what do you want to do with it?"
- Make confrontation friendly and empathic.
- Reframe apathy as an attempt to avoid hurt, hassles, and difficulties.

In summary, do not avoid dealing with reluctance and resistance, but do not reinforce these processes either. Work with your client's unwillingness, and become inventive in finding ways of dealing with it. But be realistic.

## PSYCHOLOGICAL DEFENSES: THE SHADOW SIDE OF RELUCTANCE AND RESISTANCE

It is odd to talk about the shadow side of reluctance and resistance because these realities tend to be part of the shadow side of helping in the first place. Clients don't talk openly about their reluctance and resistance, rather they demonstrate them. In many ways, they try to disguise them, call them something else. In some ways, psychological defenses are the soil from which reluctance and resistance spring.

The very concept of psychological defenses has had a long, somewhat confusing history (Cramer, 2000). Traditionally, they are treated in courses on abnormal psychology. Cramer (1998) differentiates between coping mechanisms and defenses. She sees coping as a conscious and intentional process, while defenses tend to be unconscious and unintentional. To muddy the waters, Vaillant (2000) talks about defenses as "involuntary coping mechanisms" (p. 89). Perhaps we can say that defenses play a role in coping. The "unconscious" part of this, with its ties to psychoanalytic theory, is a sore point for many. Reifying—that is, making a "thing" out of the unconscious—has been too much for some to swallow.

But today, cognitive psychologists generally agree that some mental processes go on outside of awareness. For instance, there is evidence that this is the case with decision making. Indeed, in Chapter 14, decision making is described as a less-than-rational process. Using language such as "mental processes outside of awareness" to describe defenses keeps the unconscious from sounding like some kind of black hole inside us. While everyone knows that we defend ourselves from the onslaught of reality in a variety of ways, theoreticians, researchers, and practitioners have not always been able to agree on the precise nature of these defenses. Are they good? Are they bad? Are some good and some bad? Yes. Yes. Yes.

Early psychoanalytic theory saw defenses as ways of managing instinctual drives, but some theoreticians and researchers now see them as playing a role in the maintenance of self-esteem and, generally, keeping ourselves intact (Cooper, 1998). Social psychologists today talk about defenses as processes "by which humans deceive themselves, enhance self-esteem, and foster unrealistic self-illusions" (Cramer, 2000, p. 639; see Baumeister, Dales, & Sommer, 1998). "Cognitive dissonance" has been a respected social psychology concept for years. It became almost a brand name, and there have been endless studies on it. But some now say that this term itself is just a nicer word for defenses (Paulhus, Fridhandler, & Hayes, 1997). Since dissonance is uncomfortable, people use a variety of ploys to defend themselves against it (see Chapter 10).

Cramer (2000) places defenses on a maturity continuum. High on the scale are such things as anticipation, altruism, humor, sublimation, and suppression, while delusional projection, psychotic denial, and psychotic distortion sit at the bottom. One problem with such a list is that some high-end defenses, such as anticipation, altruism, and humor, need not be "outside of awareness." Or, if you will, there are forms that are outside awareness and forms that are not. Admitting that psychologists really don't know how defenses—either mature or immature—work, Vaillant (2000) says that one possibility is that so-called mature defenses might just as well be called virtues. Going south on the maturity scale, we find (without naming them all) intellectualization and isolation, then idealization, then nonpsychotic denial and rationalization, then autistic fantasy, and finally acting out and apathetic withdrawal before hitting the bottom. Vaillant discusses the positive-psychology role of mature defenses like altruism and anticipation. There is some evidence that as clients get better, the use of immature defenses recedes and the use of mature defenses increases. Cramer (2000) cautions us on this hierarchical ranking of defenses. Using denial as an example, he notes that everyone practices denial from time to time, "but it is certainly not always pathogenic or pathological" (p. 672). It is not right, he suggests, to say that denial is always developmentally immature because sometimes it is merely adaptive.

While defenses defend people from, let's say, anxiety, they also get people in trouble. For instance, people with serious physical conditions such as diabetes and obesity use defenses to deny the seriousness of their condition. While defenses work to lower anxiety in the short term, they often put such people at risk in the longer term. Defenses, then, can keep us from doing what is in our best interests. In counseling settings, defenses can keep clients from developing insights that would help them admit and deal with their problems more creatively.

The point is that some defenses, or "adaptive mental mechanisms," can help get us into and keep us mired in our troubles, while others can help prevent us from getting into trouble and help us cope when trouble strikes. The former need to be challenged and the latter supported. This does not mean that helpers should mount an attack on every immature defense they run across. This could backfire: "By thoughtlessly challenging irritating, but partly adaptive, immature defenses, a clinician can evoke enormous anxiety and depression in a patient" (Vaillant, 1994, p. 49) and put the helping relationship itself at risk.

Immature defenses contribute a great deal to both reluctance and resistance. The problem is that clients don't show up with a list of defenses tattooed on their foreheads. Personality traits are also important, but they, too, are not always easily identifiable. Take locus of control. A client with a pronounced external locus of control might well have problems with responsibility and accountability. He or she might be looking for a great deal of guidance from you. The point is that clients can arrive on your doorstep with built-in tendencies to reluctance and resistance. This puts a burden on you to listen fully to your clients and to process what they say and do by what you know about human beings. Understanding psychological defenses can help. Understanding personality traits and their implications can help. But in the end, it is you and the client trying to do what is best for the client.

Finally, Clark (1998) has written a book describing defense mechanisms as they manifest themselves in counseling and suggests ways of helping clients challenge them. Although he deals with traditional defense mechanisms—denial, displacement, identification, isolation, projection, rationalization, reaction formation, regression, repression, and undoing—he does so with a fresh voice. The book is filled with examples of dialogue in which helpers listen, process, share highlights, probe, and challenge in the spirit of empathy as they try to help their clients face up to their defenses and what these defenses might be costing them. And that is what the next three chapters are about—challenging clients in the spirit of empathy.

# STEP I-B: I. THE NATURE OF CHALLENGING

**CHALLENGING: THE BASIC CONCEPT**
>    Helpers as sowers of discord
>    Challenge versus confrontation
>    Stop, start, continue

**BLIND SPOTS: THE TARGETS OF CHALLENGING**
>  The Nature of Blind Spots
>    Simple unawareness
>    Self-deception
>    Choosing to stay in the dark
>    Knowing but not caring
>  Specific Blind Spot Targets: Mind-Sets, Thinking, Behaving, Discrepancies, and the Behavior of Others
>    Mind-sets
>    Internal behavior, or ways of thinking
>    External behavior, or ways of acting
>    Discrepancies between thinking/saying and acting
>    Others' behavior and attitudes and their impact

**FROM BLIND SPOTS TO NEW PERSPECTIVES**
>  The Many Names of Developing New Perspectives
>  Linking New Perspectives to Action

**THE GOALS OF CHALLENGING**

**APPLICATIONS: FROM BLIND SPOTS TO NEW PERSPECTIVES TO ACTION**
>  Challenging Mind-Sets
>    Prejudices
>    Self-limiting beliefs and assumptions
>  Challenging Self-Limiting Internal Behavior
>  Challenging Self-Limiting External Behavior
>  Challenging Discrepancies

Challenging the Predictable Dishonesties of Everyday Life
 Invite clients to challenge substantial distortions
 Challenge games, tricks, and smoke screens
 Challenge excuses
Challenging Strengths and Unused Resources
Challenging Behaviors Within Helping Sessions
 Invite clients to own their problems and unused opportunities
 Invite clients to state their problems as solvable
 Invite clients to move on to the next needed stage or step of the helping process

Under the rubric "constructivism," theorists have rediscovered that both cultures and individuals, to a greater or lesser extent, construct the realities that guide their actions (see Borgen, 1992, pp. 120–121; Mahoney, 1991; Mahoney & Patterson, 1992, pp. 671–674; Neimeyer, 1993, 2000; Neimeyer & Mahoney, 1995). This view does not deny that there is a "reality out there," nor does it deny that this reality affects us. If each culture and each human being were totally constructivist, this would lead to chaos and eliminate the ability to communicate: "We're all mad here." But cultures over time develop their own "world views" and act on them. For instance, democratic world views and socialist world views differ and lead to quite different governmental policies. Within all cultures, as we saw in Chapter 3, individuals over time create subsets of these world views—that is, their own personal world views. These personal cultures drive behavior.

The tendency to construct reality or perceive the world in an individualist way produces a great deal of diversity among individuals and constitutes a key challenge for helpers. The construction of reality can contribute both to cultural and individual richness and to the creation of social and individual problems. If the reality I "construct" gets me into trouble or keeps me there, then some kind of "deconstruction" is called for. If the reality I "construct" shortchanges me because it ignores my resources and opportunities for growth and development, then some kind of reconstruction is called for. Some mental constructions are self-limiting. Challenging means, in part, helping clients explore their constructions and the actions that follow from them. Then counselors can help them reconstruct their views of themselves and their worlds in more self-enhancing, other-enhancing, and community-enhancing ways. Community here means the groups in which clients have membership: family, friends, teams, work groups, church, neighborhood, and so forth.

## CHALLENGING: THE BASIC CONCEPT

Effective helpers are not only understanders (listening, processing, sharing empathic highlights) and clarifiers (probing, summarizing) but also reality testers (challenging). Writers who have emphasized helping as a social-influence process (see Chapter 3) have always seen some form of challenge as central to helping (Bernard, 1991; Dorn, 1984, 1986; Strong & Claiborn, 1982).

**Helpers as sowers of discord.**  Martin (1994) put it well when he suggested that the helping dialogue may add the most value when it is perceived by clients as relevant, helpful, interested, and *"somehow inconsistent (discordant) with their current theories of themselves and their circumstances"* [emphasis added] (pp. 53–54). The same point is made by Trevino (1996) in the context of cross-cultural counseling:

> Certain patterns of congruency and discrepancy . . . between client and counselor facilitate change. There is a significant body of research suggesting that congruency between counselor and client enhances the therapeutic relationship, whereas discrepancy between the two facilitates change. In a review of the literature on this topic, Claiborn (1982, p. 446) concluded that the presentation of discrepant points of view contributes to positive outcomes by changing "the way the client construes problems and considers solutions." (p. 203)

Challenge adds that discordant note. Weinrach (1995, 1996), in touting the virtues of Rational Emotive Behavior Therapy (REBT), which incorporates challenge into its core, suggested that some members of the helping professions object to REBT because it is a "tough-minded therapy for a tender-minded profession" (1995, p. 296).

**Challenge versus confrontation.** Note that the word *challenge* is used here rather than the more hard-hitting term *confrontation*, a term with even more of an edge. Most people see both confronting and being confronted as unpleasant experiences. But, at least in principle, they more readily buy the softer, gentler option of challenge. This does not completely rule out confrontation, but why bring out the big guns when they are not needed? More about this later.

**Stop, start, continue.** Challenge is basically simple. Ideally, it is invitational. Challenge is about inviting clients to stop, start, and keep going.

- *Stop.* Identify and *stop* engaging in activities that keep them mired down in their problem situations or that keep them from identifying and developing resources and opportunities.
- *Start.* Identify and *start* engaging in activities that would either prevent them from getting into trouble in the first place or in activities that will either get them out of trouble or help them develop resources and opportunities.
- *Keep going.* Identify and *continue* activities that manage problems or develop opportunities, especially if the client gives signs of flagging.

The basic framework of challenge and the target of challenging is found here in Chapter 10. Methods for helping clients challenge themselves are reviewed in Chapter 11. Finally, how to go about challenging—that is, the wisdom of challenging—is the focus of Chapter 12. As usual, extensive exercises in challenging are found in Chapters 10 through 12 of *Exercises in Helping Skills*.

## BLIND SPOTS: THE TARGETS OF CHALLENGING

Just what is it that clients should start, stop, or continue doing? There are five broad targets: mind-sets, dysfunctional ways of thinking, self-limiting ways of acting, discrepancies between thinking/saying and doing, and failure to understand and deal with the behavior of others. These targets are called "blind spots." The task of the counselor is twofold: first to help clients identify blind spots, second, and more important, to help clients transform blind spots into *new perspectives* that lead to problem-managing and opportunity-developing action.

### The Nature of Blind Spots

Dysfunctional ways of thinking and acting are called blind spots because, in the end, clients don't seem to see, understand, realize, or appreciate how they are doing themselves in. Put more philosophically, clients fail to see how some of the realities they construct for themselves are self-limiting.

We all have blind spots. We all have areas of naïveté. From one perspective, life can be seen as a process of erasing blind spots and moving from naïveté to becoming more and more socially and emotionally competent. Since we are all "constructivist" to one degree or another, our personal "realities" are bound to run up against other realities. So we live in a conflict-filled world. And we are "blind" in different ways and to different degrees.

**Simple unawareness.** There are things that clients are simply not aware of. Becoming aware of them would help them know themselves better and both cope with problems and develop opportunities. Casper doesn't know that he has an acerbic communication style. If asked, he would describe himself as "assertive." But he is dissatisfied with his social life. Katya doesn't know that she is power hungry. She never thinks about such things. If asked, she would say that she is as "ambitious as anyone else." Serge was surprised when fellow members of a self-help group called him "talented." He was brought up in a family that prized modesty and "humility." He never thought about himself as talented, creative, and resourceful and had little idea how this lack of awareness had narrowed his life.

**Self-deception.** Clients are like the rest of us. There are things they would rather not know because, if they knew them, they would be challenged to change their behavior in some way. So they'd rather stay in the dark. Goleman—who, as we saw in Chapter 3, now writes extensively about "emotional intelligence"—early on wrote a book called *Vital Lies, Simple Truths* (1985) on the psychology of self-deception. Self-deception is incompatible with the social-emotional maturity Goleman describes in *Emotional Intelligence* (1995). Yet, as Eduardo Giannetti points out in *Lies We Live By: The Art of Self-Deception* (1997), it is ubiquitous. "How," he marvels, "do we carry out such feats as believing in what we don't believe in, lying to ourselves and believing the lie . . . ?" (p. viii). The defenses mentioned in Chapter 9 offer some degree of explanation. Because there are degrees of defenses, there are degrees of self-deception. Psychotic defenses represent the extreme, but the more moderate defenses such as rationalization are also forms of self-deception.

**Choosing to stay in the dark.** Choosing to stay in the dark is a form of evasiveness with oneself: "I could find out, but not now." Lots of people, when they have physical symptoms such as pain in their guts, avoid thinking about it. Finding out that the pain indicates something serious could be uncomfortable. Finding out that the pain is a symptom of a life-threatening illness could be terrifying. Clients often choose to stay in the dark. Ilia, recently released from jail, knows that she should clarify the conditions of her probation but chooses not to. When asked about the conditions, she says, "I don't know. No one really explained them to me." She knows that if she gets caught violating any of the conditions, she could go back to jail. But she puts that out of her mind. When clients are being vague or evasive with their helpers, they may well not want to know. And it is likely they don't want you to know either.

**Knowing but not caring.** Clients sometimes know that their thinking and acting is getting them into trouble or keeping them there, but they don't seem to care. Or they know what they should do to get out of trouble, but they don't lift a finger. We can use the term *blind spot*, at least in an extended sense, to describe this kind of behavior because clients don't seem to fully understand or appreciate the degree to which they are choosing their own misery. Or the degree to which they are turning their backs on a better future. For instance, Tanel says he knows that he annoys his wife when he nags her to get a job even though they have two young children at home, but he keeps on nagging her anyway. This creates a great deal of tension, but he continues to focus on his wife's reluctance to get a job rather than on his own nagging.

So, as you can see, the term *blind spot*, as used here, is somewhat elastic. We are unaware, we deceive ourselves, we don't want to know, we ignore, we don't care, or we know but not fully; that is, we do not fully understand the implications or the consequences of what we know. But it's a good term. It has great face validity. People know what you mean as soon as you use the term.

## Specific Blind Spot Targets: Mind-Sets, Thinking, Behaving, Discrepancies, and the Behavior of Others

Blind spots are mind-sets, internal and external behavior, or discrepancies that, as we have just seen, we are unaware of or choose to ignore in one way or another. They become *key* blind spots when they affect our lives substantially or have the potential to do so. We can also become "blind" with respect to others' behavior toward us.

**Mind-sets.** The term *mind-sets* here refers to more or less permanent states of mind, including assumptions, attitudes, beliefs, bias, convictions, inclinations, norms, outlook, unexamined perceptions of self/others/the world, preconceptions, prejudices, reactions, and values. They are blind spots if we are unaware or choose to be unaware of them, their implications, or their consequences. Mind-sets drive external behavior. Or at least leak out into external behavior. Gilda has a rosy outlook on life. In her conversations, she tends to emphasize the positive. You can usually count on her being in a good mood. Geoffrey, on the other hand, tends to see the dark side of things. Some have used the term *curmudgeon* to describe him. If you work for him, you know he's going to be looking for your faults. Even his friends often have to put up with him.

Mind-sets make a difference. For instance, I may grow up with a prejudice. I never question it. It's like furniture. It's just there. It may or may not influence my behavior. But it does have downside potential. On the other hand, I might have an "inquisitive mind." I might not know the origin of my sense of wonder. It's just there. It may or may not influence my behavior. It probably has a lot of upside potential.

Mind-sets can decrease or increase options. If I see the world as a hostile place, this sets me on edge and quite often restricts my options to contact people, make friends, and move around the world. On the other hand, if I see myself as a decent and caring person who has a lot to offer to the world, I increase my options. I portray myself positively in interviews. I willingly volunteer my help when it is needed.

Some clients have mind-sets that would do them a world of good were they to let these attitudes have some impact on their external behavior. Astra cares a great deal about her students and has thought of some very inventive ways of teaching, but she keeps them all inside and comes across as a very difficult person because she doesn't want students to run riot over her. A self-limiting mind-set stands in the way of creative action. Astra needs to see that expressing concern for her students and putting in practice some of her good ideas can contribute to instead of stand in the way of good discipline.

**Internal behavior, or ways of thinking.** Internally, we daydream, pray, ruminate on things, believe, make decisions, formulate plans, make judgments, question motives, approve of self and others, disapprove of self and others, wonder, value, imagine, ponder, think through, create standards, fashion norms, mull over, worry, panic, ignore, forgive, rehearse—we *do* all sorts of things. These are behaviors, not just things that happen to us.

Like mind-sets, internal behavior is inside clients' heads—you can't immediately see it. They have to tell you about it (self-disclosure). Or you have to ask them what's going on inside (probing). Or you may infer what's going on internally from clients' external behavior (making hunches, guessing, interpreting). Clients engage in ways of thinking that do them little good. For instance, Clarence, a worker in a bank, thinks, after mulling it over, that both the spirit and the skills of the "new economy" have passed him by and that he is now stuck in a career he does not like. Clarence can be challenged to see that the world is not passing him by. Rather he is letting the world pass by. Clients also fail to engage in internal behavior that would do them a world of good, like problem solving.

**External behavior, or ways of acting.** This is the stuff people can see or could see if they were looking. For some clients, their behavior constitutes trouble. When Achilles is with women at work, he engages in behavior that others, including the courts, see as sexual harassment. Other clients engage in behaviors that get them into trouble. Jake is not an alcoholic, but when he has a couple of drinks, he tends to get mean and argumentative. This leads to problems with both family and friends. Some clients engage in behaviors that keep them mired in their problems. Clarence, a self-doubting and overly cautious person, is very deferential around his manager. He does not realize that his manager interprets his deference as a "lack of ambition." Clarence's behavior keeps him stuck in a job he hates. But the manager says nothing since she would rather have him "obedient" than "too aggressive."

Clients also fail to make choices and engage in behaviors that would help them cope with problems or develop opportunities. When Clarence is offered an opportunity to update his skills, he turns it down. He also refuses a promotion, saying to himself, "I don't want to get in over my head." His self-defeating external behavior is based on a self-defeating way of thinking about himself. His manager sees him bypass opportunities but says nothing to him. She says to herself, "Anyway, I've got a hard-working drone. That's something these days."

**Discrepancies between thinking/saying and acting.** Good thinking doesn't always get translated into good behavior. That is, there are often discrepancies between what clients think/say and what they do. Clarence knows that the longer he hesitates, the more likely he will be stuck in a dead-end career. He even says this to his wife. But he still does not act. And the window of opportunity continues to close. Astra says that she cares for her students but, because of her fears, she fails to use creative teaching methods that would enhance their learning.

**Others' behavior and attitudes and their impact.** Our blind spots are not limited to our own thinking and acting. We also fail to notice others. For instance, parents often fail to notice signs that their teenage children have started to use drugs. And if at times we are blind to others and their needs—often those closest to us—we are also blind to their attitudes toward us and the impact of their behavior toward us. Sandra interpreted her husband's more relaxed attitude to her in sexual matters as a sign that "he was finally coming to his senses." So she was shocked when she learned, by accident, that he was having an affair. Sometimes we are unaware.

Sometimes we deceive ourselves. Sometimes we'd rather not know. Sometimes we know and just don't care but are blind to the implications of our not caring.

In practice, these five categories are often mixed together. Minerva believes that the world is filled with dishonest people (a self-limiting mind-set). Whenever she meets someone new, she views that person's behavior through this lens and thinks that he or she is guilty until proved innocent (an internal behavior). Therefore, when she meets someone new, she is defensive and often questions that person's actions (external actions). She also realizes that when she meets someone new, she expects some kind of initial trust from them, even though she doesn't give trust to others (a discrepancy). She fails to see that even her closest friends are becoming uncomfortable around her (other people's behavior). Minerva's blind spots cover the full range.

Helping clients deal with blind spots is one of the most important things you can do as a helper. For instance, if Lester has a prejudice and isn't aware or doesn't admit to it, he has a blind spot. If he has a prejudice, knows it (though he probably does not refer to his attitude as a prejudice), and fosters it, then he's engaging in a potentially dysfunctional bit of behavior. If he has a prejudice, fosters it, and lets it spill over into the way he deals with people, Lester has the full dysfunctional package. Contrast this with Bernice. She has been unaware that she is prejudiced, becomes aware of her prejudice, refuses to act on it, tries to get rid of it as part of her internal furniture, and even learns something about herself and the world as she does all this. She is dealing with her prejudice creatively. She has turned a problem into an opportunity. Helping clients handle dysfunctional blind spots can prevent damage from occurring, limit damage already done, and turn problems into opportunities.

## FROM BLIND SPOTS TO NEW PERSPECTIVES

We do our clients a disservice if all that we do is help them identify and explore self-limiting blind spots. The positive-psychology part of challenging is helping clients transform blind spots into new perspectives and helping them translate these new perspectives into more constructive patterns of both internal and external behavior.

### The Many Names of Developing New Perspectives

There are many upbeat names for this process of transforming blind spots into new perspectives. They include seeing things more clearly, getting the picture, getting insights, developing new perspectives, spelling out implications, transforming perceptions, developing new frames of reference, looking for meaning, shifting perceptions, seeing the bigger picture, developing different angles, seeing things in context, context breaking, rethinking, getting a more objective view, interpreting, overcoming blind spots, second-level learning, double-loop learning (Argyris, 1999), thinking creatively, reconceptualizing, discovery, having an "ah-ha" experience, developing a new outlook, questioning assumptions, getting rid of distortions, relabeling, and making connections. You get the idea. Some terms used to describe this process are *framebreaking, framebending,* and *reframing.* All these terms imply some kind of cognitive restructuring, developing understanding, or awareness that is needed to identify and manage problems and opportunities. Developing new perspectives, while painful at times, ultimately tends to be prized by clients.

In the following example, the client, Leslie, a fairly religious 83-year-old woman, is a resident of a nursing home. She is talking to one of the aides:

LESLIE: I've become so lazy and self-centered. I can sit around for hours and just reminisce . . . letting myself think of all the good things of the past—you know, the old country and all that. Sometimes a whole morning can go by.

AIDE: I'm not sure what's so self-indulgent about that.

LESLIE: Well, it's in the past and all about myself. . . . I don't know if it's right.

AIDE: The way you talk about it, the reminiscing sounds almost like a kind of meditation for you.

LESLIE: You mean like a prayer?

AIDE: Well, yes . . . like a prayer.

The client sees reminiscing as laziness. The aide helps her develop a new, more positive perspective. Reminiscing as a kind of meditation on life, or even a prayer—Leslie is comfortable with that. With that new slant, Leslie feels that she can "indulge" herself.

Effective helpers assume that clients have the resources to see the world in a less distorted way and to act on what they see. Another way of putting it is that skilled counselors help clients move from what the Alcoholics Anonymous movement calls "stinkin' thinkin'" to healthy thinking. And from healthy thinking to healthy acting. Carla is facing menopause. She is lumbered with the outmoded view of menopause as a "deficiency disease." A counselor helps her see it as a natural developmental stage of life. Although it indicates the ending of one phase, it also opens up new life-stage possibilities. Looking forward to those possibilities rather than looking back at what she's lost helps Carla a lot.

It's important to note that identifying blind spots is not always the same as developing a new perspective. When Sandra, mentioned earlier, found out that her husband was having an affair, she discovered a fact. She was blind to how serious her husband was about achieving a more satisfactory sexual life and had deceived herself by telling herself that he was finally "coming to his senses." A new perspective for her was a step further: "I now realize more fully that all aspects of marriage are two-way streets and that we have to come up with an arrangement about sexual behavior that is satisfactory to both of us." This might sound pedestrian, but it is a new perspective for her, which, if acted on, could help save the marriage.

## Linking New Perspectives to Action

Although new perspectives are important, they are not magic. Overstressing insight and self-understanding—that is, new perspectives—can actually stand in the way of action instead of paving the way for it. Unfortunately, the search for insight can too easily become a goal in itself. Much more attention has been devoted to developing insight than to linking insight to action. The former is often presented as "sexy," whereas the latter is work. I do not mean to imply that achieving useful insights into oneself and one's world is not hard work and often painful. But the pain needs to be turned into gain. Constructive behavioral change leading to valued outcomes is required. Consider this case:

Ned, a manager who thought that he was a leader, comes to realize that despite a company "empowerment" program, he still keeps making all the decisions in his team. This slows down the work and makes the department less efficient than it might be. In a leadership training program, he gets "360" feedback—that is, feedback from a survey filled in by his boss, his direct reports, his peers, and himself. Once he gets over the shock of reading the ratings, he sits down with a coach and talks about what constructive changes he needs to make in his managerial style: more delegating, getting and acting on suggestions from members of the team, and providing constructive feedback for them. Ned and the coach work together on drawing up an implementation plan.

Effective counselors, as they help their clients develop insights into themselves and their behavior, maintain a whole-model approach. Insights are not just to be relished. They are to be acted on. Clients should be able to say, "I now see much more clearly how I am putting myself in jeopardy, but *I can do something about it.*" For instance, Joanna, a woman in her late fifties, has her annual physical exam. The doctor discovers that she has a higher-than-average risk for stroke and communicates this information to Joanna.

JOANNA: What puts me at risk?

DOCTOR: A few things. The biggest factor is your smoking. Next is your cholesterol level.

JOANNA: So, if I stop smoking and get my diet under control . . .

DOCTOR: It won't make you live forever, but it will lower your risk for disease.

JOANNA: I don't want to live forever. And I don't want to be too prudent. That would take all the fun out of life. . . . But I don't want to become sick and incapacitated either.

Joanna has been ignoring the ways in which she is putting her health at risk. That her risk level is higher than average is the basic part of the new perspective. Joanna herself adds the "I can do something about it" part.

Let's put what has been outlined in the last few sections into a single case. Saul, a counseling-psychology trainee, is helped to see how his hesitancy to participate actively in the training group makes him an observer or detractor rather than a contributor to the group (a blind spot). His reluctance to participate is discussed frequently and soaks up the time and energy of the group. He fails to see that the other members of the group are becoming exasperated with him (another blind spot). Saul has been justifying his behavior to himself by telling himself that he has been "learning a lot." In a one-to-one session, the group leader helps Saul deal with his blind spots. Saul comes to realize that even though he subscribes to the theory of active participation in groups, he does not apply the theory to his own life (discrepancy).

Saul's perspective moves from observer-learner to contributor-learner. By reviewing what the other members of the group do well, he forms a concrete picture of what a contributor does (a new perspective that sits on the edge of action). He develops an action program that helps him become a contributor. For instance, as soon as he catches himself sinking into the observer role, he immediately formulates some contribution he can make to the group (translation of new perspective into new internal behavior). He makes his contribution when it seems appropriate (translation of new perspective into new external behavior). Further, he takes time before the group to think of contributions he might make (new internal behavior) and makes them whenever they are appropriate (new external behavior). All of this serves to eliminate the discrepancy between what he thinks makes groups effective and the way he has been participating in the training group.

To say that not all instances of challenge are this neat and turn out this well is, of course, an understatement.

## THE GOALS OF CHALLENGING

Let's pause a moment and review the goals of challenging. The goals of challenging are derived from all that we have seen up to this point. The overall goal of challenging is to help clients do some reality testing and invest what they learn from this in their futures. If the focus is on the ways clients are "mired down," the goal goes something like this:

> *Invite clients to challenge themselves to change ways of thinking and acting that keep them mired in problem situations and prevent them from identifying and developing opportunities.*

A parallel goal in the spirit of positive psychology is more upbeat. It deals with new perspectives and translating these new perspectives into new ways of acting. It goes something like this:

> *Become partners with your clients in helping them challenge themselves to find possibilities in their problems, to discover unused resources, both internal and external, to invest these resources in the problems and opportunities of their lives, to spell out possibilities for a better future, to find ways of making that future a reality, and to commit themselves to the actions needed to make it all happen.*

This might sound like overkill. But it contains the spirit of your role as "catalyst for a better future." Your job is not just to help clients manage problems and develop opportunities. Your task is to help clients shape a better future for themselves. Challenge should not focus just on client dysfunctioning but on the "possible self" that every client is. Putting both these statements together produces a simpler overriding principle:

> *Help clients identify and replace unproductive thinking and behaving with productive thinking and behaving.*

This "replacing" can take place at any stage or step of the helping process. For instance, helpers who do no more than listen to clients' stories and help them clarify them are, according to Kiesler (1988), "hooked" by their clients. After all, clients are experts in their own stories. Helpers "unhook" themselves by challenging clients' stories and the quality of their participation in the helping process itself. They unhook themselves further by challenging clients to move from story to possibilities to goals to action.

## APPLICATIONS: FROM BLIND SPOTS TO NEW PERSPECTIVES TO ACTION

Since challenge does not exist for its own sake, your job is to help clients challenge themselves whenever, in your estimation, it would add value. Once the skill of challenging becomes second nature to you, you will find yourself routinely inviting clients to take a "second look" at what their mind-sets are and what they are thinking, saying, or doing at every stage and step of the helping process. What follows are further examples of some of the common situations calling for challenge. Clients move from blind spots to new perspectives to action. Given the complexity of human thinking and acting, there is bound to be some overlap among these categories.

## Challenging Mind-Sets

The principle: Invite clients to transform outmoded, self-limiting mind-sets and perspectives into self-liberating and self-enhancing new perspectives that are pregnant with action. Here are examples dealing with two kinds of mind-sets—prejudices and self-limiting assumptions.

**Prejudices.** Candace is having a great deal of trouble with a colleague at work who happens to be Jewish. As she grew up, she picked up the idea that "Jews are treacherous businessmen." The counselor invites her to rethink this prejudicial stereotype. Her colleague may or may not be treacherous, but it's not because he's Jewish. Separating the individual from the prejudicial stereotype helps Candace think more clearly. But it is now clear that Candace has two problems: her troubled relationship with her colleague at work and her prejudice. She has discovered that her problem with her colleague is not his Jewishness, but she may also need to discover that often in troubled relationships, both parties contribute to the mess. She realizes that she has to go back to the drawing board and examine her behavior and the degree to which it has been affected by her prejudice.

**Self-limiting beliefs and assumptions.** Albert Ellis is the founder of Rational-Emotional-Behavior Therapy (REBT) (see Dryden, 1995; Dryden, Neenan, & Yankura, 1999; Ellis, 1999; Ellis & Harper, 1998; Ellis & MacLaren, 1998). Ellis claims that one of the most useful interventions helpers can make is to challenge clients' irrational beliefs (see also Lazarus, Lazarus, & Fay, 1993). Because clients in some way "talk themselves into" these dysfunctional beliefs, they engage in a form of self-talk. Some of the common beliefs that Ellis believes get in the way of effective living are these:

- *Being liked and loved.* I must always be loved and approved by the significant people in my life.
- *Being competent.* I must always, in all situations, demonstrate competence, and I must be both talented and competent in some important area of life.
- *Having one's own way.* I must have my way, and my plans must always work out.
- *Being hurt.* People who do anything wrong, especially those who harm me, are evil and should be blamed and punished.
- *Being danger free.* If anything or any situation is dangerous in any way, I must be anxious and upset about it. I should not have to face dangerous situations.
- *Being problemless.* Things should not go wrong in life, and if by chance they do, there should be quick and easy solutions.
- *Being a victim.* Other people and outside forces are responsible for any misery I experience. No one should ever take advantage of me.
- *Avoiding.* It is easier to avoid facing life's difficulties than to develop self-discipline; making demands of myself should not be necessary.
- *Tyranny of the past.* What I did in the past, and especially what happened to me in the past, determines how I act and feel today.

- *Passivity.* I can be happy by avoiding, by being passive, by being uncommitted, and by just enjoying myself.

I am sure that you could add to the list. Ellis suggests that if any of these beliefs are violated in a person's life, he or she tends to see the experience as terrible, awful, even catastrophic. "People pick on me. I hate it. It shouldn't happen. Isn't it awful!" Such "catastrophizing," Ellis says, gets clients nowhere. It is unfortunate to be picked on, but it's not the end of the world. Moreover, clients can often do something about the issues over which they catastrophize.

Take Allison. She is talking with a counselor about how miserable she has been since she suffered a financial reversal in the stock market. She works from the assumption that she should be problemless: "These things shouldn't happen to me!" She also works from the assumption that markets should always go up and there should never be any "corrections." She catastrophizes about her losses in the market: "Look at the state this has left me in!" In reality, while she has suffered some losses, she is still in good financial shape. All of this is an opportunity for her to rethink her approach to wealth and happiness and to come to grips with the latter-day turbulence in financial markets. Learning needs to take the place of catastrophizing.

In the past, REBT has been faulted, at least in principle, for criticizing clients' deeper beliefs, even religious beliefs. For instance, Pepito, a sales rep for a large company, believes that bribery is wrong even though it may be condoned in the countries he visits. However, challenging him to get rid of such an "outmoded" point of view is not what is meant by challenging dysfunctional beliefs. There is a huge difference between the kinds of beliefs outlined above and religious and other deeply held beliefs. And, to be fair, REBT, like other approaches to therapy, has developed significantly over the years (Ellis, 1997, 1999a). And others (Johnson, Ridley, & Nielsen, 2000; Nielsen, Johnson, & Ridley, 2000) have pointed out how the principles of REBT are congruent with and even support religious beliefs, a point of view seconded by Ellis himself (Ellis, 2000).

Jot down some of your own potentially dysfunctional mind-sets. What new perspectives might replace them? How would this change your behavior?

## Challenging Self-Limiting Internal Behavior

The principle: Invite clients to replace self-limiting and self-defeating internal behavior with more creative thinking that translates into action. The ways in which internal behavior can be self-limiting are legion. We'll consider a few examples.

John daydreams a lot, seeing himself as some kind of hero whom others admire. Thinking about unrealistic success in his social life has taken the place of working for actual success. The new perspective: Daydreaming is not all that bad. It's how you use it. With the help of a counselor, John does not stop daydreaming, but he switches its focus. He daydreams about what a fuller social life might look like. This provides him with some practical strategies for expanding and enriching his interactions with others. He begins to try these out.

Nadia, when given a project, immediately thinks of why it won't work. This annoys her colleagues. With the help of a supervisor, she sees how self-limiting such behavior is. The new perspective: The opposite approach can add value. So she gets into the habit of first trying to see what value the project or program will add to the company and what she might do to make it more realistic. Only after doing some of this "internal work" does Nadia engage her colleagues in conversation about the project. She discusses projects in a much more balanced way.

Roberto is dissatisfied with certain aspects of his marriage with Maria. Here is his story:

> Before Roberto and Maria got married, they talked a great deal about the cultural difficulties they might face. He still veered toward the Hispanic culture, whereas she had become quite "Anglo" in her attitudes and behavior. She was especially worried about norms relating to the role of women. Roberto said he would enjoy being married to someone with a "pioneer" spirit. They thought that they had things worked out. That was then. Now Maria has broken through a number of cultural taboos. She has put herself through college, gotten a job, developed it into a career, and assumed the role of both mother and co-breadwinner. She makes more money than Roberto. His woes include losing face in the community, feeling belittled by his wife's success, and being forced into an overly "democratic" marriage.

If Roberto is going to manage the conflict between himself and his wife better, he needs to challenge himself to change the way he thinks. Here are some internal behaviors he needs to start, stop, and continue:

- *Stop* telling himself that Maria is the one with the problem, stop seeing her as the offending party when conflicts arise, and stop telling himself there is no hope for the relationship. He needs to become a co-owner of both the problems and the opportunities of the marriage.

- *Start* thinking of Maria as an equal in the relationship, start understanding her point of view, and start imagining what an improved relationship with her might look like. He needs a new communication style with his wife.

- *Continue* to take stock of the ways he contributes to their difficulties and increase the number of times he tells himself to let Maria live her life as fully as he wants to live his own. He needs to identify what he does best in the marriage—he is not a dud as a communicator—and reinforce these behaviors.

That is, Roberto has to get his head straight, mobilize his resources, and turn his reconstructed thinking into action. Of course, given that any relationship is a two-way street and that both parties need to change, Maria has to develop some new perspectives and change some of the ways she thinks and acts.

The way we talk to ourselves internally has a distinct impact on the way we act. Self-talk here refers to the thematic messages we send ourselves. Some people keep saying to themselves, "You're no good." Others send a completely different message: "You're special; you're exempt from the ordinary rules." Worriers engage in a great deal of dysfunctional self-talk. The theme is that things are going to go wrong. Of course, self-talk can also be upbeat. One client learned to say to himself, "When others poke fun at you in a good-natured way, join in, enjoy it, and defuse it. That works much better than getting angry."

The following example deals with Kris, a graduate student in a counseling-psychology program. He is having his biweekly review session with his instructor-trainer. Training in counseling skills takes place in small groups. The issue is his proficiency in communication skills.

KRIS: I'm surprised to hear that you don't think that my communication skills are improving. I thought they were.

TRAINER: Kris, you know that's not what the feedback forms from your fellow group members are saying.

KRIS: Well, they're not all doing that well themselves.

TRAINER: I have a hunch about you. See if there's any truth in it. You're very bright. That's no secret. There could be a person inside you saying something like this: "This stuff is really easy. You already have most of these skills. And what you lack now can be made up by personality and smarts." Comment?

KRIS: Ouch!

TRAINER: Come on.

KRIS: I said "ouch" because there's something to what you said. I know I can be cocky, but I don't like being called on it.

TRAINER: How much time do you spend practicing these skills in your everyday life?

KRIS: I think you know the answer to that. I'm also hearing that I'm kidding myself if I think that I'm getting away with it. . . . Okay. I've got the message.

.The heart of this exchange is the trainer's hunch about Kris's self-talk, an internal behavior, and he hits the mark. The new perspective for Kris is this: "I'm not as smart as I thought. I'm not getting away with it. Exempting myself from practicing these skills means that I'm shortchanging myself in this program." His trainer challenges Kris to live up to the provision of the training contract that states: "Make communication skills second nature through practice in day-to-day interactions." Until now, Kris thought that such a program was beneath him.

For some clients, developing new perspectives and changing their internal behavior does the trick. It certainly works for Bella, a woman whose husband died two years ago. She is suffering from depression, not incapacitating, but still miserable.

BELLA: You know, I stopped wearing black a year ago. But . . .

THERAPIST: But you're still wearing black inside?

In a flash Bella had it. Not magic, but now she had the metaphor she needed. She knew that she could stop wearing black "inside." With just a couple of more sessions with the therapist, Bella found the peace she was looking for.

## Challenging Self-Limiting External Behavior

The principle: Invite clients to identify, challenge, and change self-defeating external behaviors. For many clients, it is their external behavior that gets them into and keeps them mired down in trouble. The ultimate payoff lies in changing external rather than internal behaviors. Ryan is having trouble relating to his college classmates. He is aggressive, hogs conversations, tries to get his own way when events are being planned, and criticizes others freely. One of his friends, after a couple of drinks, gets very angry with Ryan and tells him off. *Self-centered, arrogant,* and *pushy* are the kinds of words she uses. Ryan goes into a funk and finally talks things through with the dorm prefect. Once he is helped to see how self-defeating his behavior is, Ryan works with one of the counselors in the student services center to do something about his communication style. They discuss ways of being proactive and assertive rather than aggressive. His "edge" has too much of an edge about it. He takes a course in interpersonal communication and finds plenty of incentives to invest what he learns in his interactions with his classmates, in his part-time job, and at home.

The following example deals with challenging external behavior in a counselor-training program. Trainees A and B are feedback partners; that is, outside group sessions, they meet and comment on the quality of each other's participation in the group with a view to helping each other improve their performance. In this conversation, they focus on each other's behavior in what is called the counseling program's lifestyle group. In a lifestyle group trainees talk about issues in their own lives that might stand in the way of becoming effective helpers and about opportunities or unused resources they might develop to become more effective.

TRAINEE A:  You are very insightful. When you share highlights in practice counseling sessions, you almost always hit the mark. It's obvious that you listen well and understand others. You could add a lot of value if you used this skill during the lifestyle group discussions.

TRAINEE B:  Somehow I see the practice sessions as real and the lifestyle group discussions as, well, not phony, but . . . well, pedestrian. They don't do much for me.

TRAINEE A:  So as far as you are concerned, the lifestyle groups don't add much value. . . . Well, you participate a lot.

TRAINEE B:  That's so I don't get bored. I've got to do something.

TRAINEE A:  Well, in the group you mainly challenge what others say. Sometimes there's even a bit of an edge in your voice when you speak. Also, you don't say much about yourself.

TRAINEE B:  I challenge because I don't like to listen to nonsense. Nor do I particularly like talking about myself in the group. . . . It's phony.

TRAINEE A:  I'm not sure what's phony about discussing issues that can stand in the way of our effectiveness as counselors. Maybe it's personally phony for you.

TRAINEE B:  Well . . . there are a couple of things that might affect my effectiveness, but I don't think I'm ready to talk about them in an open group. After all, we're amateurs. Maybe I mean that talking about things when you're not ready to talk is phony.

TRAINEE A:  So the group might not be the right place for discussing some personal issues. . . . What forum do you think would be the best for you? I'm not sure whether you've found one yet.

TRAINEE B:  To tell you the truth, I don't talk about some things with anyone. I don't trust any of the trainers enough to talk to them one-on-one, at least not yet. . . . I talk to myself (laughs).

TRAINEE A:  So talking about some issues can be tough. . . . I'm a bit confused. In the practice counseling sessions, when students use real issues, you give the impression that you expect them to talk about real issues. And the other trainees seem to trust you. And I don't get any hint that you would betray that trust. You talk to everyone in a caring way. But you do challenge. You're no pushover.

TRAINEE B (pauses):  Why don't you just say it? I don't do what I expect others to do.

TRAINEE A:  Well. . .

TRAINEE B (interrupting):  I apologize. I'm doing it right here. I'm getting angry at you because you're speaking the truth.

In this exchange, Trainee A challenges Trainee B by describing B's behavior and by pointing out some skills and resources B has but does not use; that is, he challenges some of B's actions and some of his unused resources. The ultimate focus is on opportunity, not problem.

One way of helping clients challenge both internal and external actions is to help them explore the consequences of their actions. Let's return to Roberto. He has made some "mild" attempts at sabotaging his wife's career. He refers to his actions as "delaying tactics."

HELPER: It might be helpful to see where all of this is leading.

ROBERTO: What do you mean?

HELPER: I mean let's review the impact your "delaying tactics" have had on Maria and your marriage. And then let's review where these tactics are most likely to lead, ultimately.

ROBERTO: Well, I can tell you one thing. She's become even more stubborn.

Through their discussion Roberto discovers not only that his sabotage is not working for him but also that it is actually working against him. His campaign is headed in the wrong direction.

Roberto needs to start, continue, increase, and stop certain external behaviors if he is going to do his part in developing a better relationship with his wife. Here are some possibilities:

- *Stop* criticizing her in front of others, stop creating crises at home and assigning the blame for them to her, and stop making fun of her business friends.
- *Start* activities that will help him develop his own career, start sharing his feelings with her instead of just expressing them in negative ways, start engaging in mutual decision making, and start taking more initiative in household chores and child care.
- *Continue* visiting her parents with her and increase the number of times he goes to business-related functions with her.

With the counselor's help, Roberto begins to challenge himself to develop and implement a set of possibilities that would help him keep his relationship with Maria on an even keel while they work through the issue of Maria's career.

## Challenging Discrepancies

We don't always do what we say we're going to do. Just review last year's New Year's resolutions. Different kinds of discrepancies keep clients mired in their problem situations. There are discrepancies between

- what clients think or feel and what they say;
- what they say and what they do;
- their views of themselves and the views that others have of them;
- what they are and what they intimate that they wish to be;
- their stated goals and their behavior;
- their expressed values and their actual behavior.

For instance, a helper might challenge the following discrepancies that take place outside the counseling sessions.

- Tom sees himself as witty; his friends see him as biting.
- Minerva says that physical fitness is important, but she overeats and underexercises.
- George says he loves his wife and family, but he is seeing another woman and stays away from home a great deal.

- Clarissa, unemployed for several months, wants a job, but she doesn't want to participate in a retraining program.

Let's use the example of Clarissa to illustrate how the discrepancy between talking and acting can be challenged. Clarissa has just told the counselor that she has decided against joining the retraining program.

COUNSELOR: I thought that the retraining program would be just the kind of thing you've been looking for.

CLARISSA: Well . . . I don't know if it's the kind of thing I'd like to do. . . . The work would be so different from my last job. . . . And it's a long program.

COUNSELOR: So you feel the fit isn't good.

CLARISSA: Yeah.

COUNSELOR: Clarissa, you seemed so enthusiastic when you first talked about the program. . . . (gently) What's going on?

CLARISSA (pauses): You know, I've gotten a bit lazy. . . . I don't like being out of work, but I've gotten used to it.

The counselor sees a discrepancy between what Clarissa is saying and what she is doing. She is actually letting herself slip into a "culture of unemployment." Now that the discrepancy is out in the open, they can work together on how she wants to shape her future.

Here's another example of a discrepancy. Alicia, a woman who experiences herself as unattractive, is a member of a counseling group.

GROUP COUNSELOR: You say that you're unattractive, and yet you have talked about how you get asked out a lot. I don't find you unattractive myself. And, if I'm not mistaken, I see people here react to you in ways that say they like you. Help me pull all this together.

ALICIA: Hmm. . . . What you've just said helps me clarify what I mean. First of all, I'm no raving beauty, and when others find me attractive, I think they mean that they find me intellectually interesting, a caring person, and things like that. . . . At times I wish I were more physically attractive, though I feel ashamed when I say things like that. The real issue is that much of the time I *feel* unattractive. And sometimes I feel most unattractive at the very moment people are telling me directly or indirectly that they find me attractive.

GROUP MEMBER: So you don't really think that you actually are an unattractive person, but you've gotten into the habit of telling yourself that you are. . . . It sounds like a bad habit.

ALICIA: It is a lousy habit. If I look at my early home life and my experiences in grammar school and high school, I could probably give you the long, sad story of how it happened.

GROUP COUNSELOR: If I hear you right, you believe you'd be better off expanding your definition of attractive to include things like being intellectually interesting, having an engaging personality. . . . But there's been some kind of block to doing so.

ALICIA: Actually, there's no serious reason for me to keep apologizing for myself. The past is the past. That was then and this is now. . . . I need practice.

Since the counselor's experience of Alicia is so different from Alicia's experience of herself, he invites her to explore this discrepancy in the group. Her self-exploration clarifies the issue greatly. The way she thinks about herself (unattractive) blocks her appreciation of resources she has (being caring and intellectually stimulating) and most likely leads to the underuse of these resources.

Finally, in a touching scene from a videotaped counseling session between Carl Rogers and a client (Rogers, Perls, & Ellis, 1965), both of these dimensions are illustrated in one of his responses to the client. He is talking to a divorced woman who is having intimate relations with a man. When her daughter asks her point blank if she is having relations with the man, she says no. At one point, Rogers says to her, "So you would like to have an honest, open relationship with your daughter, and you'd like her to trust you completely—even though you feel that you have to lie to her, at least about the affair." The woman says, "Yes, that's it!" Roger pauses for a moment and then says gently, "Well, that's a tall order, isn't it?" He captures and communicates the client's point of view, but he also captures—and communicates—the dissonance.

## Challenging the Predictable Dishonesties of Everyday Life

The "predictable dishonesties of everyday life" are the distortions, evasions, games, tricks, excuse making, and smoke screens that keep clients stuck in their problem situations. All of us have ways of defending ourselves from ourselves, from others, and from the world. We all have our little dishonesties. But they are two-edged swords. Although lies, whether white or not, may help us cope with difficulties, especially unexpected difficulties, in our interactions with others, they come with a price tag, especially if they become a preferred coping strategy. Blaming others for our misfortunes helps us save face, but it disrupts interpersonal relationships and prevents us from developing a healthy sense of self-responsibility. The purpose of helping clients challenge themselves with respect to the dishonesties of everyday life, whether they take place in the helping sessions or are more widespread patterns of behavior, is not to strip clients of their defenses, which in some cases could be dangerous, but to help them cope with their inner and outer worlds more creatively.

**Invite clients to challenge substantial distortions.**  Some clients would rather not see the world as it is—it is too painful or demanding—and therefore distort it in various ways. The distortions are self-serving. Here are some examples:

- At work, Arnie is afraid of his supervisor and therefore sees her as aloof, whereas in reality she is a caring person. He is working out of past fears rather than current realities.

- Edna sees her counselor in some kind of divine role and therefore makes unwarranted demands on him.

- Nancy sees getting her own way with her friends as a measure of how much they really like her.

Let's take a little closer look at Nancy, who is married to Milan. There are some bumps in their marriage.

> Nancy and Milan come from different cultures. They fought a great deal in the early years of their marriage, but then things settled down. Now, squabbles have broken out about the best way to bring up their children. Milan is not convinced that counseling is a good idea, so the counselor is talking to Nancy alone. She has forbidden her 12-year-old son to bicycle to school because she doesn't want "his picture to end up on a milk carton." Milan thought that she was being extremely overprotective. One day, he stalks out of the house, yelling back at her, "Why don't you just keep him locked in his room!"

Nancy and her counselor have a session not long after this incident. Nancy defends her approach to her son.

NANCY: Milan's just too permissive. Now that Jan is entering his teenage years, he needs more guidance, not less. Let's face it, the world we live in is dangerous.

COUNSELOR: So from your point of view, this is not the time for letting your guard down. . . . Of course, I'm also making the assumption that Milan is not indifferent to Jan's welfare.

NANCY: Of course not! Good grief, he cares as much as I do. We just disagree on how to do it. "Safe, not sorry" is my philosophy.

Hopefully, this gets rid of an implied distortion: "I'm interested in my son's welfare, but his father isn't." They continue their dialogue.

COUNSELOR: Let's widen the discussion a bit. What other issues do you and Milan disagree on?

NANCY: Well, we used to disagree a lot. But we've put that behind us, it would seem. He leaves a lot of the home decisions to me.

COUNSELOR: I'm not sure whether you both decided that you should make the decisions at home or if it just happened that way.

NANCY (slowly): I suppose it just happened that way. . . . I don't really know.

COUNSELOR: I'm curious because he seems to be annoyed that you're the one making the decisions about how to bring up your son. . . .

NANCY (pausing): Like he wants to reassert himself. Take over again.

Another distortion. Perhaps Nancy feels the counselor is getting too close to a sensitive issue that she thought was resolved long ago.

COUNSELOR: You got a bit annoyed when I asked whether Milan was as committed to the kids as you. . . . Since he cares as much as you do about his son, I'm wondering what the disagreement is really about.

NANCY: Like he's drawing a line in the sand, taking a stand on this one? Or what?

COUNSELOR (caringly): I don't want to guess what's going through Milan's mind. . . . Maybe we could we try once more to get him to come with you.

The counselor has a hunch that the problem is as much about power and getting one's own way as it is about bringing up children. Nancy does seem to have trouble with her own "little dishonesties."

**Challenge games, tricks, and smoke screens.** If clients are comfortable with their delusions and profit by them, they will obviously try to keep them. If they are rewarded for playing games, inside the counseling sessions or outside, they will continue a game approach to life (see Berne, 1964). Consider some examples:

- Kennard plays the "Yes, but . . ." game. He gets his therapist to recommend some things he might do to control his anger. He then points out why each recommendation will not work. When the therapist calls this game, Kennard says, "Well, I didn't think you guys were supposed to tell clients what to do." A more savvy helper might have sniffed out Kennard's tendency to play games much earlier.

- Dora makes herself appear helpless and needy when she is with her friends, but when they come to her aid, she is angry with them for treating her like a child. When she tries this in an early session, her counselor invites her to drop her "helpless and needy" routine.

In these cases, a new perspective might be that they can get what they want without exposing themselves to the downside of playing games.

The number of games we can play to avoid the work involved in squarely facing the tasks of life is endless. Clients who are fearful of changing will attempt to lay down smoke screens to hide from the helper the ways in which they fail to face up to life. Such clients use communication in order not to communicate (see Argyris, 1999; Beier & Young, 1984). Therefore, helpers do well if they establish an atmosphere that discourages clients from playing games. An attitude of "nonsense is challenged here" should pervade the helping sessions.

**Challenge excuses.** Snyder, Higgins, and Stucky (1983) examined excuse-making behavior in depth. Excuse making, of course, is universal, part of the fabric of everyday life (see also Halleck, 1988; Higginson, 1999; Snyder & Higgins, 1988; Yun, 1998). Like games and distortions, it has its positive uses in life. Even if it were possible, there is no real reason for setting up a world without myths. On the other hand, avoidance behavior and excuse making contribute a great deal to the "psychopathology of the average." Clients routinely provide excuses for why they did something "bad," why they didn't do something "good," and why they can't do something they need to do. Roberto tells the helper that he has engaged in benign attempts to sabotage his wife's career "for her own good" because she would "get hurt" in the Anglo world. The counselor helps him explore the alternative hypothesis that "he is not ready" for the changes in style that his wife's career and behavior were demanding from him.

This is only skimming the surface of the games, evasions, tricks, distortions, excuses, rationalizations, and subterfuges resorted to by clients (together with the rest of the population). Skilled helpers are caring and empathic, but they do not let themselves be conned. That helps no one.

## Challenging Strengths and Unused Resources

Part of positive psychology is helping clients get in touch with unexploited opportunities and unused or underused resources. Some of these resources are client based—for instance, talents and abilities—and some are external—such as social support. Gerry Sexton (1999), focusing on internal resources, outlines what it means to be a self-directed learner. Clients often have self-healing resources that are unused or underused. According to Sexton, self-directed learners use the resources available to them in the following ways:

- They work with an underlying sense of *purpose*. Therefore, when you help clients get in touch with what they want life to be, you are helping them become learners. For clients, this means goals—short-term, medium-term, long-term.

- They have a *dream* that they refuse to surrender. As Sexton notes, "Dreams create direction." Challenging clients to develop their dreams (see Chapter 14) is one way of helping them get on the road toward self-directed learning.

- They focus on their *gifts*. Therefore, helping clients get in touch with their gifts is essential if they are to become learners in the midst of their problem situations and unused opportunities. If we bring clients to the threshold where they can see their invisible assets, they'll do whatever else it takes.

- They see themselves as *volunteers* rather than victims. "They respect the external forces in their lives but refuse to be controlled by them." Counselors help clients "volunteer" for their own causes. Even in the worst problem situations, there are things clients can control.

- They act *despite their fears*. Helping clients initiate action—even small steps—helps them drive out their fears. Fear and misgivings cannot be entirely banished from life, but we can help clients find motivating forces in their lives that enable them to move beyond fear to action.

- They thrive on *interdependence*. Counselors help clients identify or establish the kinds of relationships they need to support their dreams. Neither isolation nor exaggerated independence does the trick. As Sexton says, "Success is impossible without interconnections."

There is no one set of rules to identify and put in motion opportunities, strengths, and unused resources. Helpers need a self-healing mind-set. Effective helpers are subliminally asking themselves this question as they partner with their clients: How can I help this client unleash and marshal both internal and external resources? They are there. They need to be mined.

In a backhanded way, helping clients identify self-limiting blind spots by exploring cognitive perspectives, dysfunctional behavior, and life-limiting external behavior can also be a search for resources. Driscoll (1984), as we have seen, points out that helpers can show clients that their "irrationalities" actually make sense. Instead of forcing clients to see how stupidly they are thinking and acting, helpers can challenge them to find the logic embedded even in seemingly dysfunctional ideas and behaviors. Then clients can use that logic as a resource to manage problem situations instead of perpetuating them. A psychiatrist friend of mine helped a client see the "beauty," as it were, of a very carefully constructed self-defense system. The client, through a series of mental gymnastics and external behaviors, was cocooning himself from real life. My friend helped the client see how inventive he had been and how powerful the system that he had created was. He went on to help him redirect that power into more productive channels.

## Challenging Behaviors Within Helping Sessions

How clients talk about their problems and unused opportunities has a lot to do with the "feel" of helping sessions. Three sets of behavior are reviewed here. The first set deals with the ownership of problems and unused opportunities, the second with problem or opportunity definition, and the third with movement within the helping model.

**Invite clients to own their problems and unused opportunities.**  It is all too common for clients to refuse to take responsibility for their problems and unused opportunities. Instead, they create a whole list of outside forces and other people who are to blame. Therefore, clients need to challenge themselves or be challenged to own their problem situations. Here is the experience of one counselor who had responsibility for about 150 young men in a youth prison within the confines of a larger central prison:

I believe I interviewed each of the inmates. And I did learn from them. What I learned had little resemblance to what I had found when I read their files containing personal histories, including the description of their crimes. What I learned when I talked with them was that they didn't belong there. With almost universal consistency, they reported a "reason" for their incarceration that had little to do with their own behavior. An inmate explained with perfect sincerity that he was there because the judge who sentenced him had also previously sentenced his brother, who looked very similar; the moment this inmate walked into the courtroom he knew he would be sentenced. Another explained with equal sincerity that he was in prison because his court-appointed lawyer wasn't really interested in helping him. (Miller, 1984, pp. 67–68)

This is perhaps an extreme form of a common phenomenon. But we don't have to go behind prison walls to find this lack of ownership. Lack of ownership is a common blind spot.

Carkhuff (1987) talks about helping clients' own problem situations and unused opportunities in terms of "personalizing." Take the case of a client who feels that her business partner has been pulling a fast one on her. He's made a deal on his own. She's alarmed, but she hasn't done anything about it so far. Let's say that three different helpers, A, B, and C, share a highlight. Consider how they differ.

HIGHLIGHT A:  You feel angry because he unilaterally made the decision to close the deal on his terms.

HIGHLIGHT B:  You're angry because your legitimate interests were ignored.

HIGHLIGHT C:  You're furious because you were ignored, your interests were not taken into consideration, maybe you were even financially victimized, and you let him get away with it.

These three statements become progressively more personal. Personalizing means helping clients understand that in some situations, they may have some responsibility for creating or at least perpetuating their problem situations. Statement C does precisely that. It is about ownership.

Not only problems but also opportunities need to be seized and owned by clients. As Wheeler and Janis (1980) note, "Opportunities usually do not knock very loudly, and missing a golden opportunity can be just as unfortunate as missing a red-alert warning" (p. 18). Consider Tess and her brother Josh.

Tess and Josh's father died years ago. When their mother died about a year ago, she left a small country cottage to both of them. They have been fighting over its use. Some of the fighting has been quite bitter. They have never been that close, but until now they have just had squabbles, not all-out war. Without admitting it, Tess has been shocked by her angry and bellicose behavior—but not shocked enough to do anything about it. When Josh had a heart attack, Tess knew that this was an opportunity to do something about their relationship. But she kept putting it off. She realized that the longer she put it off, the harder it would be to do something about it. Here is a part of one counseling session:

TESS:  I thought that this was going to be our chance to patch things up, but he hasn't said anything.

COUNSELOR:  So, nothing from his camp. . . . What about yours?

TESS:  I think it might already be too late. We're falling right back into our old patterns.

COUNSELOR:  I didn't think that's what you wanted.

TESS (angrily):  Of course that's not what I wanted! . . . That's the way it is.

COUNSELOR:  Tess, if someone put a gun to your head and said, "Make this work or I'll shoot," what would you do?

TESS (after along pause):  You mean it's up to me. . . .

COUNSELOR:  If you mean that I'm assigning this to you as a task, then no. If you mean that you can still seize the opportunity no matter what Josh does, well. . . .

TESS:  I think I'm really angry with myself. . . . I know deep down it's up to me, no matter what Josh does. And I keep putting it off.

Josh, of course, is not in the room. So it's about Tess's ownership of the opportunity. And she senses, now more deeply, that "missing a golden opportunity can be just as unfortunate as missing a red-alert warning."

**Invite clients to state their problems as solvable.** Jay Haley (1976, p. 9) says that if "therapy is to end properly, it must begin properly—by negotiating a solvable problem." Or exploring a realistic opportunity, someone might add. It is not uncommon for clients to state problems as unsolvable. This justifies a "poor-me" attitude and a failure to act.

UNSOLVABLE PROBLEM:  In sum, my life is miserable now because of my past. My parents were indifferent to me and at times even unjustly hostile. If only they had been more loving, I wouldn't be in this mess. I am the failed product of an unhappy environment.

Of course, clients will not use this rather stilted language, but the message is common enough. The point is that the past cannot be changed. As we have seen, clients can change their attitudes about the past and deal with the present consequences of the past. Therefore, when a client defines the problem exclusively as a result of the past, the problem cannot be solved. "You certainly had it rough in the past and are still suffering from the consequences now" might be the kind of response that such a statement is designed to elicit. The client needs to move beyond such a point of view.

A solvable or manageable problem is one that clients can do something about. Consider a different version of the foregoing unsolvable problem

SOLVABLE PROBLEM:  Over the years, I've been blaming my parents for my misery. I still spend a great deal of time feeling sorry for myself. As a result, I sit around and do nothing. I don't make friends, I don't involve myself in the community, I don't take any constructive steps to get a decent job.

This message is quite different from that of the previous client. The problem is now open to being managed because it is stated almost entirely as something the client does or fails to do. The client can stop wasting her time blaming her parents, since she cannot change them; she can increase her self-esteem through constructive action and therefore stop feeling sorry for herself; and she can develop the interpersonal skills and courage she needs to enter more creatively into relationships with others.

This does not mean that all problems are solvable by the direct action of the client. A teenager may be miserable because his self-centered parents are constantly squabbling and seem indifferent to him. He certainly can't solve the problem by making them less self-centered, stopping them from fighting, or getting them to care for him more. But he can be helped to find ways to cope with his home situation more effectively by developing fuller social opportunities outside the home. This could mean helping him develop new perspectives on himself and family life and challenging him to act both internally and externally in his own behalf.

**Invite clients to move on to the next needed stage or step of the helping process.** We have touched on this already in discussing probing. There is no reason to keep going "'round the mulberry bush" with clients. You can help clients challenge themselves to

- clarify problem situations by describing specific experiences, behaviors, and feelings when they are being vague or evasive;
- talk about issues—problems, opportunities, goals, commitment, strategies, plans, actions—when they are reluctant to do so;
- develop new perspectives on themselves, others, and the world when they prefer to cling to distortions;
- review possibilities, critique them, develop goals, and commit themselves to reasonable agendas when they would rather continue wallowing in their problems;
- search for ways of getting what they want, instead of just talking about what they would prefer;
- spell out specific plans instead of taking a scattered, hit-or-miss approach to change;
- persevere in the implementation of plans when they are tempted to give up;
- review what is and what is not working in their pursuit of change "out there."

In sum, counselors can help clients challenge themselves to engage more effectively in all the stages and steps of problem management during the sessions themselves and in the changes they are pursuing in everyday life. Box 10-1 reviews the kinds of questions counselors can help their clients ask themselves to uncover blind spots.

**Box 10-1    Questions to Uncover Blind Spots**

These are the kinds of questions you can help clients ask themselves in order to develop new perspectives, change internal behavior, and change external behavior.

- What problems am I avoiding?
- What opportunities am I ignoring?
- What's really going on?
- What am I overlooking?
- What do I refuse to see?
- What don't I want to do?
- What unverified assumptions am I making?
- What am I failing to factor in?
- How am I being dishonest with myself?
- What's underneath the rocks?
- If others were honest with me, what would they tell me?

# STEP I-B: II. SPECIFIC CHALLENGING SKILLS

ADVANCED EMPATHIC HIGHLIGHTS: THE MESSAGE BEHIND THE MESSAGE
    Help clients make the implied explicit
    Help clients identify themes in their stories
    Help clients make connections they may be missing
    Share educated hunches based on empathic understanding

INFORMATION SHARING: FROM NEW PERSPECTIVES TO ACTION

HELPER SELF-DISCLOSURE
    Include helper self-disclosure in the contract
    Make sure that your disclosures are appropriate
    Be careful of your timing
    Keep your disclosures selective and focused
    Don't disclose too frequently
    Do not burden already overburdened clients
    Remain flexible

IMMEDIACY: DIRECT, MUTUAL TALK
  Types of Immediacy in Helping and Principles for Using Them
    Overall relationship immediacy
    Event-focused immediacy
    Self-involving statements
  Situations Calling for Immediacy

USING SUGGESTIONS AND RECOMMENDATIONS

CONFRONTATION

ENCOURAGEMENT

EVALUATION QUESTIONS FOR STEP I-B:
THE USE OF SPECIFIC CHALLENGING SKILLS

Helpers can challenge clients to develop new perspectives and to change their internal and external behaviors in a number of ways. The following are discussed and illustrated in this chapter: (a) sharing advanced empathic highlights, (b) sharing information, (c) helper self-disclosure, (d) immediacy, (e) suggestions and recommendations, (f) confrontation, and (g) encouragement. It has already been noted that both probing and summarizing—and even sharing accurate highlights—can challenge clients to rethink their attitudes and behavior.

## ADVANCED EMPATHIC HIGHLIGHTS: THE MESSAGE BEHIND THE MESSAGE

Chapter 3 outlined the characteristics of empathy as a value, and Chapter 6 discussed the communication skill of sharing basic empathic highlights. However, as skilled helpers listen intently to clients, they often see clearly what clients only half see and barely hint at. In Chapter 5, this was called "listening for the slant" or the "spin" clients give to their messages. This deeper kind of empathic listening involves "sensing meanings of which the client is scarcely aware" (Rogers, 1980, p. 142) or, in broader terms, listening to and grasping the "message behind the message."

One way of challenging a client is to share with him or her your understanding of the message behind the message. For instance, Gordon talks about and expresses his anger with his wife, but as he talks, the helper hears not just anger but also hurt. Gordon can talk with relative ease about his anger but not as easily about his feelings of hurt. When you share basic empathic highlights—provided, of course, that you are accurate—clients recognize themselves almost immediately: "Yes, that's what I meant." However, because advanced empathic highlights dig a bit deeper, clients might not immediately recognize themselves in helpers' responses. Or, as seen in the last chapter, they might experience a bit of disequilibrium. That's what makes sharing advanced empathic highlights a form of challenge. For instance, the helper says something like this to Gordon: "It's pretty obvious that you really get steamed when she acts like that. But I thought I sensed, mixed in with the anger, a bit of hurt." At that, Gordon looks down and pauses. He finally says, "She can still get to me. She certainly can." This appreciably broadens or deepens the discussion of the problem situation.

Here are some questions helpers can ask themselves to probe a bit deeper as they listen to clients:

- What is this person only half saying?
- What is this person hinting at?
- What is this person saying in a confused way?
- What covert message is behind the explicit message?

Note that advanced empathic listening and processing focuses on what the client is actually saying or at least expressing, however tentatively or confusedly. It is not an interpretation of what the client is saying. Sharing advanced empathic highlights is not an attempt to "psych the client out."

Advanced empathic listening in the hands of skilled helpers focuses not just on the problematic dimensions of clients' behavior but also on unused opportunities and resources. Effective helpers listen for the resources that are buried deeply in clients and often have been forgotten by them. Consider the following example. A soldier who has been thinking seriously about making the army his career has been talking to a chaplain about his failure to be promoted. He has seen service in both Bosnia and Kosovo and has performed very well. As he talks, it becomes fairly evident that part of the problem is that he is so quiet and unassuming that it is easy for his superiors to ignore him.

SOLDIER: I don't know what's going on. I work hard, but I keep getting passed over when promotion time comes along. I think I work as hard as anyone else, and I work efficiently, but all of my efforts seem to go down the drain. I'm not as flashy as some others, but I'm just as substantial.

CHAPLAIN A: You feel it's quite unfair to do the kind of work that merits a promotion and still not get it.

SOLDIER: Yeah. . . . I suppose there's nothing I can do but wait it out. [A long silence ensues.]

Chaplain A tries to understand the client from the client's frame of reference. He deals with the client's feelings and the experience underlying those feelings. In responding, he shares a basic empathic highlight. But the client merely retreats more into himself. Here's a different approach:

CHAPLAIN B: It's depressing to put out so much effort and still get passed by. . . . Tell me more about this "not as flashy" bit. What in your style might make it easy for others not to notice you, even when you're doing a good job?

SOLDIER: You mean I'm so unassuming that I could get lost in the shuffle? Or maybe it's the guys who make more noise, the squeaky wheels my dad called them, who get noticed. . . . I guess I've never really thought of selling myself. It's not my style."

From the context, from the discussion of the problem situation, from the client's manner and tone of voice, Chaplain B picks up a theme that the client states in passing in the phrase "not as flashy." That is, the client is so unassuming that his best efforts go unnoticed. They go on to discuss how he might "market himself" in a way that is consistent with his values. Advanced empathic highlights can take a number of forms. Here are some of them.

**Help clients make the implied explicit.** The most basic form of an advanced highlight involves helping clients give fuller expression to what they are implying rather than saying directly. In the following example, the client has been discussing ways of getting back in touch with his wife after a recent divorce, but when he speaks about doing so, he expresses very little enthusiasm:

CLIENT (somewhat hesitatingly): I could wait to hear from her. But I suppose there's nothing wrong with calling her up and asking her how she's getting along.

COUNSELOR A: It seems that there's nothing wrong with taking the initiative to contact her. After all, you'd like to find out if she's doing okay.

CLIENT (somewhat drearily): Yeah, I suppose I could.

Counselor A's response might have been fine at an earlier stage of the helping process, but it misses the mark here, and the client grinds to a halt.

COUNSELOR B: You've been talking about getting in touch with her, but unless I'm mistaken, I don't hear a great deal of enthusiasm in your voice.

CLIENT: To be honest, I don't really want to talk to her. But I feel guilty—guilty about the divorce, guilty about her going out on her own. Frankly, all I'm doing is I'm taking care of her all over again. And that's one of the reasons we got divorced. I had a need to take care of her, and she let me do it. That was the story of our marriage. I don't want to do that anymore.

COUNSELOR B: What would a better way of going about all this be?

CLIENT: I need to get on with my life and let her get on with hers. Neither of us is helpless. [His voice brightens.] For instance, I've been thinking of quitting my job and starting a business with a friend of mine—helping small businesses use the Internet to improve their businesses—you know, Web sites, marketing, all of that.

Counselor B bases her response not only on the client's immediately preceding remark but also on the entire context of his story. Her response hits the mark, and the client moves forward. As with basic highlights, there is no such thing as a good advanced highlight in itself. Does the response help the client clarify the issue more fully so that he or she might begin to see the need to act differently?

**Help clients identify themes in their stories.** When clients tell their stories, you'll notice certain themes emerge. Thematic material might refer to feelings (hurt, depression, anxiety), to behavior (controlling others, avoiding intimacy, blaming others, overwork), to experiences (being a victim, being seduced, being punished, being ignored, being picked on), or some combination of these. Once you see a self-defeating theme or pattern emerging from your discussions, you can share your perception and help the client check it out.

In the following example, a counseling-psychology trainee is talking with his supervisor. The trainee has four clients. In the past week, she has seen each of them for the third time. This dialogue takes place in the middle of a supervisory session.

SUPERVISOR: You've had a third session with each of your four clients this past week. Even though you're at different stages with each because each started in a different place, you have a feeling, if I understand what you've been saying, that you're "going 'round the mulberry bush" a bit with a couple of them.

TRAINEE: Yes, I'm grinding my wheels. I don't have a sense of movement.

SUPERVISOR: Any thoughts on what's going on?

TRAINEE: Well, they seem willing enough. And I think I've been very good at listening and sharing highlights. It keeps them talking.

SUPERVISOR: But this doesn't seem to be enough to get them moving forward. I tell you what. Let's listen to one of the tapes.

They listen to a segment of one of the sessions. The trainee turns off the recorder.

TRAINEE: Oh, now I see what I'm doing! It's all sharing highlights with a few uh-huhs. And I thought I was being pushy. But this is as far from pushy as you can get.

SUPERVISOR: So what's missing?

TRAINEE: There are very few probes and nothing close to summaries or mild challenges. Certainly some probes would have given much more focus and direction to the session.

SUPERVISOR: Let me role-play the client as well as I can and see how you might redo the session.

They then spend about 15 minutes in a role-playing session. The trainee mingles probes with highlights, and the result is quite different.

SUPERVISOR: How close did you get to challenging, even mild challenging?

TRAINEE: I didn't get there at all. . . . You know, I think that I see probes as challenges. . . . The thread through all of this is playing it safe. I think I'm playing it safe because I don't want to damage the client. But now it's clear that I'm afraid to push.

The theme that the supervisor helps the trainee bring to the surface is a fear of "being pushy," which explains his "playing it safe" behavior. Notice how the supervisor creates a climate that helps the trainee challenge himself. The trainee identifies the theme in terms such as "playing it safe" and "afraid to push."

**Help clients make connections they may be missing.** Clients often tell their stories in terms of experiences, behaviors, and emotions in a hit-or-miss way. The counselor's job, then, is to help clients make the kinds of connections that provide the insights or perspectives they need to move forward in the helping process.

- Cymae's counselor helps her see that she is having difficulty developing strategies for her chosen goals because she is only halfheartedly committed to her goals. They revisit the goals she has set for herself.

- A managerial coach helps Finnbar relate the trouble he is having with a strong woman supervisor to carefully hidden sexist attitudes that leak out into his behavior.

- A therapist helps Joanna see the link between her ingratiating style and her inability to influence her colleagues at work.

- A supervisor helps Dieter see that the persistent anxiety he feels when working with clients is related to the perfectionistic standards he has set for himself.

The following client has a full-time job and is finishing the last two courses he needs for his college degree. His father has recently had a stroke and is incapacitated. The client talks about being progressively more anxious and tired in recent weeks. He visits his father regularly. He meets frequently with his mother, his two sisters, and his two brothers to discuss how to manage the family crisis. Under stress, fault lines in family relationships appear. He has deadlines for turning in papers for current courses.

JOHN: I don't know why I'm so tired all the time. And edgy. I'm supposed to be the calm one. I wonder if it's something physical. You know, what's happened to Dad and all that. I never even think about my health.

COUNSELOR: A lot has happened in the past few weeks. Work. School. Your dad's stroke. Juggling schedules.

JOHN (interrupting): But that's what I'm good at. Working hard. Juggling schedules. I've done that all the time. And I've never gotten so tired and edgy before.

COUNSELOR: Add now, with your dad's illness. . . .

JOHN: You know, I could handle that, too. If I were the only one, you know, just me and Mom, I bet I could do it.

COUNSELOR: All right, so besides your dad's illness, what's so new?

JOHN (slowly): Well, I hate to say it. . . . It's the squabbling. We usually get on pretty well. We all like getting together. But the meetings about Dad, they can be awful. I keep thinking about them at work. And the other evening, when I was trying to write a paper for school, I was still ticked off at my older sister.

COUNSELOR: So the family stuff is getting at you no matter what you're doing.

JOHN: I'm just not used to all that. I thought we'd rally together. You know, get support from one another. Sometimes it's just the opposite.

John handles the normal stress of everyday life quite well. But the "family stuff" is acting like a multiplier. They go on to discuss what the family dynamics are like and what John can do to cope with them.

**Share educated hunches based on empathic understanding.** As you listen to clients, thoughtfully process what they say and put it in context. Hunches about the messages behind the messages or the stories behind the stories will naturally begin to form. Then share the hunches that you feel might add value. The more mature and socially competent you become and the more experience you have helping others, the more "educated" your hunches become. Here are some examples:

• Hunches can help clients see the bigger picture. A counselor is talking with a client who is having trouble with his perfectionism. The client also mentions problems with his brother-in-law, whom his wife enjoys having over. The two men often argue, and sometimes the arguments have an edge to them. At one point the client describes his brother-in-law as "a guy who can never get anything right." Later the counselor says, "We started out by talking about perfectionism in terms of the inordinate demands you place on yourself. I wonder whether it could be 'spreading' a bit. You should be perfect. But so should everyone else." They go on to discuss the ways his perfectionism is interfering with his social life.

• Hunches can help clients see what they are expressing indirectly or merely implying. A counselor is talking to a client who feels that a friend has let her down: "I think I might also be hearing you say that you are more than disappointed—perhaps even betrayed." Since the client has been making every effort to avoid her friend, "betrayal" rings truer than "let down." She is finally in touch with the depth of her feelings.

• Hunches can help clients draw logical conclusions from what they are saying. A manager is having a discussion with one of his team members who has expressed, in a rather tentative way, some reservations about one of the team's projects: "If I stitch together everything that you've said about the project, it sounds as if you are saying that it was ill advised in the first place and probably should be shut down. I know that might sound drastic and you've never put it in those words. But if that's how you feel, we should discuss it in more detail."

• Hunches can help clients open up areas they are only hinting at. A school counselor is talking to a senior in high school: "You've brought up sexual matters a number of times, but you haven't pursued them. My guess is that sex is a pretty important area for you but perhaps pretty touchy, too."

• Hunches can help clients see things they may be overlooking. A counselor is talking to a client who probably has only six months to live. The man is unmarried and has never made a will. He has some money but has expressed indifference to money matters. "I'm financially lazy" is his theme. He adds, "I'm ready to die." Later in the session, the counselor says, "I wonder if your financial laziness has spread a bit. For instance, you live alone and, if I'm not mistaken, you haven't given anyone power of attorney in health matters either. That could mean that how you die will be in the hands of the doctors." This helps the client begin to rethink how he wants to die. They even discuss finances. He may not be a slave to money, but whatever money he has could go to a good cause.

• Hunches can help clients identify themes. A counselor is talking to a woman who has been abused by her husband but has hesitated to make use of the services available to women who face abuse. After listening to her story and getting some background, the counselor says: "If I'm not mistaken, you've mentioned in a couple of different ways how difficult it can be for you to stick up for your rights. For instance. . . ." He gives a couple of examples. They go on to discuss how she might become more assertive without being pushy.

• Hunches can help clients take fuller ownership of partially owned experiences, behaviors, feelings, points of view, and decisions. A counselor is talking to a client who is experiencing a lot of pain in a physical rehabilitation program following an automobile accident. She keeps focusing on how difficult the program is. At one point, the counselor says, "You sound as if you have already decided to quit. Or I might be overstating the case. . . ." This helps the client enormously. She has been thinking of quitting, but she has been afraid to discuss it. They go on to discuss her wanting to give up and her dread of giving up. When the counselor finds out that she has never even mentioned the pain to the members of the rehabilitation staff, they discuss strategies for coping with the pain, including talking directly with the staff about the pain.

Like all responses, hunches should be based on your understanding of your clients. If your clients were to ask you where your hunches come from—"What makes you think that?"—you should be able to identify the experiential and behavioral clues on which they are based. Of course, sharing advanced empathic highlights is not license to draw inferences from clients' history, experiences, or behavior at will (see MacDonald, 1996). Nor is it license to load clients with interpretations that are more deeply rooted in your favorite psychological theories than in the realities of the client's world. Sharing advanced empathic highlights constructively depends on social competence and emotional intelligence.

## INFORMATION SHARING: FROM NEW PERSPECTIVES TO ACTION

Sometimes clients are unable to explore their problems fully and proceed to action because they lack information of one kind or another. Information can help clients at any stage or step of the helping process. For instance, in Stage I, it helps many clients to know that they are not the first to try to cope with a particular problem. In Stage II, information can help clients further clarify or identify possibilities and set goals. In the implementation stage, information on commonly experienced obstacles can help clients cope and persevere.

The skill or strategy of information sharing is included under challenging skills because it helps clients develop new perspectives on their problems or shows them how to act. It includes both giving information and correcting misinformation. In some cases, the information can prove to be quite confirming and supportive. For instance, a parent who feels responsible following the death of a newborn baby may experience some relief through an understanding of the warning signs of Sudden Infant Death Syndrome. This information does not "solve" the problem, but the parent's new perspective can help him or her handle self-blame. In other cases, the information may be painful and challenging.

In some cases, the new perspectives clients gain from information sharing can be both comforting and painful. Consider the following example:

> Adrian was a college student of modest intellect. He made it through school because he worked very hard. In his senior year, he learned that a number of his friends were going on to graduate school. He, too, applied to a number of graduate programs in psychology. He came to see a counselor in the student services center after being rejected by all the schools to which he had applied. In the interview, it soon became clear to the counselor that Adrian thought that many, perhaps even most, college students went on to graduate school. After all, most of his closest friends had been accepted in one graduate school or another. The counselor shared with him the statistics of what could be called the educational pyramid—the decreasing percentage of students attending school at higher levels. Adrian did not realize that just finishing college made him part of an elite group. Nor was he completely aware of the extremely competitive nature of the graduate programs in psychology to which he had applied. He found much of this relieving but then found himself suddenly faced with what to do now that he was finishing school. Up to this point, he had not thought much about it. He felt disconcerted by the sudden need to look at the world of work.

Giving information is especially useful when it is needed to manage a problem or when lack of accurate information either is one of the principal causes of a problem situation or is making an existing problem worse.

In some medical settings, doctors team up with counselors to give clients messages that are hard to hear and to provide them with information needed to make difficult decisions. For instance, Lester, a 54-year-old accountant, has been given a series of diagnostic tests for possible prostate cancer. He finds out that he does have cancer, but now he faces the formidable task of choosing what to do about it. The doctor sits down and talks with him, laying out the alternatives. Since there are many different options, including doing nothing, the doctor also describes the pluses and minuses of each option. Later, Lester has a discussion with a counselor. The counselor helps Lester cope with the news, process the information, and begin the process of making a decision.

There are some cautions helpers should observe in giving information. When information is challenging, or even shocking, be tactful and help the client handle the disequilibrium that comes with the news. Do not overwhelm the client with information. Make sure that the information you provide is clear and relevant to the client's problem situation. Don't let the client go away with a misunderstanding of the information. Be supportive; help the client process the information. Finally, be sure not to confuse information giving with advice giving; the latter is seldom useful. Professional guidance is not to be confused with telling clients what to do. Neither the doctor nor the counselor tells Lester which treatment to choose. But Lester needs further help with the burden of choosing.

## HELPER SELF-DISCLOSURE

A third skill of challenging involves the ability of helpers to constructively share some of their own experiences, behaviors, and feelings with clients (Edwards & Murdoch, 1994; Hendrick, 1990; Knox, Hess, Petersen, & Hill, 1997; Mathews, 1988; Simon, 1988; Stricker & Fisher, 1990; Watkins, 1990; Weiner, 1983). In one sense, counselors cannot help but disclose themselves: "The counselor communicates his or her characteristics to the client in every look, movement, emotional response, and sound, as well as with every word" (Strong & Claiborne, 1982, p. 173). This is the kind of indirect disclosure that goes on all the time. Effective helpers, as they tune in, listen, process, and respond, try to track and manage the impressions they are making on clients.

Here, however, it is a question of direct self-disclosure. Research into direct helper self-disclosure has led to mixed and even contradictory conclusions. Some researchers have discovered that helper self-disclosure can frighten clients or make them see helpers as less well adjusted. Or helper self-disclosure, instead of helping, might place another burden on clients. Other studies have suggested that helper self-disclosure is appreciated by clients. Some clients see self-disclosing helpers as down-to-earth and honest.

Direct self-disclosure on the part of helpers can serve as a form of modeling. Self-help groups such as Alcoholics Anonymous use such modeling extensively. This helps new members get an idea of what to talk about and find the courage to do so. It is the group's way of saying, "You can talk here without being judged and getting hurt." Even in one-to-one counseling dealing with alcohol and drug addiction, extensive helper self-disclosure is the norm.

> Beth is a counselor in a drug rehabilitation program. She herself was an addict for a number of years but "kicked the habit" with the help of the agency where she is now a counselor. It is clear to all addicts in the program that the counselors there were once addicts themselves and are not only rehabilitated but also intensely interested in helping others rid themselves of drugs and develop a kind of lifestyle that helps them stay drug-free. Beth freely shares her experiences, both of being a drug user and of her rather agonizing journey to freedom, whenever she thinks that doing so can help a client.

Other things being equal, ex-addicts often make excellent helpers in programs like this. They know from the inside the games addicts play. Sharing their experience is central to their style of counseling and is accepted by their clients. It helps clients develop both new perspectives and new possibilities for action. Such self-disclosure is challenging. It puts pressure on clients to talk about themselves more openly or in a more focused way.

Helper self-disclosure is challenging for at least two reasons. First, it is a form of intimacy and, for some clients, intimacy is not easy to handle. Therefore, helpers need to know precisely why they are divulging information about themselves. Second, the message to the client is, indirectly, a challenging "You can do it, too," because revelations on the part of helpers, even when they deal with past failures, often center on problem situations they have overcome or opportunities they have seized. However, done well, such disclosures can be very encouraging for clients.

In the following example, Rick has had a number of sessions with his client Tim, who has had a rather tumultuous adolescence. For instance, he fell into the "wrong crowd" and got in trouble with the police a few times. His parents were shocked, and his relationship with them became very strained. One of Tim's themes is how the past can remain a burden for a long time. Rick decides to share some of his own experiences.

RICK: In my junior year in high school, I was expelled for stealing. I thought that it was the end of the world. My Catholic family took it as the ultimate disgrace. We even moved to a different neighborhood in the city.

TIM: What did it do to you?

Rick briefly tells his story, a story that includes setbacks not unlike Tim's. But Rick, with the help of a very wise and understanding uncle, was able to put the past behind him. He does not overdramatize his story. In fact, his story makes it clear that developmental crises are normal. How we interpret and manage them is the critical issue.

Since current research does not give us definitive answers, we need to stick to common sense. Helper self-disclosure is at present not a science but an art. Here are some guidelines for using it.

**Include helper self-disclosure in the contract.**  In self-help groups and in the counseling of addicts by ex-addicts, helper self-disclosure is an explicit part of the contract. If you don't want your disclosures to surprise your clients, let them know that you may talk about your own experiences when they seem relevant to clients' concerns. At some time toward the beginning of the counseling process, you might say something like this: "From time to time, I might share with you some of my own life experiences if I think they might help." Although Rick did not include it in the contract, he saw that he could weave it almost seamlessly into his dialogue with Tim.

**Make sure that your disclosures are appropriate.**  Sharing yourself is appropriate if it helps clients achieve treatment goals. Don't disclose more than is necessary. Helper self-disclosure that is exhibitionistic is obviously inappropriate. Self-disclosure on the part of helpers should be a natural part of the helping process, not a gambit. Rick's self-disclosure helps Tim get a different view of the "bad things" that happen in the past. Rick's "we" helps Tim see that he is not the only person who has had problems in the past.

**Be careful of your timing.**  Timing is critical. Common sense tells us that premature helper self-disclosure can turn clients off (Goodyear and Shumate, 1996). Rick did not share anything about himself immediately. He waited for a few sessions. However, once he saw a natural opening, he moved in.

**Keep your disclosures selective and focused.**  Don't distract clients with rambling stories about yourself. In the following example, the helper is talking to a first-year graduate student in a clinical-psychology program. The client is discouraged and depressed by the amount of work he has to do. The counselor wants to "help" him by sharing his own experience of graduate school.

COUNSELOR: Listening to you brings me right back to my own days in graduate school. I don't think that I was ever busier in my life. I also believe that the most depressing moments of my life took place then. On any number of occasions, I wanted to throw in the towel. I remember once toward the end of my third year when . . .

It may be that selective bits of this counselor's experience in graduate school would be useful in helping the student get a better conceptual and emotional grasp of her problems, but he has wandered off into the kind of reminiscing that meets his needs rather than the client's. In contrast, Rick's disclosure was selective and focused.

**Don't disclose too frequently.**  Helper self-disclosure is inappropriate if it is too frequent. Some research (Murphy & Strong, 1972) suggests that if helpers disclose themselves too frequently, clients tend to see them as phony and suspect that they have hidden motives. If Rick had continued to share his experiences whenever he saw a parallel with Tim's, Tim might have wondered who was helper and who was client.

**Do not burden already overburdened clients.**  One novice helper thought that he would help make a client who was sharing some sexual problems more comfortable by sharing some of his own sexual experiences. After all, he saw his own sexual development as not too different from the client's. However, the client reacted by saying: "Hey, don't tell me your problems. I'm having a hard enough time dealing with my own." This novice counselor shared too much of himself too soon. He was caught up in his own willingness to disclose rather than its potential usefulness to the client.

**Remain flexible.**  Take each client separately. Adapt your disclosures to differences in clients and situations. When asked directly, clients say that they want helpers to disclose themselves (see Hendrick, 1988), but this does not mean that every client in every situation wants it or would benefit from it. Even though Rick's disclosure to Tim was natural, it was made as a result of an explicit decision on his part.

## IMMEDIACY: DIRECT, MUTUAL TALK

Many, if not most, clients who seek help have trouble with interpersonal relationships either as a primary or a secondary concern. Some of the difficulties clients have in their day-to-day relationships are also reflected in their relationships with their helpers. For instance, if they are compliant outside, they are often compliant in the helping process. If they become aggressive and angry with authority figures outside, they often do the same with helpers. Therefore, the client's interpersonal style can be examined, at least in part, through an examination of his or her relationship with the helper. If counseling takes place in a group, then the opportunity is even greater. The package of skills enabling helpers to explore their relationships with their clients or vice versa or enabling clients to do the same with fellow group members and the group leader has been called "immediacy" by Robert Carkhuff (see Carkhuff 1969a, 1969b; Carkhuff & Anthony, 1979; see also Hill & O'Brien, 1999, Chapter 16). As you can see, it is an important tool for monitoring and managing the working alliance, a tool or kind of interaction that can be used by either client or helper.

## Types of Immediacy in Helping and
## Principles for Using Them

Three kinds of immediacy are reviewed here. First is the immediacy that focuses on the overall relationship: "How are you and I doing?" Second is the immediacy that focuses on some particular event in a session: "What's going on between you and me right now?" Third is self-involving statements—present tense, personal responses to the client (see Robitschek & McCarthy, 1991): "I like it when you challenge me. It shows spunk." Let's take a look at each type of immediacy.

**Overall relationship immediacy.** General relationship immediacy refers to your ability to discuss with a client where you stand in your overall relationship with him or her and vice versa. The focus is not on a particular incident but on the way the relationship itself has developed and how it is helping or standing in the way of progress. In the following example, the helper is a 44-year-old woman working as a counselor for a large company. She is talking to a 36-year-old man she has been seeing once every other week for about two months. One of his principal problems is his relationship with his supervisor, who is also a woman.

COUNSELOR: We seem to have developed a good relationship here. I feel we respect each other. I have been able to make demands on you, and you have made demands on me. There has been a great deal of give-and-take in our relationship. You've gotten angry with me, and I've gotten impatient with you at times, but we've worked it out. I'm wondering what our relationship has that is missing in your relationship with your supervisor.

CLIENT: Well, for one thing, you listen to me, and I don't think she does. On the other hand, I listen pretty carefully to you, but I don't think I listen to her at all, and she probably knows it. I think she's dumb, and I guess I "say" that to her in a number of ways, even without using the words. She knows how I feel.

The review of the relationship helps the client focus more specifically on his relationship with his supervisor.

Here is another example. Norman, a 38-year-old trainer in a counselor-training program, is talking to Weijun, 25, one of the trainees.

NORMAN: Weijun, I'm a bit bothered about some of the things that are going on between you and me. When you talk to me, I get the feeling that you are being very careful. You talk slowly. And you seem to be choosing your words, sometimes to the point that what you are saying sounds almost prepared. You have never challenged me on anything in the group. When you talk most intimately about yourself, you seem to avoid looking at me. I find myself giving you less feedback than I give others. I've even found myself putting off talking to you about all this. Perhaps some of this is my own imagining, but I want to check it out with you.

WEIJUN: I've been afraid to talk about all this, so I keep putting it off, too. I'm glad that you've brought it up. A lot of it has to do with how I relate to people in authority, even though you don't come across as an "authority figure." You don't act the way an authority figure is supposed to act.

In this case, cultural differences are part of the problem. For Weijun, giving direct feedback to someone in authority is not natural. However, he does go on to talk to Norman about his misgivings. He thinks that Norman's interventions in the training group are too "unorganized" and that he plays favorites. He has not wanted to bring it up because he fears that his position in the program will be jeopardized. But now that Norman has made the overture, he accepts the challenge.

The interaction can, of course, be initiated by the client, though many clients, for obvious reasons, would hesitate to do so. Who wants to take on the leader? In the following example, Cheryl is the mother of a boy, Bobby, who has been sexually molested by Luke, a minister, though not the minister of the church the family attends. She comes to talk with the minister of her church. They have been talking for a few minutes.

CHERYL: As you can imagine, all sorts of feelings swirled around inside me as I thought of having this session with you.

MINISTER: Given the circumstances, that probably doesn't surprise either of us. I wasn't sure you wanted to talk to me at all.

CHERYL: I've always respected you, but now, well, it's like someone dumped garbage on you, even though you haven't done anything. I know it's unfair, but something inside keeps saying, "He's one of them." Not a molester, but . . . someone who's supposed to be above all that.

MINISTER: I've got my own kind of turmoil. I want to apologize, but I haven't done anything. I want to apologize for the church, but you're as much a part of the church as I am. And apologizing for Luke sounds hollow. . . . Like you, I'm in pain . . . because of Bobby, because of the misery that's engulfing all of us.

CHERYL: I didn't think I'd see you this vulnerable. I thought I'd find you caring and . . . rational. I don't think of you as vulnerable. It makes me back down . . . from being so angry.

MINISTER: Vulnerable . . . hmm. . . . You used the phrase "above all that." I don't know who's above all that. Certainly not me. But I know that doesn't make me a predator either. . . . I'm just me.

This dialogue helps clear the air for both of them. Two things conspire to make this interaction possible. The minister makes it clear that he is open to being challenged and to engaging in a you-me conversation. And Cheryl is a gutsy person to begin with. Even if she had not been gutsy, the minister could have used his communication skills at least to offer this kind of conversation.

**Event-focused immediacy.**    Here-and-now immediacy refers to your ability to discuss with clients what is happening between the two of you in the here and now of any given transaction. It is not the entire relationship that is being considered but rather the specific interaction or incident. In the following example, the helper, a 43-year-old woman, is a counselor in a neighborhood human services center. Agnes, a 49-year-old woman who was recently widowed, has been talking about her loneliness. Agnes seems to have withdrawn quite a bit, and the interaction has bogged down.

COUNSELOR: I'd like to stop a moment and take a look at what's happening right now between you and me.

AGNES: I'm not sure what you mean.

COUNSELOR: Well, our conversation today started out quite lively, and now it seems rather subdued. I've noticed that the muscles in my shoulders have become tense. I sometimes tense up that way when I feel that I might have said something wrong. It could be just me, but I sense that things are a bit strained between us right now.

AGNES (hesitatingly): Well, a little. . . .

Agnes goes on to say how she resented one of the helper's remarks early in the session. She thought that the counselor had intimated that she was lazy. Agnes knows that she isn't lazy. They discuss the incident, clear it up, and move on.

The purpose of event-focused or here-and-now immediacy is to strengthen the working alliance. Research has shown that too much support can actually weaken the working alliance (see Kivlighan, 1990). The relationship needs some fiber. Immediacy is a way of balancing support with challenge on both sides (Kivlighan & Schmitz, 1992; Tryon & Kane, 1993).

**Self-involving statements.** Such statements can be either positive or negative in tone. Let's start with the positive.

HELPER: I like the way you interrupt me, Adler. You don't do it often, but it seems that it's your way of saying, "Hey, this is supposed to be a dialogue."

This self-involving remark is also a challenging statement, because the implication is "Keep it up." Clients tend to appreciate positive self-involving statements: "During the initial interview, the support and encouragement offered through the counselor's positive self-involving statements may be especially important because they put clients at ease and allay their anxiety about beginning counseling" (Watkins & Schneider, 1989, p. 345).

Negative self-involving statements are much more directly challenging in tone. Carl Rogers, the dean of client-centered therapy, recounts the following incident:

I am quite certain even before I stopped carrying individual counseling cases, I was doing more and more of what I would call confrontation. That is, confrontation of the other person with my feelings. . . . For example, I recall a client with whom I began to realize I felt bored every time he came in. I had a hard time staying awake during the hour, and that was not like me at all. Because it was a persisting feeling, I realized I would have to share it with him. I had to confront him with my feeling and that really caused a conflict in his role as a client. . . . So with a good deal of difficulty and some embarrassment, I said to him, "I don't understand it myself, but when you start talking on and on about your problems in what seems to me a flat tone of voice, I find myself getting very bored." This was quite a jolt to him and he looked very unhappy. Then he began to talk about the way he talked and gradually he came to understand one of the reasons for the way he presented himself verbally. He said, "You know, I think the reason I talk in such an uninteresting way is because I don't think I have ever expected anyone to really hear me." . . . We got along much better after that because I could remind him that I heard the same flatness in his voice I used to hear. (See Landreth, 1984, p. 323)

Rogers's self-involving statement, genuine but quite challenging, helped the client move forward. But there is another point of view. Someone once said, "Boredom is a self-indictment." Rogers was bored because he restricted himself to sharing empathic highlights with clients. On principle, he did not ordinarily use probing, summaries, and challenging lest he rob his clients of responsibility. He was a master at understanding clients and sharing highlights. Without doubt he helped many clients. This story also reveals the direction in which Rogers was moving toward the end of his career—adding the "spice" of probing and challenging to his interactions with clients.

## Situations Calling for Immediacy

Part of skilled helping—and, more generally, social intelligence—is knowing when to use any given communication skill. Immediacy can be useful in the following situations:

- *Lack of direction.* When a session is directionless and it seems that no progress is being made: "Let's take a time-out. I feel that we're not clicking. If we're stuck, I'm wondering what I'm doing, what we're doing to stay stuck." Of course, either helper or client could initiate this.

- *Tension.* When there is tension between helper and client: "We seem to be getting on each other's nerves. It might be helpful to stop a moment and see if we can clear the air."

- *Trust.* When trust seems to be an issue: "I see your hesitancy to talk, and I'm not sure whether it's related to me or not. You're talking about pretty sensitive issues. It might still be hard for you to trust me."

- *Diversity.* When diversity, some kind of "social distance," or widely differing interpersonal styles between client and helper seem to be getting in the way: "You're older, I'm younger. I'm an introvert, and there's more of the extrovert in you. I don't know whether this has anything to do with the fact that we seem to be stumbling along here. Or at least that's how it seems to me."

- *Dependency.* When dependency seems to be interfering with the helping process: "You don't seem willing to explore an issue until I give you permission to do so or urge you to do so. And I seem to have let myself slip into the role of permission giver and urger."

- *Counterdependency.* When counterdependency is blocking the helping relationship: "It seems that we're letting this session turn into a kind of struggle between you and me. And, if I'm not mistaken, both of us would like to win."

- *Attraction.* When attraction is sidetracking either helper or client: "I think we've liked each other from the start. Now I'm wondering whether that might be getting in the way of the work we're doing here." Care is needed in this situation. Talking about attraction can increase it. Someone once described romantic moments as "when we are alone together and the topic is only us."

Immediacy, both in counseling and in everyday life, is a difficult, demanding skill. It is difficult, first of all, because the helper—or the client—needs to be aware of what is happening in the relationship without becoming preoccupied with it. A helper should not use immediacy as an opportunity to "psych out" the client. Second, immediacy demands both social intelligence and social courage. It is not always easy to bring up relationship issues. The helper needs backbone. The client needs backbone. But it's worth the effort. Immediacy can help both counselor and client move beyond a variety of relationship obstacles. It is also a learning opportunity for clients. If helpers use immediacy well, clients can see its value and learn how to apply it to their own sticky relationships.

## USING SUGGESTIONS AND RECOMMENDATIONS

This section begins with a few imperatives: Don't tell clients what to do. Don't try to take over their lives. Let clients make their own decisions. These imperatives flow from the values of respect and empowerment. Does this mean, however, that suggestions and recommendations are absolutely forbidden? Never say never. It was mentioned earlier that there is a natural tension between helpers' desire to have their clients manage their lives better and respecting their freedom. If helpers build strong, respectful relationships with their clients, then "stronger" interventions sometimes make sense. In this context, suggestions and recommendations can stimulate clients to "get off the dime" and do something. Helpers move from counseling mode to guidance role. Research has shown that clients will generally go along with recommendations from helpers if those recommendations are clearly related to the problem situations, challenge clients' strengths, and are not too difficult (Conoley, Padula, Payton, & Daniels, 1994). Effective helpers can provide suggestions, recommendations, and even directives without robbing clients of their autonomy or integrity.

Here is a classic example of this from Cummings's (1979, 2000) work with addicts. Substance abusers came to him because they were hurting in many ways. He used every communication skill available to listen to and understand their plight.

> During the first half of the first session the therapist must listen very intently. Then, somewhere in mid-session, using all the rigorous training, therapeutic acumen, and the third, fourth, fifth, and sixth ears, the therapist discerns some unresolved wish, some long-gone dream that is still residing deep in that human being, and then the therapist pulls it out and ignites the client with a desire to somehow look at that dream again. This is not easy, because if the right nerve is not touched, the therapist loses the client. (1979, p. 1123)

So Cummings shared both basic and advanced empathic highlights to let clients know that he understood their plight and their longings. The addicts came knowing how to play every game in the book. But Cummings knew all their games, too. Toward the end of the first session, he told them they could have a second session—which they invariably wanted—only when they were "clean." The time of the second session depended on the withdrawal period for the kind of substance they were abusing. They screamed, shouted "foul," tried to play games, but he remained adamant. The directive "Get clean, then return" was part of the therapeutic process. And the vast majority did return. Clean.

In the hands of the socially competent helper, the use of suggestions, advice, and directives is an adjunct to the rest of the process. One manager was arguing with a consultant in a coaching and counseling session about a possible change in his communication style with the members of his team. Finally, the consultant said, "Just try it. Then we'll talk about it." The relationship was a good one. Giving the directive was one way of breaking through the manager's pugnacious communication style. He tried the change, liked it, and worked at incorporating it into his style.

Suggestions, advice, and directives need not always be taken literally. They can act as stimuli to get clients to come up with their own packages. One client said something like this to her helper: "You told me to let my teenage son have his say instead of constantly interrupting and arguing with him. What I did was make a contract with him. I told him that I would listen carefully to what he said and even summarize it and give it back to him. But he had to do the same for me. At least now we have some mutual monologuing going on. And we avoid our usual shouting matches. My hope is to find a way to turn it into dialogue."

In daily human interactions, people feel free to give one another advice. It goes on all the time. But helpers must proceed with caution. Suggestions, advice, and directives are not for novices. It takes a great deal of experience with clients and a great deal of savvy to know when they might work.

## CONFRONTATION

What about clients who keep dragging their feet? Some clients who don't want to change or don't want to pay the price of changing simply terminate the helping relationship. However, those who stay stretch across a continuum from mildly to extremely reluctant and resistant. Or they may be collaborative on some issues but reluctant when it comes to others. For instance, Hester is quite willing to work on career development but very reluctant to work on improving relationships, even though relationship building is part of the career package. "That's my private world," she says of her relationships.

If *inviting* clients to challenge themselves is at one end of the continuum, what's at the other? Where does respecting clients' rights to be themselves stop and placing demands on them to live more fully begin? Different helpers answer these questions differently. Therefore, helpers differ, both theoretically and personally, in their willingness to confront. "Traumatic confrontation" (one wonders about this choice of words) is a cognitive behavior modification technique that involves challenging youths to face up to and change dysfunctional behavior (Lowenstein, 1993). For example, a 12-year-old boy who had become involved in criminal activity after the disappearance of his father was confronted about his behavior. At first he denied everything but then decided to face up to the situation.

When all is said and done, there is a place in helping for interventions strong enough to merit the term *confrontation*. Confrontation, as intimated earlier, means challenging clients to develop new perspectives and to change both internal and external behavior even when they show reluctance and resistance to doing so. When helpers confront, they "make the case" for more effective living. Confrontation does not involve "do this or else" ultimatums. More often it is a way of making sure that clients understand what it means not to change—that is, making sure they understand the consequences of persisting in dysfunctional patterns of behavior or of refusing to adopt new behaviors.

Both advice giving and confrontation require high levels of social intelligence and social competence on the part of the helper. They are not for everyone and, as suggested earlier, can go wrong in the hands of novices. In most cases, they should be used sparingly. And when helpers do judge that advice giving and confrontation might well serve the interests of their clients, they should be guided by the values outlined in Chapter 3.

## ENCOURAGEMENT

This chapter ends on a positive-psychology note. You may not have noticed, but in the last two chapters, some form of the word *encouragement* has been used only a couple of times. If the whole purpose of challenging is to help clients move forward, and if encouragement (sugar) works as well as challenge (vinegar), why don't we hear more about encouragement? The sugar-vinegar analogy is not exactly right, because many clients find challenge both refreshing and stimulating. Challenge certainly does not preclude encouragement. After all, encouragement is a form of support, and research shows that support is one of the main ingredients in successful therapy (Beutler, 2000, p. 1004).

Miller and Rollnick (1991; Rollnick & Miller, 1995) introduced an approach to helping called "motivational interviewing." A simple Internet search on motivational interviewing reveals an extensive literature, including theory, research, and case studies (for instance, Borsari & Carey, 2000; Colby et al., 1998; Dench & Bennett, 2000; Baer, Kivlahan, & Donovan, 1999). Their original work focused on helping clients deal with addictive behavior, but their methodology over the years has been adapted to a much wider range of human problems. Much of the literature highlights the main elements of a problem-management approach. The values of respect, empathy, self-empowerment, and self-healing are emphasized (see Chapter 3). The spirit of encouragement rather than confrontation pervades the approach. Typically, clients (for instance, pregnant women who are smokers or users of alcohol) receive personal feedback on their problem areas—for instance, on how smoking has affected their lungs. There are discussions of personal responsibility, and advice on ways of managing the problem situations is offered. Clients are encouraged to find the motives, incentives, or levers of change that make sense for them and to use the change options that they feel are a best fit. Intrinsic motives— that is, motives that clients have internalized for themselves ("I want to be free")— rather than extrinsic motives ("I'll get in trouble if I don't change") are emphasized. Clients are also given help on identifying obstacles to change and ways of overcoming them. Empathy, both as a value and as a form of communication (empathic highlights), is used extensively.

Common sense suggests that direct realistic encouragement be included in any set of helping skills. Like most of the skills we have been discussing, encouragement can be used at any stage or step of the helping process. Clients can be encouraged to identify and talk about their problems and unused opportunities, to review possibilities for a better future, to set goals, to engage in actions that will help them achieve their goals, and to overcome the inevitable obstacles that stand in their way. Effective encouragement is not patronizing. It is not sympathy, nor does it rob the client of autonomy. It respects the client's self-healing abilities. It is a fully human nudge in the right direction.

# Evaluation Questions for Step I-B:
# The Use of Specific Challenging Skills

How effectively have I developed the communication skills that serve the process of challenging?

- *Sharing advanced empathic highlights:* How well do I share hunches with clients about their experiences, behaviors, and feelings to help them move beyond blind spots and develop needed new perspectives?

- *Information sharing:* How well do I give clients needed information or assist them in their search for it to help them see problem situations in a new light and to provide a basis for action?

- *Helper self-disclosure:* How well do I share my own experience with clients as a way of modeling nondefensive self-disclosure and helping them move beyond blind spots?

- *Immediacy:* How well do I discuss aspects of my relationship with my clients to improve the working alliance?

- *Suggestions and recommendations:* How well do I point out ways in which clients can more effectively manage problems and develop opportunities, or move productively through the stages and steps of the helping process?

- *Confrontation:* How well do I use a solid relationship with clients to challenge them more forcefully when they show signs of reluctance?

- *Encouragement*: How well do I encourage clients, at any step of the helping process, to find within themselves the desire to move forward?

# STEP I-B: III. THE WISDOM OF CHALLENGING

GUIDELINES FOR EFFECTIVE CHALLENGING
   Keep the Goals of Challenging in Mind
   Encourage Self-Challenge
   Earn the Right to Challenge
   Be Tentative but Not Apologetic in the Way You Challenge Clients
   Challenge Unused Strengths More Than Weaknesses
   Build on Clients' Successes
   Be Specific in Your Challenges
   Respect Clients' Values
   Deal Honestly, Caringly, and Creatively with Client Defensiveness

LINKING CHALLENGE TO ACTION

THE SHADOW SIDE OF CHALLENGING
   Clients' Shadow-Side Responses to Challenge: Dealing with Dissonance
      Discredit challengers
      Persuade challengers to change their views
      Devalue the issue
      Seek support elsewhere for the views being challenged
      Cooperate in the helping session, but then do nothing about it outside
   Challenge and the Shadow Side of Helpers
      The "MUM effect"
      Excuses for not challenging
      Helpers' blind spots

EVALUATION QUESTIONS FOR STEP I-B:
THE PROCESS AND WISDOM OF CHALLENGING

# GUIDELINES FOR EFFECTIVE CHALLENGING

All challenges should be permeated by the spirit of the client-helper relationship values discussed in Chapter 3; that is, they should be caring (not power games or put-downs), genuine (not tricks or games), and designed to increase the self-responsibility of the client (not expressions of helper control). They should also serve the stages and steps of the helping process, moving it forward for the purpose of constructive change (not an endless search for insight). Empathy should permeate every kind of challenge. Clearly, challenging well is not a skill that comes automatically. It needs to be learned and practiced. The following principles constitute some basic guidelines.

## Keep the Goals of Challenging in Mind

Challenge must be integrated into the entire helping process. Keep in mind that the goal is to help clients develop the kinds of alternative perspectives, internal behavior, and external actions needed to get on with the stages and steps of the helping process. To what degree do the new perspectives developed lead to problem-managing and opportunity-developing action? Are the insights relevant to real problems and opportunities rather than merely dramatic? Are the calls to action solution-focused?

## Encourage Self-Challenge

Invite clients to challenge themselves, and give them ample opportunity to do so. You can provide clients with probes and structures that help them engage in self-challenge. In the following excerpt, the counselor is talking to a man who has discussed at length his son's ingratitude. There has been something cathartic about his complaints, but it is time to move on.

COUNSELOR: People often have blind spots in their relationships with others, especially in close relationships. Picture your son sitting with some counselor. He is talking about his relationship with you. What's he saying?

CLIENT: Well, I don't know. . . . I guess I don't think about that very much. . . . Hmm. . . . He'd probably say . . . well, that he loves me . . . [pauses]. And then he might say that since his mother died, I have never really let him be himself. I've done too much to influence the direction of his life rather than let him fashion it the way he wanted. . . . Hmm. He'd say that he loves me but he has always resented my "interference."

COUNSELOR: So both love for you and resentment for all that control.

CLIENT: And he'd be right. I'm still doing it. Not with him. He won't let me. But with lots of others. Especially in my business.

The counselor provides a structure that enables the client to challenge himself with respect to his son and then apply what he learns to other settings of life. Would that all clients responded so easily! Alternatively, the counselor might have asked this client to list three things he thinks he does right and three things he thinks he should reconsider in his relationship with his son. The point is to be inventive with the probes and structures you provide clients to help them challenge themselves.

## Earn the Right to Challenge

Berenson and Mitchell (1974) maintain that some helpers don't have the right to challenge others because they are not doing a good job keeping their own houses in order. Here are some of the factors that earn you the right to challenge:

- *Develop a working relationship.* Challenge only if you have spent time and effort building a relationship with your client. If your rapport is poor or you have allowed your relationship with the client to stagnate, challenge yourself to deal with the relationship more creatively.

- *Make sure you understand the client.* Empathy drives everything. Effective challenge flows from accurate understanding. Only when you see the world through the client's eyes can you begin to see what he or she is failing to see.

- *Be open to challenge yourself.* Hesitate to challenge unless you are open to being challenged. If you are defensive in the counseling relationship or in your relationships with supervisors or in your everyday life, your challenges might ring hollow. Model the kind of nondefensive attitudes and behavior that you would like to see in your clients.

- *Work on your own life.* How important is constructive change in your own life? Berenson and Mitchell claim that only people who are striving to live fully according to their value systems have the right to challenge others, for only such persons are potential sources of human nourishment for others.

In summary, ask yourself, "To what degree am I the kind of person from whom clients would be willing to accept challenges?"

## Be Tentative but Not Apologetic in the Way You Challenge Clients

Tentative challenges are generally viewed more positively than strong, direct challenges (see Jones & Gelso, 1988). The principle is this: Challenge clients in such a way that they are more likely to respond than to react. The same challenging message can be delivered in such a way as to invite the cooperation or arouse the resistance of the client. Deliver challenges tentatively, as hunches that are open to review and discussion rather than as accusations. Challenging is certainly not an opportunity to browbeat clients or put them in their place.

On the other hand, challenges that are delivered with too many qualifications—either verbally or through the helper's tone of voice—sound apologetic and can be easily dismissed by clients. I once worked in a career-development center. As I listened to one client, it soon became evident that one reason he was getting nowhere was that he engaged in so much self-pity. When I shared this observation with him, I overqualified it. These are not my exact words, but it must have sounded something like this:

HELPER: Has it ever, at least in some small way, struck you that one possible reason for not getting ahead, at least as much as you would like, could be that at times you tend to engage in a little bit of self-pity?

I still remember his response. He paused, looked me in the eye, and said, "A little bit of self-pity?" When he paused again, I said to myself, "I've been too harsh!" He continued, "I *wallow* in self-pity." We moved on to explore what he might do to move beyond self-pity to constructive change.

## Challenge Unused Strengths More Than Weaknesses

Berenson and Mitchell (1974), taking a page from positive psychology, found that successful helpers tend to challenge clients' strengths rather than their weaknesses. What we talk about sets the tone for helping. Individuals who focus on their failures find it difficult to change their behavior. Clients who dwell too much on their shortcomings tend to belittle their achievements, withhold rewards from themselves when they do achieve, and live with anxiety. All this tends to undermine performance (see Bandura, 1986, p. 339).

Challenging strengths is a positive-psychology approach. It means pointing out to clients the assets and resources they have but fail to use. In the following example, the helper is having a one-to-one session with a woman who is a member of a self-help group in a rape crisis center. She is very good at helping others but is always down on herself.

COUNSELOR: Ann, in the group sessions, you provide a great deal of support for the other women. You have an amazing ability to spot a person in trouble and provide an encouraging word. And when one of the women wants to give up, you are the first to challenge her, gently and forcibly at the same time. But when Ann is dealing with Ann . . .

ANN: I know where you're headed. . . . I know I'm a better giver than receiver. I'm not sure why that is. . . . Or that it even matters. I wonder if the others see this in me and then begin thinking that I'm just doing it. If I'm not both demanding *and* caring to myself. . . .You know, I've been this way for a long time. I think I've got some bad habits when it comes to dealing with myself. I'm so fearful of being self-indulgent.

The counselor helps her place a demand on herself to use her rather substantial resources in her dealings with herself. Since she isn't self-indulgent, it's time to take a second look, a different kind of look, at her fear. At least now it's on the table.

Adverse life experiences can be sources of strength. For instance, McMillen, Zuravin, and Rideout (1995) studied adult perceptions of the benefits from experiencing sexual abuse as children. Almost half the adults reported some kind of benefit, including increased knowledge of child sexual abuse, protecting other children from abuse, learning how to protect themselves from others, and developing a strong personality—without, of course, discounting the horror of child abuse. Counselors, therefore, can help clients "mine" benefits from adverse experiences, putting to practical use the age-old dictum that "good things can come from evil things."

## Build on Clients' Successes

Effective helpers do not urge clients to place too many demands on themselves all at once. Rather, they help clients place reasonable demands on themselves and, in the process, help them appreciate and celebrate their successes. In the following example, the client is a boy in a detention center who is rather passive and allows the other boys to push him around. Recently, however, he has made a few halfhearted attempts to stick up for his rights in the counseling group. The counselor is talking to him alone after a group meeting.

COUNSELOR A: You're still not standing up for your own rights the way you need to. You said a couple of things in there, but you still let them push you around.

This counselor emphasizes the negative and browbeats the client. The following counselor takes a different tack.

COUNSELOR B: Here's what I've noticed. In the group meetings, you have begun to speak up. You say what you want to say, even though you don't say it really forcefully. And I get the feeling that you feel good about that. You've got some power. Now the challenge is to find ways of using it more effectively.

CLIENT: I didn't think anyone noticed. You think it was a good start?

COUNSELOR B: Certainly. . . . But what you think is more important. And it sounds like you're proud of what you did.

The second counselor emphasizes the client's success, however modest, and goes on to provide some encouragement to do even better.

## Be Specific in Your Challenges

Specific challenges hit the mark. Vague challenges get lost. Clients don't know what to do about them. Statements such as "You need to pull yourself together and get on with it" may satisfy some helper need, such as venting frustration, but they do little for clients. Specific statements, on the other hand, can hit the mark. In the following example, the client is experiencing a great deal of stress both at home and at work:

HELPER: You say that you really want to spend more time at home with the kids and you really enjoy it when you do, but you keep taking on new assignments at work, like the Eclipse project, that will add to your travel schedule. Maybe it would be helpful to talk a bit more about work-life balance.

CLIENT: Boy, there's that phrase! Work-life balance. The company talks a lot about it, but nothing much happens. I'm not sure there's anyone at work who's got the work-life balance right.

HELPER: You know what they say about career: "If you're not in charge of your career, no one is." It sounds like the same is true with work-life balance.

CLIENT: I hadn't thought about it like that. . . . But I'm afraid you're right. It's right where it belongs, I suppose—on my shoulders. I've been waiting for my family and my company to figure it out for me.

Some helpers avoid clarity and specificity because they feel that they're being too intrusive. Helping has to be intrusive to make a difference.

## Respect Clients' Values

Challenge clients to clarify their values and to make reasonable choices based on them. Be wary of using challenging, even indirectly, to force clients to accept your values. This violates the empowerment value discussed in Chapter 3. In the following example, the client is a 21-year-old woman who has curtailed her social life, her education, and her career to take care of her elderly mother who is suffering from incipient Alzheimer's.

CLIENT: I admit that juggling work, home, and school is a real challenge. I keep feeling that I'm not doing justice to any of them.

COUNSELOR A: You have every right to have a life of your own. Why not get your mother into a nursing facility? You can still visit her regularly. Then get on with life. That's probably what she wants anyway.

Challenging clients to clarify their values is, of course, legitimate. But this counselor does little to help the client clarify her values. She makes suggestions without finding out why the client is doing what she's doing. Counselor B takes a different approach.

COUNSELOR B: You're trying to juggle four very important areas of your life: caring for your mother, school, work, and social life. That's a tough assignment. It might help to explore what's driving you in all this. Maybe I could help you lay out the values that drive your behavior. Then you could ask yourself about your priorities.

CLIENT: I've never thought about values as things that drive what I do. I thought we had values and just, well, did them. Help me take a look.

This counselor challenges her gently to find out what she really wants. Helpers can assist clients to explore the consequences of the values they hold, but that is not the same as questioning them.

## Deal Honestly, Caringly, and Creatively with Client Defensiveness

Do not be surprised when clients react strongly to being challenged, even when you're trying to help them respond rather than react. If they react negatively, follow the principles outlined in Chapter 9 on reluctance and resistance. Help them share and work through their emotions. If they seem to "clam up," try to find out what's going on inside. In the following example, the helper has just delivered a brief summary of the main points of the problem situation he and the client have been discussing. He has gently pointed out the self-destructive nature of some of the client's behaviors.

HELPER: I'm not sure how all this sounds to you.

CLIENT: I thought you were on my side. Now you sound like all the others. And I'm paying you to talk like this to me!

Even though the helper is tentative in his challenge, the client still reacts defensively. Here are two different approaches to the client's defensiveness:

HELPER A: All I've done is summarize what you have been saying about yourself. And you know you're doing yourself in. Let's look at each point we've been discussing and see if this isn't the case.

This helper takes a defensive, judicial approach. He's about to assemble the evidence. This could well lead to an argument rather than further dialogue. Helper B backs up a bit.

HELPER B: So I'm sounding harsh and unfair to you. . . . Kind of dumping on you . . . Let's back up.

This helper backs off without saying that her summary was wrong. She is giving the client some space. It may be that the client needs time to think about what the helper has said. Helper B tries to get into a constructive dialogue with the angered client.

The principles just outlined are, of course, guidelines, not absolute prescriptions. In the long run, use your common sense. The more flexible you are, the more likely you are to add value to your clients' search for solutions.

## LINKING CHALLENGE TO ACTION

More and more theoreticians and practitioners are stressing the need to link insight to problem-managing and opportunity-developing action. Wachtel (1989) puts it succinctly: "There is good reason to think that the really crucial insights are those closely linked to new actions in daily life and, moreover, that insights are as much a product of new experience as their cause" (p. 18).

Is challenge enough to stimulate problem-managing action? For some, yes. A few well-placed challenges might be all that some clients need to move to constructive action. Once challenged, they are off to the races. Others may well have the resources to manage their lives better but not the will. They know what they need to do but are not doing it. A few nudges in the right direction help them overcome their inertia. I have had many one-session encounters that included a bit of listening and thoughtful processing, some sharing of highlights, and a brief challenge that produced a new perspective that sent the client off on some useful course of action. On the other hand, if helping clients challenge themselves to develop new perspectives leads to one profound insight after another but no action and behavioral change, then once more we are whistling in the wind. Helping becomes an intellectual game rather than a serious attempt to partner with clients in their search for a better life.

## THE SHADOW SIDE OF CHALLENGING

Challenging, because of its very nature, has a strong shadow side. While we hate being confronted, we often dislike being challenged, even when it is done well. If helpers are very effective in challenging, they might well sense some reluctance in their clients' responses. If they are poor at it, they are more likely to experience resistance. Inviting clients to challenge themselves and helping them do it to themselves is your best bet.

### Clients' Shadow-Side Responses to Challenge:
### Dealing with Dissonance

Even when challenge is a response to a client's plea to be helped to live more effectively, it can precipitate some degree of disorganization in the client. Different writers refer to this experience under different names, including crisis, disorganization, a sense of inadequacy, disequilibrium, and beneficial uncertainty (Beier & Young, 1984). As the last of these terms implies, counseling-precipitated crises can be beneficial for the client. Whether they are or not depends, to a great extent, on the skill of the helper.

Even when an invitation to self-challenge or a direct challenge is accurate and delivered caringly, some clients still dodge and weave. Cognitive-dissonance theory (Festinger, 1957; Draycott & Dabbs, 1998) gives us some insight into the dynamics of this. Since dissonance (discomfort, crisis, disequilibrium) is an uncomfortable state, the client will try to get rid of it (see Adler & Towne, 1999, pp. 400–405). Following are five of the most typical ways clients experiencing dissonance attempt to rid themselves of their discomfort. Let's examine them briefly as they apply to being challenged. Since these responses are forms of resistance, review the ways in which resistance can be avoided and managed (Chapter 9) and see how they might apply to the following examples.

**Discredit challengers.** The client might confront the helper whose challenges are getting too close for comfort. Some attempt is made to point out that the helper is no better than anyone else. In the following example, the client has been discussing her marital problems and has been invited by the helper to take a look at her own behavior rather than giving example after example of what her spouse does wrong.

CLIENT: It's easy for you to sit there and suggest that I be more responsible in my marriage. You've never had to experience the misery in which I live. You've never experienced his brutality. You probably have one of those nice middle-class marriages.

Counterattack is a common strategy for coping with challenge. You may even have to field some sarcastic barbs: "Thanks for coming down out of your ivory tower to tell me how to shape up." How do you think the "nice middle-class marriage" counterattack should be handled?

**Persuade challengers to change their views.** In this approach, clients reason with their helpers. They urge them to see what they have said as misinterpretations and to revise their views. In the following example, the client has been talking about the way she blows up at her husband whenever he makes "some stupid mistake." The counselor has invited her to explore the consequences of this pattern of behavior.

CLIENT: I'm not so sure that my anger at home isn't called for. I think that it's a way in which I'm asserting my own identity. If I were to lie down and let others do what they want, I would become a doormat at home. And, as you have helped me see, assertiveness should be part of my style. I think you see me as a fairly reasonable person. I don't get angry here with you because there is no reason to.

Sometimes a client like this will lead an unwary counselor into an argument about who's really right. How would you respond to this client?

**Devalue the issue.** This is a form of rationalization. A client who is being invited to challenge himself about his sarcasm points out that he is rarely sarcastic, that "poking fun at others" is just that, good-natured fun, that everyone does it, and that it is a very minor part of his life not worth spending time on. The client has a right to devalue a topic if it really isn't important. The counselor has to be sensitive enough to discover which issues are important and which are not. How would you handle the devaluing this client engages in?

**Seek support elsewhere for the views being challenged.** Some clients, once challenged, go out and gather testimonials supporting their views. In extreme cases, clients leave one counselor and go to another because they feel they aren't being understood. They try to find helpers who will agree with them. More commonly, the client remains with the same counselor and offers evidence from others contesting the helper's point of view.

CLIENT: I asked my wife about my sarcasm. She said she doesn't mind it at all. And she thinks that my friends see it as humor and as a part of my style.

This is an indirect way of telling the counselor she is wrong. The counselor might well be wrong, but if the client's sarcasm is really dysfunctional in his interpersonal life, the counselor should find some way of pressing the issue. What would you do in this case?

**Cooperate in the helping session, but then do nothing about it outside.** The client can agree with the counselor as a way of dismissing an issue. However, the purpose of challenging is not to get the client's agreement but to develop new perspectives that lead to constructive action. Consider this client:

CLIENT: To tell you the truth, I am pretty lazy and manipulative. And I'm not even very clever about it.

In truth, this client is lazy and manipulative. She has given little evidence so far that she will do anything about it. If it were up to you to break through this screen of "honesty," what would you do?

## Challenge and the Shadow Side of Helpers

Sol Garfield introduces a set of four articles on "the therapist as a neglected variable in psychotherapy research" in an issue of *Clinical Psychology: Science and Practice* (Spring, 1997). Helpers have many characteristics and engage in many behaviors that add value to clients' efforts to manage their lives better. But their behaviors in helping sessions can also be part of the shadow side of helping. Two shadow-side areas are addressed here: the reluctance of some helpers to challenge clients or invite them to challenge themselves, and the blind spots helpers themselves have. These phenomena are hardly restricted to helping situations. Consider the workplace. When supervisors give feedback to the members of their teams, they are engaging in a form of challenge. They challenge their direct reports—or better, they invite their direct reports to challenge themselves—to correct their mistakes and to use their ingenuity to find ways to improve their performance. This is the theory. However, *all* the companies I have worked for as a consultant have said that giving feedback is an area where their managers need a great deal of improvement.

**The "MUM effect."** Initially, some counselor trainees are quite reluctant to help clients challenge themselves. They become victims of what has been called the "MUM effect"—the tendency to withhold bad news even when it is in the other's interest to hear it (Rosen & Tesser, 1970, 1971; Tesser & Rosen, 1972; Tesser, Rosen, & Batchelor, 1972; Tesser, Rosen, & Tesser, 1971). In ancient times, the person who bore bad news to the king was sometimes killed. That obviously led to a certain reluctance on the part of messengers to bring such news. Bad news—and, by extension, the kind of "bad news" that is involved in any kind of invitation to self-challenge—arouses negative feelings in the challenger, no matter how he or she thinks the receiver will react. If you are comfortable with the supportive dimensions of the helping process but uncomfortable with helping as a social-influence process, you could fall victim to the MUM effect and become less effective than you might otherwise be.

**Excuses for not challenging.** Reluctance to challenge is not a bad starting position. In my estimation, it is far better than being too eager to challenge. However, all helping, even the most client-centered, involves social influence. It is important for you to understand your reluctance (or eagerness) to challenge—that is, to challenge yourself on the issue of challenging and on the very notion of helping as a social-influence process. When trainees examine how they feel about challenging others, here are some of the things they discover:

- I am just not used to challenging others. My interpersonal style has had a lot of the live-and-let-live in it. I have misgivings about intruding into other people's lives.
- If I challenge others, then I open myself to being challenged. I may be hurt, or I may find out things about myself that I would rather not know.
- I might find out that I like challenging others and that the floodgates will open and my negative feelings about others will flow out. I have some fears that deep down I am a very angry person.
- I am afraid that I will hurt others, damage them in some way or other. I have been hurt or I have seen others hurt by heavy-handed confrontations.
- I am afraid that I will delve too deeply into others and find that they have problems that I cannot help them handle. The helping process will get out of hand.
- If I challenge others, they will no longer like me. I want my clients to like me.

Vestiges of this kind of thinking persist long after trainees move out into the field as helpers. Helpers, cure yourselves! Managers are also bedeviled by the MUM effect and come up with their own sets of excuses for not giving feedback. Of course, being willing to challenge responsibly is one thing; having the skills and wisdom to do so is another. One reason managers fail to give feedback is that they do not have the communication, coaching, and counseling skills needed to do it well.

**Helpers' blind spots.**  There is interesting literature on the humanity and flaws of helpers (Kottler, 1993; Kottler & Blau, 1989; Pope & Tabachnick, 1994; Wood, Klein, Cross, Lammers, & Elliot, 1985; Yalom, 1989) that can be of enormous help to both beginners—because prevention is infinitely better than cure—and to old-timers—because you *can* teach old dogs new tricks. More recently, Kottler (2000) has given trainees and novices an upbeat view of what passion and commitment in the helping professions look like.

Because helpers are as human as their clients, they too can have blind spots that detract from their ability to help. There is some evidence that adopting a resource-collaborator role is difficult for some helpers (see Bohart & Tallman, 1999, pp. 263–266). For instance, in one study (Atkinson, Worthington, Dana, & Good, 1991), counselors were almost unanimous in their preference for a "feeling" approach to counseling, whereas the majority of male clients preferred either a "thinking" or an "acting" orientation. This is not a sign of collaboration. Of course, other helpers lose control of their part of the helping dialogue. It does not dawn on them that the client has taken charge of the helping process and is manipulating them.

Helpers sometimes prevent clients from moving forward by playing the "insight" game, a perverted form of what has been discussed in this chapter. They help clients develop one insight after another without linking those insights to problem-managing action. Many of the insights end up being oriented to the theories of the helper rather than to the problems of the client. There is an unsurfaced assumption that insight will ultimately cure.

One of the critical responsibilities of supervisors is to help counselors identify their blind spots and learn from them. Once out of training, skilled helpers use different forums or methodologies to continue this process, especially with difficult cases. They take counsel with themselves, asking, "What am I missing here?" They take counsel with colleagues. Without becoming self-obsessed, they scrutinize and challenge themselves and the role they play in the helping relationship. Or, more simply, throughout their careers, they continue to learn about themselves, their clients, and their profession.

## Evaluation Questions for Step I-B: The Process and Wisdom of Challenging

How well do I do each of the following as I try to help my clients?

### The Process of Challenging

- Help clients become aware of their blind spots in their mind-sets, thinking, and acting, and help them develop new perspectives leading to more constructive behaviors
- Use challenging seamlessly whenever it is needed in the helping process
- Keep in mind the goals of challenging—that is, helping clients move beyond blinds spots to more effective mind-sets and change both internal and external patterns of behavior that keep them mired in problems and ineffective in developing unused opportunities
- Help clients participate fully in the helping process
- Help clients own their problems
- Help clients state problems as solvable and opportunities as doable
- Help clients correct faulty interpretations of their experiences, actions, and feelings
- Help clients identify and move beyond the predictable dishonesties of life
- Help clients spot opportunities
- Help clients get in touch with unused resources
- Help clients link new insights to problem-managing action
- Help clients develop a sense of self-efficacy
- Help clients generally move beyond discussion and inertia to action

*(continued)*

*(continued)*

## The Wisdom of Challenging

- Invite clients to challenge themselves
- Earn the right to challenge by
  - developing an effective working alliance with clients
  - working at seeing clients' points of view
  - being open to challenge myself
  - managing problems and developing opportunities in my own life
- Be tactful and tentative in challenging without being insipid or apologetic
- Be specific, developing challenges that hit the mark
- Challenge clients' strengths rather than their weaknesses
- Don't ask clients to do too much too quickly
- Invite clients to clarify and act on their own values

## The Shadow Side of Challenging

- Identify the games my clients attempt to play with me without becoming cynical in the process
- Become comfortable with the social-influence dimension of the helping role, with the kind of "intrusiveness" that goes with helping
- Incorporate challenging into my counseling style without becoming a confrontation specialist
- Develop the assertiveness needed to overcome the MUM effect
- Challenge the excuses I give myself for failing to challenge clients
- Come to grips with my own imperfections and blind spots both as a helper and as a "private citizen"

# STEP I-C: LEVERAGE— HELPING CLIENTS WORK ON THE RIGHT THINGS

THE ECONOMICS OF HELPING

SCREENING: THE INITIAL SEARCH FOR LEVERAGE

LEVERAGE: WORKING ON ISSUES THAT MAKE A DIFFERENCE

SOME PRINCIPLES OF LEVERAGE

    If there is a crisis, first help the client manage the crisis

    Begin with the problem that seems to be causing the most pain

    Begin with issues the client sees as important and is willing to work on

    Begin with some manageable subproblem of a larger problem situation

    Begin with a problem that, if handled, will lead to some kind of general improvement in the client's condition

    Focus on a problem for which benefits will outweigh the costs

FOCUS AND LEVERAGE: THE LAZARUS TECHNIQUE

STEP I-C AND ACTION

THE SHADOW SIDE OF STEP I-C

EVALUATION QUESTIONS FOR STEP I-C

# THE ECONOMICS OF HELPING

Helping is expensive, both financially and psychologically. It should not be undertaken lightly. Therefore, a word is in order about the economics of helping, which relates to all the stages and steps of the helping process and not just Step I-C. The issue is raised here because Step I-C introduces the notion of *leverage*. The leverage question is, How can we help our clients get the most out of the helping process? Helpers need to ask themselves, "Am I adding value through each of my interactions with this client?" Clients need to be helped to ask themselves, "Am I spending my time well? Do the decisions I am making have the potential of adding value to my life?" The question here is not "Does helping help?" but "Is helping working in this situation? Is it worth it?"

It's important that helpers determine whether the client is ready to invest in constructive change or not. And to what degree. Change requires work on the part of clients. If they do not have the incentives to do the work, they might begin and then trail off. If this happens, it's a waste of resources. For instance, Helmut and Gretchen are mildly dissatisfied with their marriage and seem to be looking for a "psychological pill" that will magically make things better. There seem to be few incentives for the work required to reinvent the marriage. On the other hand, Beth is an intelligent but bored empty-nester. Her husband travels a great deal, and she has too much time on her hands. She's quite dissatisfied with her current lifestyle. She has plenty of incentives to deal with her malaise and to identify and develop opportunities for a fuller life. Leverage comes from having a good client-helper relationship, working on the right problems and opportunities, choosing the right goals, and pursuing these goals through the right strategies—all ending up in constructive change.

# SCREENING: THE INITIAL SEARCH FOR LEVERAGE

Relatively little is said in the literature about screening—that is, about deciding whether any given problem situation or opportunity deserves attention. The reasons are obvious. Helpers-to-be are rightly urged to take their clients and their clients' concerns seriously. They are also urged to adopt an optimistic attitude, an attitude of hope, about their clients. Finally, they are schooled to take their profession seriously and are convinced that their services can make a difference in the lives of clients. For those and other reasons, the first impulse of the average counselor is to try to help clients no matter what the problem situation might be.

There is something very laudable about that. It is rewarding to see helpers prize people and express interest in their concerns. It is rewarding to see helpers put aside the almost instinctive tendency to evaluate and judge others and to offer their services to clients just because they are human beings. However, like other professions, helping can suffer from the "law of the instrument." A child given a hammer soon discovers that almost everything needs hammering. Helpers, once equipped with the models, methods, and skills of the helping process, can see all human problems as needing their attention. In fact, in many cases, counseling may be a useful intervention and yet be a luxury whose expense cannot be justified. The problem-severity formula discussed in Chapter 8 is a useful tool for screening.

Under the term *differential therapeutics*, Frances, Clarkin, and Perry (1984) discuss ways of fitting different kinds of treatment to different kinds of clients. They also discuss the conditions under which "no treatment" is the best option. In the no-treatment category, they include clients who have a history of treatment failure or who seem to get worse from treatment, such as

- criminals trying to avoid or diminish punishment by claiming to be suffering from psychiatric conditions—"We may do a disservice to society, the legal system, the offenders, and ourselves if we are too willing to treat problems for which no effective treatment is available" (p. 227);
- patients with malingering or fictitious illness;
- chronic nonresponders to treatment;
- clients likely to improve on their own;
- healthy clients with minor chronic problems;
- reluctant and resistant clients who refuse treatment.

Although a decision needs to be made in each case, and although some might dispute some of the categories proposed, the possibility of no treatment deserves serious attention.

The no-treatment or no-further-treatment option can do a number of useful things: interrupt helping sessions that are going nowhere or are actually destructive; keep both client and helper from wasting time, effort, and money; delay help until the client is ready to do the work required for constructive change; provide a "breather" period that allows clients to consolidate gains from previous treatments; provide clients with the opportunity to discover what they can do without treatment; keep helpers and clients from playing games with themselves and one another; and provide motivation for the client to find help in his or her daily life. However, a decision on the part of helping professionals not to treat or to discontinue treatment that is proving fruitless is countercultural and therefore difficult to make.

It goes without saying that screening can be done in a heavy-handed way. Statements such as the following are not useful:

- "Your concerns are actually not that serious."
- "You should be able to work that through without help."
- "I don't have time for problems as simple as that."

Whether such sentiments are expressed or implied, they obviously indicate a lack of respect and constitute a caricature of the screening process.

Practitioners in the helping professions are not alone in grappling with the economics of treatment. Doctors face clients day in and day out with problems that run the gamut from life-threatening to inconsequential. Statistics suggest that more than half of the people who come to doctors have nothing physically wrong with them. Doctors consequently have to find ways to screen patients' complaints. I am sure that the best physicians find ways to do so that preserve the dignity of their patients.

Effective helpers, because they are empathic, pick up clues relating to clients' commitment, but they don't jump to conclusions. They test the waters in various ways. If clients' problems seem inconsequential, they probe for more substantive issues. If clients seem reluctant, resistant, and unwilling to work, they challenge clients' attitudes and help them work through their resistance. But in both cases, they realize that there may come a time, and it may come fairly quickly, to judge that further effort is uncalled for because of lack of results. It is better, however, to help clients make such decisions themselves or to challenge them to do so. In the end, helpers sometimes have to stop seeing certain clients, but their way of doing so should reflect their basic counseling values.

## LEVERAGE: WORKING ON ISSUES THAT MAKE A DIFFERENCE

Clients often need help to get a handle on complex problem situations. A 41-year-old depressed man with a failing marriage, a boring, run-of-the-mill job, deteriorating interpersonal relationships, health concerns, and a drinking problem cannot work on everything at once. Priorities need to be set. The blunt questions go something like this: Where is the biggest payoff? Where should the limited resources of both client and helper be invested? Where to start?

> Andrea, a woman in her mid-thirties, is referred to a neighborhood mental-health clinic by a social worker. During her first visit, she pours out a story of woe both historical and current: brutal parents, teenage drug abuse, a poor marriage, unemployment, poverty, and the like. Andrea is so taken up with getting it all out that the helper can do little more than sit and listen.

Where is Andrea to start? How can the time, effort, and money invested in helping provide a reasonable return to her? What are the economics of helping in Andrea's case?

## SOME PRINCIPLES OF LEVERAGE

The following principles of leverage—a reasonable return on the investment of the client's, the helper's, and third-party resources—serve as guidelines for choosing issues to be worked on. These principles overlap; more than one can apply at the same time. The first three principles focus on client priorities. You can first help the client cope and then help him or her move on.

- If there is a crisis, first help the client manage the crisis.
- Begin with the problem that seems to be causing the most pain for the client.
- Begin with issues the client sees as important and is willing to work on.
- Begin with some manageable subproblem of a larger problem situation.
- Begin with a problem that, if handled, will lead to some kind of general improvement in the client's condition.
- Focus on a problem for which the benefits will outweigh the costs.

Underlying all these principles is an attempt to make clients' initial experience of the helping process rewarding so that they will have the incentives they need to continue to work. Examples of uses and abuses of these principles follow.

**If there is a crisis, first help the client manage the crisis.** Although crisis intervention is sometimes seen as a special form of counseling (Baldwin, 1980; Janosik, 1984), it can also be seen as a rapid application of the three stages of the helping process to the most distressing aspects of a crisis situation.

PRINCIPLE VIOLATED: Zachary, a student near the end of the second year of a four-year doctoral program in counseling, gets drunk one night and is accused of sexual harassment by a student whom he met at a party. Knowing that he has never sexually harassed anyone, he seeks the counsel of a faculty member whom he trusts. The faculty member asks him many questions about his past, his relationship with women, how he feels about the program, and so on. Zachary becomes more and more agitated and then explodes: "Why are you asking me all these silly questions?" He stalks out and goes to a fellow student's house.

PRINCIPLE USED: Seeing his agitation, his friend says, "Good grief, Zach, you look terrible! Come in. What's going on?" He listens to Zachary's account of what has happened, interrupting very little, merely putting in a word here and there to let his friend know he is with him. He sits with Zachary when he falls silent or cries a bit, and then slowly and reassuringly "talks Zach down," engaging in an easy dialogue that gives his friend an opportunity gradually to find his composure again. Zach's student friend has a friend in the university's student services department. They call him up, go over, and have a counseling and strategy session on the next steps in dealing with the harassment charges.

The friend's instincts are much better than those of the faculty member. He does what he can to defuse the immediate crisis and helps Zach take the next crisis-management step.

**Begin with the problem that seems to be causing the most pain.** Clients often come for help because they are hurting even though they are not in crisis. Their hurt, then, becomes a point of leverage. Their pain also makes them vulnerable. If it is evident that they are open to influence because of their pain, seize the opportunity, but move cautiously. Their pain may also make them demanding. They can't understand why you cannot help them get rid of it immediately. This kind of impatience may put you off, but it, too, needs to be understood. Such clients are like patients in the emergency room, each seeing himself or herself as needing immediate attention. Their demands for immediate relief may well signal a self-centeredness that is part of their character and therefore part of the broader problem situation. It may be that their pain is, in your eyes, self-inflicted, and ultimately you may have to challenge them on it. But pain, whether self-inflicted or not, is still pain. Part of your respect for clients is your respect for them as vulnerable.

PRINCIPLE VIOLATED: Rob, a man in his mid-twenties, comes to a counselor in great distress because his wife has just left him. The counselor's first impression is that Rob is an impulsive, self-centered person with whom it would be difficult to live. The counselor immediately challenges him to take a look at his interpersonal style and the ways in which he alienates others. Rob seems to listen, but he does not return.

PRINCIPLE USED: Rob goes to a second counselor, who also sees a number of clues indicating self-centeredness and a lack of maturity and discipline. However, she listens carefully to his story, even though it is one-sided. She explores with him the incident that precipitated his wife's leaving. Instead of adding to his pain by making him come to grips with his selfishness, she focuses on what he wants for the future, especially the immediate future. Of course, Rob thinks that his wife's return is the most important part of a better future. She says, "I assume she would have to be comfortable with returning." He says, "Of course." She asks, "What could you do on your part to help make her more inclined to want to return?" His pain provides the incentive for working with the counselor on how he needs to change even in the short term to get what he wants.

The second helper does not use pain as a club. However guilty he might be, Rob doesn't need his nose rubbed in his pain. The helper uses his pain as a point of leverage. What is Rob willing to do to rid himself of his pain and create the future he says he wants?

**Begin with issues the client sees as important and is willing to work on.** The frame of reference of the client is a point of leverage. Given the client's story, you may think that he or she has not chosen the most important issues for initial consideration. However, helping clients work on issues that are important in their eyes sends an important message: "Your interests are important to me."

PRINCIPLE VIOLATED: A woman comes to a counselor complaining about her relationship with her boss. She believes that he is sexist. Male colleagues not as talented as she get the best assignments, and a couple of them are promoted. After listening to her story, the counselor has her explore her family background for "context." After listening, he believes that she probably has some leftover developmental issues with her father and an older brother that affect her attitude toward older men. He pursues this line of thinking with her. She is confused and feels put down. When she does not return for a second interview, the counselor says to himself that his hypothesis has been confirmed.

PRINCIPLE USED: The woman seeks out a lawyer who deals with equal opportunity cases. The lawyer, older and not only smart but wise, listens carefully to her story and probes for missing details. Then he gives her a snapshot of what such cases involve if they go to litigation. Against that background, he helps her explore what she really wants. Is it more respect? more pay? a promotion? a different kind of boss? a better use of her talents? a job in a company that does not discriminate? Once she names her preferences, he discusses with her options for getting what she wants. Litigation is not one of them.

The first helper substituted his own agenda for hers. He turned out to be somewhat sexist himself. The second helper accepted her agenda and helped her broaden it. He saw no leverage in litigation, but he did suggest that her problem situation could be an opportunity to reset her career. He knew there were plenty of firms eager to employ people of her caliber.

**Begin with some manageable subproblem of a larger problem situation.** Large, complicated problem situations often remain vague and unmanageable. Dividing a problem into manageable bits can provide leverage. Most larger problems can be broken down into smaller, more manageable subproblems.

PRINCIPLE VIOLATED: Aaron and Ruth, in their mid-fifties, have a 25-year-old son living at home who has been diagnosed as schizophrenic. Aaron is a manager in a manufacturing concern that is in crisis. His wife has a history of panic attacks and chronic anxiety. All these problems have placed a great deal of strain on the marriage. They feel guilty about their son's illness. The son has become quite abusive at home and has been stigmatized by people in the neighborhood for his "odd" behavior. Aaron and Ruth have also been stigmatized for "bringing him up wrong." They have been seeing a counselor who specializes in a "systems" approach to such problems. They are confused by his "everything is related to everything else" approach. They are looking for relief but are exposed to more and more complexity. They finally come to the conclusion that they do not have the internal resources to deal with the enormity of the problem situation and drop out.

PRINCIPLE USED: A couple of weeks pass before they screw up their courage to make contact with a psychiatrist, Fiona, whom a family friend has recommended highly. Although Fiona understands the systemic complexity of the problem situation, she also understands their need for some respite. She first sees the son and prescribes some antipsychotic medication. His odd and abusive behavior is greatly reduced. Even though Aaron and his wife are not actively religious, she also arranges a meeting with a rabbi from the community known for his ecumenical activism. The rabbi puts them in touch with an ecumenical group of Jews and Christians who are committed to developing a "city that cares" by starting up neighborhood groups. This is a positive-psychology approach. Involvement with one of these groups helps diminish their sense of stigma. They still have many concerns, but they now have some relief and better access to both internal and community resources to help them with those concerns.

The psychiatrist helps them target two manageable subproblems: the son's behavior and the couple's sense of isolation from the community. Some immediate relief puts them in a much better position to tackle their problems in the long term.

**Begin with a problem that, if handled, will lead to some kind of general improvement in the client's condition.** Some problems, when addressed, yield results beyond what might be expected. This is the spread effect.

PRINCIPLE VIOLATED: Jeff, a single carpenter in his late twenties, comes to a community mental-health center with a variety of complaints, including insomnia, light drug use, feelings of alienation from his family, a variety of psychosomatic complaints, and temptations toward exhibitionism. He also has an intense fear of dogs, something that occasionally affects his work. The counselor sees this last problem as one that can be managed through the application of a manualized behavior modification methodology. He and Jeff spend a fair amount of time in the desensitization of the phobia. Jeff's fear of dogs abates quite a bit and many of his symptoms diminish. But gradually the old symptoms reemerge. His phobia, though significant, is not related closely enough to his primary concerns to involve any kind of significant spread effect.

PRINCIPLE USED: Predictably, Jeff's major problems reemerge. One day he becomes disoriented and bangs a number of cars near his job site with his hammer. He is overheard saying, "I'll get even with you." He is admitted briefly to a general hospital with a psychiatric ward. The immediate crisis is managed quickly and effectively. During his brief stay, he talks with a psychiatric social worker. He feels good about the interaction, and they agree to have a few sessions after he is discharged. In their talks, the focus turns to his isolation. This has a great deal to do with his lack of self-esteem: "Who would want to be my friend?" The social worker believes that helping Jeff to get back into community may well help with other problems. He has managed this problem by staying away from close relationships with both men and women. They discuss ways in which he can begin to socialize. Instead of focusing on the origins of Jeff's feelings of isolation, the social worker, taking an opportunity-development approach, helps him involve himself in mini-experiments in socialization. As Jeff begins to get involved with others, his symptoms begin to abate.

The second helper finds leverage in Jeff's lack of human contact. The mini-experiments in socialization reveal some underlying problems. One of the reasons that others shy away from him is his self-centered and abrasive interpersonal style. When he repeats his question, "Who would want to be my friend?" the helper responds, "I bet a lot of people would . . . if you were a bit more concerned about them and a bit less abrasive. You're not abrasive with me. We get along fine. What's going on?"

**Focus on a problem for which the benefits will outweigh the costs.** This is not an excuse for not tackling difficult problems. If you demand a great deal of work from both yourself and the client, then basic laws of both economics and behavior suggest that there will be some kind of reasonable payoff for both of you.

PRINCIPLE VIOLATED: Margaret discovers to her horror that Hector, her husband, has been found to be HIV-positive. Tests reveal that she and her recently born son have not contracted the disease. This helps cushion the shock. But she has difficulty with her relationship with Hector. He claims that he picked up the virus from a "dirty needle." But she didn't even know that he had ever used drugs. The counselor focuses on the need for the "reconstruction of the marital relationship." He tells Margaret that some of this will be painful because it means looking at areas of their lives that they have never reviewed or discussed. But Margaret is looking for some practical help in reorienting herself to her husband and to family life. After three sessions she decides to stop coming.

PRINCIPLE USED: Margaret still searches for help. The doctor who is treating Hector suggests a self-help group for spouses, children, and partners of HIV-positive patients. In the sessions, Margaret learns a great deal about how to relate to someone who is HIV-positive. The meetings are very practical. The fact that Hector is not in the group helps. She begins to understand herself and her needs better. In the security of the group, she explores mistakes she has made in relating to Hector. Although she does not "reconstruct" either her relationship with Hector or her own personality, she does learn how to live more creatively with both herself and him. She realizes that there will probably be further anguish, but she also sees that she is getting better prepared to face that future.

Reconstructing both relationships and personality, even if possible, is a very costly and chancy proposition. The cost-benefit ratio is out of balance. Once more it may be a question of a helper more committed to his or her theories than the needs of clients. Margaret gets the help she needs from the group and from sessions she and Hector have with a caring psychiatric aide.

Some counselors misuse these principles or otherwise fail to spot opportunities for leverage.

- They begin with the framework of the client, but never get beyond it.
- They deal effectively with the issues the client brings up but fail to challenge clients to consider significant issues they are avoiding.
- They help clients explore and deal with problems but never help clients translate them into opportunities or move on to other unused opportunities.
- They recognize the client's pain as a source of leverage but then overfocus on the pain or allow pain to mask other dimensions of the client's problems.
- They help clients achieve small victories but fail to help clients build on their successes.
- They help clients start with small, manageable problems, but fail to help clients face more demanding problems or to see the larger opportunities embedded in these problems.
- They help clients deal with a problem or opportunity in one area of life (for instance, self-discipline in an exercise program) but fail to help them generalize what they learn to other, more difficult areas of living (for instance, self-control in interpersonal relationships).

The leverage mind-set is part of positive psychology. It is second nature in effective helpers.

## FOCUS AND LEVERAGE: THE LAZARUS TECHNIQUE

Arnold Lazarus (1976, 1981), in a film on his multimodal approach to therapy (Rogers, Shostrom, & Lazarus, 1977), uses a helpful focusing technique. The technique highlights how important language is in counseling and how important it is to ask questions that help clients point themselves in the right direction. For example, a helper asks a client to use just one word to describe her problem. She searches around a bit and then offers the word "Cloudy." Then the helper asks the client to put the word in a phrase. She says, "cloudy thinking." In the next step, the client is to come up with a simple sentence that describes her problem. She says, "My mind is cloudy and I can't think straight." The helper then asks the client to move from a simple sentence to a more extended description of the issue. She says something like this:

> When I say that my mind is cloudy and I can't think straight, I mean that this is the reaction to what my boss told me last week. I have been working very hard. And learning a lot. Over the past two years, I've received two awards for special projects. And all along I've gotten good ratings. But then someone else got the promotion I thought I was going to get. Someone I think my boss preferred personally. But if I was supposedly so good, why couldn't he find a promotion for me? . . . I was shell-shocked. . . . I've been very confused . . . in a fog. . . . I was loyal. . . . It's hit me very hard.

I'm not suggesting that you begin all sessions or any session that way, but it is a simple way to bring a session into focus. This methodology can be used at any stage or step of the helping process. For instance, clients can be asked to use one word to describe what they want.

HELPER:  Now that you have explored some of your concerns and against this background, tell me in just one word or short phrase what you want.

CLIENT:  Let's see . . . my family.

HELPER:  All right, your family. Now put that in a sentence. Describe what you want in a short sentence.

CLIENT:  I want my family and my family life back.

This client had achieved financial success in a business venture rather quickly. Success went to his head. He left his wife and three children and began living with a "flashier" woman. Some six years later, the shine was off the apple. He was successful but unhappy. He wanted his family back. He wanted family life as he once knew it again. How realistic it was and how it could be achieved was another issue. But the Lazarus technique provided focus.

There are other techniques for helping clients in their search for leverage. In self-examination therapy (SET), clients are given a 45-page booklet containing all the information needed to use it (Bowman, 1995; Bowman, Ward, Bowman, & Scogin, 1996). In an experiment using this therapy (Bowman, Scogin, Floyd, Patton, & Gist, 1997)—a therapy designed to help reduce anxiety and depression—clients used worksheets, first to record what mattered to them and then problems they were having. They then compared the two lists. If a problem that bothered them did not relate to what mattered for them, they crossed that particular problem off the list. The economics of working on it would not be good. Next they went

## Box 13-1   Leverage Questions for Step I-C

Help clients ask themselves such questions as:

- What problem or opportunity should I really be working on?
- Which issue, if faced, would make a substantial difference in my life?
- Which problem or opportunity has the greatest payoff value?
- Which issue do I have both the will and the courage to work on?
- Which problem, if managed, will take care of other problems?
- Which opportunity, if developed, will help me deal with critical problems?
- What is the best place for me to start?
- If I need to start slowly, where should I start?
- If I need a boost or a quick win, which problem or opportunity should I work on?

on to look for solutions for the problems that remained and were encouraged to try out the solutions that seemed to have the best chance of working. The point is that the therapeutic approach is built on leverage. Box 13-1 points out leverage questions that you can help clients ask themselves.

## STEP I-C AND ACTION

Like every other step of the helping model, Step I-C should act as a stimulus for client action. Helping clients identify and deal with high-leverage issues should, if done well, help them move toward the little actions that precede the formal plan for constructive change.

> One client, a man in his early thirties, discussed all sorts of problems and unused opportunities with a counselor. He skipped from one topic to another, usually settling on the issue that had caused him the most trouble the previous week. The counselor always listened attentively. At the beginning of the third session she said, "I have a hypothesis I'd like to explore with you." She went on to name what she saw as a thread that wove its way through most of the issues he discussed: a reluctance on his part to commit himself fully to anyone or anything. He was thunderstruck, because he instantly recognized the truth. In the next few weeks, he began to explore his problems and concerns from this perspective. Time after time, he saw himself withdrawing when he should have been committing himself, whether to a project or to a person. Gradually, he began to commit himself in little ways. For instance, when an intimate woman friend of his began talking about where their relationship was going, he said, "There is something in me that wants to change the subject or run away. But no. Let's begin talking about the future and what future we might have together." At least he committed himself to opening a discussion about the relationship if not to the relationship itself.

Discussing issues that make a difference can galvanize some clients into using resources that have lain dormant for years.

## The Shadow Side of Step I-C

For obvious reasons, no one knows how much time is wasted in counseling sessions discussing problems that won't make much of a difference in clients' lives. Some helpers perhaps unknowingly encourage clients to discuss problems that fit their theories and their approaches to therapy rather than the needs of clients. They fit the client to the "technology" rather than joining clients in the search for issues that will make a difference in clients' lives. And, as mentioned in Chapter 1, manualized treatments, while effective in themselves, may be too narrowly focused and leave some clients with critical unmanaged problems in living. But helping is a young profession. And we often look back and smile wistfully at the indiscretions of our youth.

Client behaviors also contribute to the shadow side of Step I-C. Some clients for whatever set of reasons don't bring up their key concerns. Effective helpers can spot clues that suggest that clients are hiding some concerns, skirting important issues, or even lying. Then, with tact, they can help clients approach these issues. But in the end counselors are not mind readers. Clients can avoid talking about their real concerns if they want to or feel compelled to. One way of handling this kind of avoidance behavior is to point out its possibility in the initial helping contract shared with the client. The contract could say something like this: "Sometimes clients don't bring up things that really bother them. Or they are uncertain whether they want to tackle a particular problem. They feel that they are not ready. They may even feel caught and lie about their problems. Since counselors are not mind readers, all this might escape the notice of the counselor. All I can say is that I am willing to help you explore any issue that you believe will make a difference in your life. I am quite aware of human foibles, both my own and those of my clients. I will do my best to deal with sensitive issues in a caring and respectful way." Choose your own words. One client told me, "I'm glad that you mentioned that no-fault part about lying. In the last couple of sessions I haven't told you the entire truth. But now I'm ready." Then we got down to business.

### Evaluation Questions for Step I-C

How well am I doing the following?

- Helping clients focus on issues that have payoff potential for them
- Maintaining a sense of movement and direction in the helping process
- Avoiding unnecessarily extending the problem identification and exploration stage
- Moving to other stages of the helping process as clients' needs dictate
- Encouraging clients to act on what they are learning

# Stage II: Helping Clients Determine What They Need and Want

Stages II and III together with the action arrow get at the heart of helping because they deal with goals, outcomes, and action. Chapter 14 is an introduction to Stage II. It focuses on the nature of solutions, the relationship between solutions and brief therapy, the importance of goal setting, and the peculiar nature of decision making.

Chapter 15, Step II-A, deals with helping clients review possibilities for a better future. Because it is concerned with the future, the chapter also focuses on the place of hope in helping.

Chapter 16, Step II-B, outlines a process for helping clients choose and shape their goals. Chapter 17, Step II-C, reviews the importance of commitment to goals and to the work clients must do to achieve those goals. It also deals with self-efficacy and helping clients find the incentives for moving forward.

# INTRODUCTION TO STAGE II: "WHAT SOLUTIONS MAKE SENSE FOR ME?" HELPING CLIENTS IDENTIFY, CHOOSE, AND SHAPE GOALS

## THE THREE STEPS OF STAGE II

Step II-A: Possibilities
Step II-B: Choices
Step II-C: Commitment

## SOLUTION-FOCUSED HELPING

Solution-Focused Therapies, Brief Therapy, and Appreciative Inquiry
Solutions Versus Solutions
The Beneficial Effects of Brief Therapy

## HELPING CLIENTS DISCOVER AND USE THEIR POWER THROUGH GOAL SETTING

Goals help clients focus their attention and action
Goals help clients mobilize their energy and effort
Goals provide incentives for clients to search for strategies to accomplish them
Clear and specific goals help clients increase persistence

## HELPING CLIENTS BECOME MORE EFFECTIVE DECISION MAKERS

Rational Decision Making
The Shadow Side of Decision Making: Choices in Everyday Life
Making Smarter Decisions

## THE THREE STEPS OF STAGE II

In many ways Stages II and III together with the action arrow are the most important parts of the helping model because they are about "solutions." It is here that counselors help clients develop and implement programs for constructive change. The payoff for identifying and clarifying both problem situations and unused opportunities lies in doing something about them. The skills needed to help clients do precisely that—engage in constructive change—are reviewed and illustrated in Stages II and III. In these stages, counselors help clients answer the following two commonsense but critical questions:

What do you want?

and

What do you have to do to get what you want?

Problems can make clients feel hemmed in and closed off. To a greater or lesser extent, they have no future, or the future they have looks troubled. But, as Gelatt (1989) notes, "The future does not exist and cannot be predicted. It must be imagined and invented" (p. 255). The steps of Stage II outline three ways in which helpers can partner with their clients in exploring and developing this better future.

**Step II-A: Possibilities.** "What possibilities do I have for a better future?" "What are some of the things I think I want?" "What about my needs?" Counselors, in helping clients ask themselves these questions, help them develop a sense of hope.

**Step II-B: Choices.** "What do I really want and need?" Here clients craft a viable change agenda from among the possible choices. Helping them design and shape the right goals is the central task of helping.

**Step II-C: Commitment.** "What am I willing to pay for what I want?" Help clients discover incentives for commitment to their change agenda. It is a further look at the economics of personal change.

Figure 14-1 highlights these three steps of the helping process. Without minimizing in any way what counselors can help their clients accomplish through Stage I skills and interventions—that is, problem and opportunity clarification; the development of new, more constructive perspectives of self, others, and the world; and the choice of high-leverage issues to work on—the real power of helping lies in helping clients set goals and move to accomplish them.

## SOLUTION-FOCUSED HELPING

O'Hanlon and Weiner-Davis (1989) claimed that a trend "away from explanations, problems, and pathology, and toward solutions, competence, and capabilities" (p. 6) was emerging in the helping professions. An earlier study showed that clients were interested in solutions to their problems and feeling better, whereas many helpers were concerned about the origin of problems and transforming them through insight (Llewelyn, 1988). Solution-focused therapies (Berg, 1994; de

**The Skilled-Helper Model**

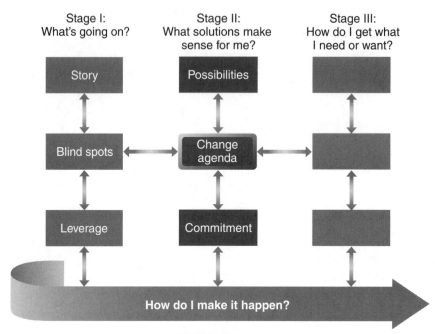

FIGURE 14-1
The Helping Model—Stage II

Shazer, 1985; Fish, 1995, 1997; Manthei, 1998; Metcalf, 1998; Miller, Hubble, & Duncan, 1996; Murphy, 1997; O'Connell, 1998; Walter & Peller, 1992; Zimmerman, Prest, & Wetzel, 1997) tackle this disconnect. Even today too many approaches to helping still focus on Stage I activities. Too many helper training programs still emphasize—or overemphasize—the exploration methods of Stage I. Communication skills are required in every approach to helping, but limiting their use to Stage I endeavors is a waste. Intensive discussion of problem situations is often based on a "working through" mentality, whereas action or "solution" approaches are based on the assumption that many problems need to be dealt with or even "transcended" rather than worked through. At any rate, the goal of helping, as stated in Chapter 1, is "problems managed," not just "problems explored and understood," and "opportunities developed," not just "opportunities identified and discussed."

## Solution-Focused Therapies, Brief Therapy, and Appreciative Inquiry

Solution-focused therapies, brief therapy (Bloom, 1997; Cade & O'Hanlon, 1993; Cooper, 1995; Matthews & Edgette, 1999; Frieman, 1997; Hoyt, 1995; Preston, 1998; Ratner, 1998), and an approach to problem management and opportunity development called "appreciative inquiry" (Cooperrider & Srivasta, 1987; Zemke,

1999) have a common philosophy and approach to helping that is similar to the approach taken in this book. Each in its own way is a problem-management and, especially, opportunity-development approach. For instance, appreciative inquiry (Zemke, 1999) has the following four stages or steps:

- **Discovery:** What gives life? Help the clients identify past and current successes, strengths, and resources.
- **Dreaming**: What might be? Help the client identify possibilities for a better future.
- **Design:** What's the ideal? Help the client design outcomes, mapping out how the client's life will look in the areas of concern.
- **Delivery:** How do I move forward? Help the client find and implement the best way to realize the future he or she has designed. Some now call this stage "destiny," meaning that the client who finds his or her own power is now in charge of his or her own destiny.

The words are different but the process is more or less the same as the problem-management and opportunity-development model.

Here is a quick overview of what these approaches have in common. You have already seen and will see more of the philosophy, spirit, and methods outlined here. The following is a bit staccato-ish. It is meant to give you the flavor of a philosophy and approach that is common to these three positive-psychology approaches to helping.

**Philosophy.** In relating with clients, focus on resources rather than deficits, on success rather than failure, on credit rather than blame, on solutions rather than problems. Use common sense. Don't let theory get in the way of helping clients.

**View of clients.** Clients are people like the rest of us. See them as people with complaints about life, not symptoms. Don't assume that they will arrive ambivalent about change and resistant to therapy. Clients have a reservoir of wisdom, learned and forgotten but still available. Clients have resources and strengths to resolve complaints. Clients will have their own view of life just as everyone else. Respect the reality they construct, even though they might have to move beyond it. In a way, clients are experts in their own lives. Help them feel competent to solve their own problems. When helpers see clients as problems to be solved, they impoverish them and take away their power.

**Dealing with the past.** There is no escape from past trauma. It did happen. However, if you help clients dwell on it, they will become captives of it. If fact, many will arrive as captives. They need to liberate themselves from the past. That said, clients should get an organized or integrated view of past bad experiences, but they should not go into origins and causation. Looking for deep, underlying causes for symptoms is a mistake. Focus on the client's ability to survive a problem situation. Getting at causes does not usually resolve a complaint. Resolving the complaint resolves it. Nonetheless, clients have more confidence and comfort in their journey to the future when they carry forward parts of the past. The best things they can bring forward are past successes, what they have done that works. If clients carry part of the past forward, it should be what is best about the past. As Bushe (1995) sees appreciative inquiry (and by extension, brief therapy and solution-focused therapy), as an attempt to generate a collective image of a new and better future by exploring the best of what is and has been.

**The role of the helper.** Helpers are consultants, catalysts, guides, facilitators, assistants. By adopting the client's world view, at least temporarily, helpers can lessen reluctance and resistance. Sharing highlights helps demonstrate your understanding of the client's world. Your job is to notice and amplify life-giving forces within the client and any sign of constructive change. Become a detective for good things. Develop an "appreciative ear." Listen to the problem, but listen even more to the opportunity buried within the problem. Use questions that inspire and encourage the client to give positive examples. Questions should stimulate dialogue. Remember that questions are not just questions, they are interventions.

**The discovery phase: helping clients explore and exploit competencies, successes, and "normal times."** Help them identify ways of thinking, behaving, and interacting that have worked in the past. And, since clients are not continually manifesting problem behavior, help them explore the times when they are free of such behavior. The "free" times point toward solutions. Help them identify what has been working during these misery-free periods and capitalize on it. How can they amplify what has been working? Have them recall successes from the past—for instance, when they have handled disagreements more creatively. When your marriage was good, what was it like? What did you do when you successfully resisted the urge to drink? Catch clients being competent and resourceful and help them take a good look at themselves at such times. Notice competencies revealed in a client's story and behavior. There are things that work in every client. Recognize and discuss these competencies because they are strengths that clients can build on. What is the client like when performing well?

**The nature of problems and how we talk about things.** Clients, like the rest of us, become what they talk about. If you always encourage them to talk about problems, they run the risk of becoming "problem people." Then helping turns into remedying pathology and deficits. What clients focus on becomes their chronic reality. Help them see their problems as "complaints." We all have complaints. In other words, help them "normalize" their problems—they are the ordinary difficulties of life. For instance, overeating is showing too much enthusiasm for the wonderful texture, taste, and comfort of food. Hyperactivity is energy that at times gets the better of us and interferes with rest, relaxation, and relationships. A perfectionist is a person who loves quality but who goes too far. Help clients see problems as external to themselves, not things that define and control their lives. Problems are intruders that get the best of us at times. Problems are complaints that bother us rather than define us. Ordinarily, you don't need to know a great deal about complaints to resolve them. So be careful about the questions you ask. They should not keep clients mired in problem talk, because problem talk can keep clients immersed in frustration, impotence, and even despair.

**Insight.** Insight is not necessary for change. Therefore, avoid insight generation. Too often, insights are about problems, not about solutions. Therefore, they encourage clients to keep defining themselves in terms of the problem. Rather, help clients generate "outcome" scenarios.

**The dreaming phase: possibilities for a better future.** The principle is this: The future we anticipate is the future we create. The helper and the client should partner in the systematic search for possibilities and potential. The client's imagination needs to be "provoked" to discover new ways of approaching life. Questions should stimulate the client to think as creatively as possible about a better future: What images capture your hopes for your future? What can you do to keep these hopeful images alive? If you no longer needed help, how would that show up in your actions? If you did the right thing for yourself this week and were filmed, what would the film's highlights show? On the assumption that you would like to move forward, what might you do to push the envelope a bit in the coming week? Appreciative inquiry aims at engaging client and helper in dialogue that leads to the development of a "textured vocabulary of hope" (Ludema, Wilmot, & Srivasta, 1997).

**The design phase: helping clients define their goals.** There is no one correct way to live one's life. Help clients actively design solutions that will turn their possibilities into opportunity-developing realities. Solutions do not have to be complex, even if the problem is complex. Rather, help clients to look for simple solutions to complex problems. Often only small changes are necessary. Nor do solutions have to take care of everything. Troubles don't have to be totally solved. Help clients find systemic solutions. A change in one part of the system can produce good results in another part of the system. Look for interventions that break up patterns of self-limiting behavior. Don't hesitate to design solutions that get rid of symptoms. Getting rid of symptoms is not shallow, useless, or dangerous. Solutions should include clients' ability to grapple with future problems on their own. Clients should leave therapy with identified tools to do so.

**The delivery phase: implementation is everything.** The pace of change will be different for each client. Some need help to ease themselves into solutions gradually. The smallest action is a step forward. On the other hand, rapid change is possible. Don't shortchange clients. Solutions often require that clients develop new ways of relating to their social environment. Where will support come from? Who needs to be engaged to make things work?

Recall that the helping model outlined in this book can be used as a browser to identify and integrate frameworks, methods, and skills in other approaches to helping. Simple Internet searches in solution-focused therapies, brief therapy, and appreciative inquiry will reveal a world of insights, frameworks, methods, and skills that you can integrate into the overall problem-management and opportunity-development framework.

**Criticisms.** There have been some criticisms of this positive-psychology approach to helping (see Zemke, 1999, and the excellent Comment section of the *American Psychologist*, January, 2001, pp. 75–90). For instance, some say that it runs the risk of being a "don't worry, just be happy" approach. That it's too pie-in-the-sky. Others say, let's get real. Change comes from dealing with problems. People are used to dealing with problems, so this approach might be too new for some practitioners and some clients.

On the other hand, traditional problem-solving approaches to helping also come in for their knocks (Zemke, 1999). Critics say that traditional approaches are painfully slow, asks the client to look back at yesterday's failures, looks for the causes of problems, rarely results in a new vision, assumes that either the helper or the client knows what should be in place and therefore leads to talk about closing gaps, places blame and therefore promotes defensiveness, and uses deficit-focused language.

However, wedding the positive-psychology approach of solution-focused therapies to a problem-management and opportunity-development approach to problem solving faces down all these criticisms. There is no question of either/or. The interplay of the two philosophies provides the most robust system. The problem-management and opportunity-development approach provides the backbone of helping. But the use of its stages and steps are dictated by client need. The solution-focused philosophy gives direction to *how* clients and helpers partner in using these stages and steps. Flexibility is the key. The arbiter is common sense and social intelligence. Do what is best for the client.

## Solutions Versus Solutions

This book offers a solution-focused approach to helping. There is, however, a semantic problem with the word *solution*. It means two distinct things. First and foremost, it means an end state—results, accomplishments. Take Pinta. Her eating was out of control. She was killing herself with food. But now she is eating moderately and has lost a lot of weight. A new approach to eating is in place. This is a solution in the end-state sense. *Solution* also means a strategy a client uses to get to the end state. For instance, Pinta joined a 12-step program for overeaters and faithfully attends the meetings. "My eating is out of control," she said at the first meeting. In the group, she learned a variety of ways to get back in control of her eating. Joining the group and using the strategies she learned in the group were solutions in the second, instrumental sense. They are activities or strategies that helped her get to the end state she now enjoys. Stage II of the helping process deals with solutions in the primary sense—end states, accomplishments, goals, outcomes. Stage III focuses on solutions in the secondary sense—means, actions, strategies.

The distinction is not inconsequential. Many approaches to problem solving confuse the two. Or they pay little attention to solutions as end states—what clients really want and need—and talk mostly about solutions in terms of the strategies clients must use to "solve" problems. They leap from problem or unused opportunity to action without linking action to outcome. The correct logic is this: Link solutions-as-goals to problem situations or unused opportunities. Pinta's problem is overeating. Her solution-as-goal is "a fitness program consisting of healthy exercise and nutritional habits consistently in place." Her solution-as-means is the 12-step group together with the self-regulation strategies it sponsors. Goals, not problems, should drive action.

## The Beneficial Effects of Brief Therapy

Brief therapies are of their very essence solution focused. If there is little time, most of it had better be spent focusing on a better future. In fact, many books and articles on brief therapy have "solution-focused" in their titles and vice versa. Research has demonstrated that brief interventions can produce substantive changes that last. And solution-focused therapists like to work with fairly well-defined goals that are realizable within a reasonable amount of time. Brief therapy can be brief but still comprehensive (Lazarus, 1997). Asay and Lambert (1999), after reviewing the research on brief therapy, drew the following conclusions:

> The beneficial effects of therapy can be achieved in short periods (5 to 10 sessions) with at least 50% of clients seen in routine clinical practice. For most clients, therapy will be brief. . . . In consequence, therapists need to organize their work to optimize outcomes within a few sessions. Therapists also need to develop and practice interventions methods that assume clients will be in therapy for fewer than 10 sessions. (p. 42)

They also found that there are three categories of clients who do poorly in brief therapy. First, poorly motivated and hostile clients: Therapists who have the skills of handling resistance and know something about "motivational interviewing" have a better chance of success with such clients. Second, clients who come with a history of poor relationships: The helper's ability to establish a collaborative working alliance that is a "just society" is very important with this type of client. Third, clients who expect to be passive recipients of a medical procedure: Helping such a client quickly find a sense of self-responsibility and agency, however deeply buried, is the helper's challenge. Strategies for doing all of this are found in the pages of this book. Of course, not all helping should be brief. Some 20% to 30% of clients require treatment lasting more than 25 sessions.

## HELPING CLIENTS DISCOVER AND USE THEIR POWER THROUGH GOAL SETTING

Goal setting, whether it is called that or not, is part of everyday life. We all do it all the time.

> Why do we formulate goals? Well, if we didn't have goals, we wouldn't do anything. No one cooks a meal, reads a book, or writes a letter without having a reason, or several reasons, for doing so. We want to get something we want through our actions or we want to prevent or avoid something we don't want. *These desires are beacons for our actions;* they tell us which way to go. When formalized into goals, they play an important role in problem solving. [emphasis added] (Dorner, 1996, p. 49)

Even not setting goals is a form of goal setting. If we don't name our goals, that does not mean that we don't have any. Instead of overt goals, we have a set of covert goals. These are our default goals. They may be enhancing or limiting. We don't like the sagging muscles and flab we see in the mirror. But not deciding to get into better shape is a decision to continue to allow fitness to drift. Since life is filled with goals—chosen goals or goals by default—it makes sense to make them work for us rather than against us.

At their best, goals mobilize our resources; they get us moving. They are a critical part of the self-regulation system. If they are the right goals for us, they get us headed in the right direction. According to Locke and Latham (1984), helping clients set goals empowers them in the following four ways.

**Goals help clients focus their attention and action.**  A counselor at a refugee center in London described Simon, a victim of torture in a Middle Eastern country, to her supervisor as aimless and minimally cooperative in exploring the meaning of his brutal experience. Her supervisor suggested that she help Simon explore possibilities for a better future. The counselor started one session by asking, "Simon, if you could have one thing you don't have, what would it be?" Simon came back immediately, "A friend." During the rest of the session, he was totally focused. What was uppermost in his mind was not the torture but the fact that he was so lonely in a foreign country. When he did talk about the torture, it was to express his fear that torture had "disfigured" him, if not physically, then psychologically, thus making him unattractive to others.

**Goals help clients mobilize their energy and effort.**  Clients who seem lethargic during the problem-exploration phase often come to life when asked to discuss possibilities for a better future. A patient in a long-term rehabilitation program who had been listless and uncooperative said to her counselor after a visit from her minister, "I've decided that God and God's creation and not pain will be the center of my life. This is what I want." That was the beginning of a new commitment to the arduous program. She collaborated more fully in exercises that helped her manage her pain. Clients with goals are less likely to engage in aimless behavior. Goal setting is not just a "head" exercise. Many clients begin engaging in constructive change after setting even broad or rudimentary goals.

**Goals provide incentives for clients to search for strategies to accomplish them.**  Setting goals, a Stage II task, leads naturally into a search for means to accomplish them, a Stage III task. Lonnie, a woman in her seventies who had been described by her friends as "going downhill fast," decided, after a heart-problem scare that proved to be a false alarm, that she wanted to live as fully as possible until she died. She searched out ingenious ways of redeveloping her social life, including remodeling her house and taking in two young women from a local college as boarders.

**Clear and specific goals help clients increase persistence.**  Not only are clients with clear and specific goals energized to do something, but they also tend to work harder and longer. An AIDS patient who said that he wanted to be reintegrated into his extended family managed, against all odds, to recover from five hospitalizations to achieve what he wanted. He did everything he could to buy the time he needed. Clients with clear and realistic goals don't give up as easily as clients with vague goals or with no goals at all.

One study (Payne, Robbins, & Dougherty, 1991) showed that high-goal-directed retirees were more outgoing, involved, resourceful, and persistent in their social settings than low-goal-directed retirees. The latter were more self-critical, dissatisfied, sulky, and self-centered. People with a sense of direction don't waste time in wishful thinking. Rather, they translate wishes into specific outcomes toward which they can work. Picture a continuum. At one end is the aimless person; at the other, a person with a keen sense of direction. Your clients may come from any point on the continuum. Taz knows that he wants to become a better supervisor but needs help in developing a program to do just that. On the other hand, Lolita, one of Taz's colleagues, doesn't even know whether this is the right job for her and does little to explore other possibilities. One client can be at different points on the continuum with respect to different issues—mature in seizing opportunities for education, for instance, but aimless in developing sexual maturity. Most of us have had directionless periods at one time or another in life.

Setting goals, whether formally or informally, provides clients with a sense of direction. People who have a sense of direction

- have a sense of purpose;
- live lives that are going somewhere;
- have self-enhancing patterns of behavior in place;
- focus on results, outcomes, and accomplishments;
- don't mistake aimless action for accomplishments;
- have a defined rather than an aimless lifestyle.

Locke and Latham (1990) pulled together years of research on the motivational value of setting goals. Although the motivational value of goal setting is incontrovertible, the number of people who disregard problem-managing and opportunity-developing goal setting and its advantages are legion. The challenge for counselors is to help clients do it well.

There is a massive amount of sophisticated theory and research on goals and goal setting (Karoly, 1999; Locke & Latham, 1990, 1994) As you can well understand, not all theory and research is easily translated into practical advice for helpers. There is also an extensive self-help literature dealing with goal setting and implementation in everyday life (for instance, D. Ellis, 1999; K. Ellis, 1998; Secunda, 1999). There is a great deal of practical wisdom to be mined from the self-help literature, and helpers-to-be would be doing themselves a disservice if they were to turn their noses up at it. Once more, balance is the answer. The best in theory and research should be used to spot the best practical advice for both helpers and clients in the popular literature.

## HELPING CLIENTS BECOME MORE EFFECTIVE DECISION MAKERS

The second overall goal for helping—outlined in Chapter 1—is to help clients, either directly or indirectly, become better problem solvers and opportunity developers in their everyday lives. Since the encouragement of client self-responsibility is a key helping value, helping clients not only make good decisions but also become better decision makers is not an amenity but a necessity. Consider this case:

> Alice's third marriage has just ended. A counselor is helping her explore the decisions she has made in developing and dissolving intimate relationships. Alice discovers that she makes poor decisions about people in general—for instance, by being too trusting too soon. Although she is horrified by all the mistakes she has made, she realizes that without these sessions, she might well make the same mistakes all over again. She must become a more aware and savvy decision maker.

Though we make decisions of greater or less magnitude every day of our lives, society has not made education in decision making a priority.

## Rational Decision Making

Decision making pervades problem management and opportunity development. One of the reasons clients get into trouble in the first place is that they make poor decisions. We need only review our own experiences to see how often the decisions we make or our failure to make decisions gets us into trouble. There are many decision points in the helping process. We have already seen a number of them. Clients must decide to come to a counseling interview in the first place; to talk about themselves; to return for a second session; to respond to the helper's empathic highlights, probes, and challenges; and to choose issues to work on. We are about to see that clients must also decide what they want, to set goals, to develop strategies, to make plans, and to implement those plans. Deciding—or letting the world decide for you—is at the heart of helping, as it is at the heart of living.

Decision making in its broadest sense is the same as problem solving. Indeed, this book could be called a decision-making approach to helping. In this chapter, however, the focus is on decision making in a narrower sense—the internal (mental) action of identifying alternatives or options and choosing from them. It is a commitment to do or to refrain from doing something:

- "I have decided to discuss my career problems but not my sexual concerns."
- "I have decided to start a new business."
- "I have decided to ask the courts to remove artificial life support from my comatose wife."
- "I have decided to get a better balance between work and home life."
- "I have decided not to undergo chemotherapy."
- "I have decided to stop putting myself down."
- "I have decided to move into a retirement home."

The commitment can be to an internal action—"I have decided to get rid of my preoccupation with my ex-wife"—or to an external action—"I have decided to confront my son about his drinking." Decision making, in the fullest sense, includes the implementation of the decision: "I made a resolution to give up smoking, and I haven't smoked for three years." "I decided that I was being too hard on myself, so I took a week off work and just enjoyed myself."

Traditionally, decision making has been presented as a rational, linear process involving information gathering, analysis, and making a choice. Here are the bare essentials of the decision-making process.

**Information gathering.**  The first rational task is to gather information related to the particular issue or concern. A patient who must decide whether to have a series of chemotherapy treatments needs some essential information: What are the treatments like? What will they accomplish? What are the side effects? What are the consequences of not having them? What would another doctor say? And so forth. There is a whole range of ways in which she might gather this information: from the Internet, reading books, talking to doctors, talking to patients who have undergone treatment or who have refused treatment. Patients today routinely mount extensive Internet searches on their medical conditions to make better informed decisions.

**Analysis.**  The next rational step is processing the information. This includes analyzing, thinking about, working with, discussing, meditating on, and immersing oneself in the information. Just as there are many ways of gathering information, so there are many ways of processing it. Effective information processing leads to a clarification and an understanding of the range of possible choices. "Now, let's see, what are the advantages and disadvantages of each of these choices?" is one way of analyzing information. Effective analysis assumes that decision makers have criteria, whether objective or subjective, for comparing alternatives. For instance, a patient wants to determine whether the weeks or months of life she will gain through a series of chemotherapy treatments will be worth the effort and discomfort.

**Making a choice.**  Finally, decision makers need to make a choice—that is, commit themselves to some internal or external action that is based on the analysis: "After thinking about it, I have decided to sue for custody of the children." As indicated earlier, the fullness of the choice includes an action: "I had my lawyer file the custody papers this morning." There are also rational "rules" that can be used to make a decision. For instance, one rule, stated as a question, deals with the consequences of the decision: "Will it get me everything I want or just part?" Values also enter the picture because, from one point of view, values are criteria for making decisions. "Should I do X or Y? Well, what are my values?" The woman suing for the custody of the children says to herself, "I value fairness. I'm not going to try to extort a lot of money for child care. I'll make reasonable demands."

Counselors help clients engage in rational decision making; that is, they help clients gather information, analyze what they find, and then base action decisions on the analysis. Although this indeed does happen, it is not the full story.

## The Shadow Side of Decision Making: Choices in Everyday Life

Thinking and reasoning are not always what they are supposed to be or seem to be in everyday life. And, when people get into trouble, thinking and reasoning can go even "further south." This means that decision making in everyday life, and in counseling, is not the straightforward, rational process just outlined. Rather, it is an ambiguous, highly complicated process with a deep shadow side (Cosier &

Schwenk, 1990; Etzioni, 1989; Gilovich, 1991; Heppner, 1989; Kaye, 1992; March, 1994; Schoemaker & Russo, 1990; Stroh & Miller, 1993; Whyte, 1991). For instance, Gati, Krausz, and Osipow (1996) discuss the messiness associated with making career decisions and list ten ways in which such decisions can be flawed. There is no such thing as the perfect career decision.

Headlee and Kalogjera (1988) found evidence that some of the roots of the shadow side of decision making begin in childhood. Some children are allowed too many choices, while others are given too few. Moreover, in the early years, distortions of choice evolve because of racial, ethnic, sexual, religious, and other prejudices. By the time the child becomes an adult, these distortions are ingrained in the decision-making process and nobody thinks about them. The sources of possible distortion are myriad. In everyday life, decision making is often confused, covert, difficult to describe, unsystematic, and, at times, quite irrational. A shadow-side analysis of decision making as it is actually practiced reveals a less-than-rational application of the three dimensions outlined in the previous section.

**Information gathering.** Information gathering should lead to a clear definition of the issues to be decided. A client trying to decide whether to pursue a divorce needs information about that entire process. However, information gathering is practically never straightforward. Decision makers, for whatever reason, are often complacent and engage in perfunctory searches. They get too much or too little information; the information they gather is inaccurate or misleading; or they cloud their search for information with emotion. In counseling, the client trying to decide whether to proceed with therapy may have already made up his or her mind and therefore may not be open to confirming or disconfirming information. Since full, unambiguous information is never available, all decisions are at risk. In fact, there is no such thing as completely objective information. All information, received by the decision maker takes on a subjective cast. A patient with prostate cancer who goes to the Internet to get some idea of what to do faces a bewildering range of opinions and options. In view of all this, Ackoff (1974) calls human problem solving "mess management" (p. 21).

> Eloise wanted to make a decision about whether to marry her partner or not. One obstacle was their conflicting careers. She didn't know whether she'd be in the same career five years from now; neither did he. Another obstacle lay in the fact that she knew little about his past. She thought it didn't matter. She liked him now. He knew that she was a nonpracticing Catholic but knew little about how her Catholicism affected her or how it would affect them in the future, especially if they had children. Since religion was not currently an issue, he did not explore it. There were many other things they did not know about themselves and each other. They eventually did marry, but the marriage lasted less than a year.

Granted, clients' stories are never complete, and information will always be partial and open to distortion. Though counselors cannot help clients make information gathering perfect, they can help them make it at least "good enough" for problem management and opportunity development.

**Processing the information.** Since it is impossible to separate the decision from the decision maker, the processing of information is as complex as the person making the decision. Factors affecting the analyzing of information include clients' feelings and emotions, their values-in-use, which often differ from their espoused values, their assumptions about "the way things work," and their level of motivation. There is no such thing as full, objective processing of gathered information. Poorly gathered information is often subjected to further mistreatment. Clients, because of their biases, focus on bits and pieces of the information they have gathered rather than seeing the full picture. Furthermore, few clients have the time or the patience to spell out all possible choices related to the issue at hand, together with the pros and cons of each. Therefore, some say that most decisions are based not on evidence but on taste: "I like it. It sounds good."

> Jamie was in a high-risk category for AIDS because of occasional drug use and sexual promiscuity. Once, when he was busted for drug use, he had to attend a couple of sessions on AIDS awareness. He listened to all the information, but he processed it poorly. These were problems for "other people." He engaged in risky sexual behavior "only occasionally." He was sure that his sexual partners were "clean." One or two "mistakes" were not going to do him in. He knew others who engaged in much riskier behavior than he and "nothing had happened to them." He'd be "more careful," though it was not clear what that meant as far as his behavior was concerned. He was in good health and "healthy people can take a lot."

Jamie distorted information and rationalized away most of the risk of his current lifestyle. He was living not on the edge but on a precipice.

Up to a point, counselors can help clients overcome inertia and biases and tackle the work of analysis. For instance, a client who says his values have "matured" but still automatically makes decisions based on his former values can be challenged to get these more mature values into his decision making. One client, trying to make a decision about a career change, kept moving toward options in the helping professions even though she had become quite interested in business. There was something in her that kept saying, "You have to choose a helping profession. Otherwise you will be a traitor." The counselor helped her see her bias. In the end, the client became a consultant, then a manager, then a senior manager. But she still had to salve her conscience by noting both to herself and to others that "running a successful business is an important contribution to society."

**Choice and execution.**   A host of strange things can happen on the way to executing a decision. Here are some of the things decision makers do:

- Skip the analysis stage and move quickly to choice. "Let's get married. Love will conquer all."
- Ignore the analysis and base the decision on something else entirely. The analysis was nothing but a sham, because the decision criteria, however covert, were already in place. Reiner goes through an extended analysis of the reasons for becoming an entrepreneur and starting his own business, but then he accepts an offer from a large firm. He ignored the fact that security was his main driver.
- Engage in what Janis and Mann (1977) called "defensive avoidance." That is, they procrastinate, attempt to shift responsibility, or rationalize delaying a choice. An elderly man says, "I know that it makes sense to sell this big house and move into a retirement village, but what do the kids want? And what if we don't like it? We might run into people we don't like. We'd better take a closer look at this."

- Confuse confidence in decision making with competence. "I know what I want. If it's prostate cancer, I'm going for surgery and get it over with."
- Panic and seize on a hastily contrived solution that gives promise of immediate relief. The choice may work in the short term but have negative long-term consequences. Tess and Lars panic and get married quickly because she is pregnant. The next two years are very rocky.
- Are swayed by a course of action that is most salient at the time or by one that comes highly recommended, even though it is not right for them. Imogene, single, gives up her child for adoption. Later she bitterly regrets her decision.
- Let enthusiasm and other emotions govern their choices. Ben is so elated to be offered a promotion that he says yes right away. Only later does he realize that he was not cut out to be a manager.
- Announce a choice to themselves or to others but then do nothing about it. Bert and Linda tell their teenage children that they want to involve them more in household decision making but plan a summer vacation without their input. The kids become even more resentful.
- Translate the decision into action only halfheartedly. Sandra, grieving over the loss of her husband, decides to renew her social life. But she often fails to return phone calls, cancels engagements, and leaves get-togethers shortly after arriving, offering what seem to others as rather lame excuses.
- Decide one thing but do another. Ted decides to turn down a job offer because it's "not for me" but ends up taking it anyway.

The fact that choices do not necessarily make life easier for oneself and others explains a great deal of the shadow side of decision making. It is clear that counselors cannot help clients avoid all the pitfalls involved in making decisions, but they can help clients minimize them.

In summary, rational, linear decision making, in its pure form, has probably never been the norm in human affairs. Decision making goes on at more than one level. There is, as it were, the rational decision-making process in the foreground and an emotional or impulsive decision-making process in the background. Gelatt (1989) called for an approach to decision making that factors in these shadow-side realities: "What is appropriate now is a decision and counseling framework that helps clients deal with change and ambiguity, accept uncertainty and inconsistency, and utilize the nonrational and intuitive side of thinking and choosing" (p. 252). Positive uncertainty means, paradoxically, being positive (comfortable and confident) in the face of uncertainty (ambiguity and doubt)—feeling both uncertain about the future and positive about the uncertainty. Stages II and III, together with an understanding of the shadow side of these two stages, provide methodologies clients can use to make decisions, explore their consequences, and act on them.

## Making Smarter Decisions

Learning how to make "good" decisions is left to chance in society. As with other skills—interpersonal communication, problem solving, parenting, and managing—

everyone thinks that decision-making skills are very important. But when asked in what forum these critical skills are to be learned, the usual reaction is a shrug of the shoulders. Once more, life itself is to be the teacher. James March (1994) discusses how decisions are actually made. In the last chapter on "decision engineering," March moves toward making suggestions on how to make "quality" or "intelligent" decisions. Hammond, Keeney, and Raiffa (1998, 1999), offer a guide to making better decisions, taking into account, of course, the shadow side of decision making. In one article (1998), they focus on hidden traps in decision making and how to handle them. Here are some of them.

**The status quo bias.** Clients often have a bias toward alternatives that perpetuate the status quo. The status quo is seen as the "safe" option even when it is not. Jeff, who has prostate cancer, is trying to make a decision about treatment. One option is "do nothing," because prostate cancer is usually slow growing, and older men who take a watch-and-wait approach often die of something else. Jeff, however, ignores the fact that he is not a good candidate for this approach. How could you help Jeff deal with his bias?

- Help him determine what his real objective is: comfort? living longer? living disease free? What do you want, Jeff?
- Help him review the alternatives in the light of what he wants. He might want a combination of things.
- Help him see whether he is choosing the status quo approach precisely because it is the status quo.
- Help him determine whether he wants to avoid the risk, pain, or trouble of choosing a non–status quo alternative. This is a different problem.
- Help him look further into the future. In Jeff's case, that might not be very far. Depending on the nature of his cancer and the likelihood of its spreading, the status quo situation might not last that long. "It's two years from now. You're looking back at this time in your life. Which decision would you have rather made?"
- Help him determine whether he is defaulting to the status quo option because it is difficult to choose from among the other alternatives. If this is the case, help him cope with the agony of making such a choice.

Amidst all this, aim your challenges at the "self-healing person" inside Jeff. Your job is not to choose for him, nor is it to talk him into anything. You can do any or all the above without robbing Jeff of any self-responsibility.

**The confirming-evidence trap.** If I have secretly—hidden more or less even from myself—decided to do or avoid doing something, I can begin looking for evidence that will confirm my choice or avoid evidence that will challenge it. Sheila, a college junior, was seeing a counselor because she was both shy and perfectionistic. She grew tired of counseling quickly because the counselor seemed to be trying to determine whether she was shy because she was perfectionistic—staying away from people gave her time to "get things done right"—or whether she was perfectionistic because she was shy—her "high standards" meant that she had very few friends. Or maybe it was something else.

A second counselor, after hearing her recapitulate her experience with the first counselor, said during their second session, "Which do you want: to be perfect or to be alone?" She remained totally self-possessed, paused, and then said, "I want to leave school. My mother is dying. She needs me. She might have six months. She might have two years. No matter. My place is at her side." It was something else. When the counselor asked, "Is this what your mother wants," Sheila replied, "That's not the issue." In a later session, the counselor, after finding out that Sheila's decision was not seconded by her mother, her father, or any of her three younger brothers, tried to come at it from various angles. Sheila was very bright. She amassed evidence supporting her decision from every source: psychology, sociology, the Bible, theology, and her commitment "to my family and myself." This sounds as if she was being battered on all sides, but this was not the case. Her family and the counselor knew that the decision was hers. No one badgered her. But everyone wanted her to make the decision for the "right" reasons. According to Hammond, Keeney, and Raiffa, what could the counselor do? Here are a few suggestions:

• Help Sheila examine all the evidence with equal vigor. She was being very intellectual about it all. She did a great job with the evidence from the human and godly sciences but attended very little to what the significant people in her life were saying.

• Get someone Sheila respects to act as devil's advocate. Sheila had friends, and there were family friends, her doctor, her minister, and so forth. Or better yet, get Sheila herself to play her own devil's advocate. Reverse roles. The helper becomes Sheila, and Sheila becomes the helper.

• Have Sheila take a closer look at her motives. What does she really want? Is there something behind her leaving school besides her mother's illness? Is it her way of being both "alone" in some sense and "perfect"? If her mother were not sick, would she still be leaving school? These questions are not meant to be probes into her "deeper" internal dynamics. Rather, it would be helpful for Sheila to know what she is doing and what is moving her to do it.

• If Sheila seeks advice from others, help her frame her questions so they don't merely invite confirmation of what she has already decided. Sheila could say, "Here's what I'm thinking of doing. Grill me on it, will you?"

It might well be that Sheila wants to leave school because she wants to be at her mother's side during these difficult times. It may have nothing to do with either isolation or perfectionism. And, even if others think that it's a lousy decision, Sheila thinks it's right for herself. After all, it is her life. There are, of course, other decision traps.

In their book, *Smart Choices: A Practical Guide to Making Better Decisions* (1999), Hammond, Keeney, and Raiffa lay out a *system* for making smart choices. The eight elements of the system highlight the eight most common and most serious errors in decision making (p. 189):

- Working on the wrong problem
- Failing to identify your key objectives
- Failing to develop a range of good, creative alternatives
- Overlooking crucial consequences of your alternatives
- Giving inadequate thought to tradeoffs
- Disregarding uncertainty
- Failing to account for your risk tolerance
- Failing to plan ahead when decisions are linked over time

As you can see, when put positively ("Working on the right problem" and so forth), these are also elements of the problem-management process, most of which are addressed in Stages II and III.

Finally, since helpers themselves are human, they do not escape the shadow side of decision making. Helpers, as you can see, make decisions throughout the helping process. Pfeiffer, Whelan, and Martin (2000), after reviewing the decision-making research, comment:

> When examined as a whole, this research suggests that people tend to preferentially attend to information, gather information, and interpret information in a manner that supports, rather than tests, their decisions about another person. Therapists may not be exempt from this tendency, particularly given the often complex and ambiguous nature of clients' problems (p. 429).

No matter how empathic you are, as a helper you will still make hypotheses about your clients throughout the helping process and base some of your decisions on these hypotheses. Your challenge is to continually test these hypotheses against the reality of your clients in the context of their lives. Theories are theories. Clients are clients.

# STEP II-A: "WHAT DO I NEED AND WANT?" POSSIBILITIES FOR A BETTER FUTURE

POSSIBILITIES FOR A BETTER FUTURE

    The Psychology of Hope

        The nature of hope

        The benefits of hope

    Possible Selves

SKILLS FOR IDENTIFYING POSSIBILITIES FOR A BETTER FUTURE

    Creativity and Helping

    Divergent Thinking

    Brainstorming: A Tool for Divergent Thinking

        Suspend your own judgment, and help clients suspend theirs

        Encourage clients to come up with as many possibilities as possible

        Help clients use one idea to stimulate others

        Help clients let themselves go and develop some "wild" possibilities

    Future-Oriented Probes

    Exemplars and Models as Sources of Possibilities

CASES FEATURING POSSIBILITIES FOR A BETTER FUTURE

    The Case of Brendan: Dying Well

    The Washington Family Case

EVALUATION QUESTIONS FOR STEP II-A

# POSSIBILITIES FOR A BETTER FUTURE

The goal of Step II-A is to help clients develop a sense of direction by exploring possibilities for a better future. I once was sitting at the counter of a late-night diner when a young man sat down next to me. The conversation drifted to the problems he was having with a friend of his. I listened for a while and then asked, "Well, if your relationship was just what you wanted it to be, what would it look like?" It took him a bit to get started, but eventually he drew a picture of the kind of relationship he could live with. Then he stopped, looked at me, and said, "You must be a professional." I believe he thought that because this was the first time in his life that anyone had ever asked him to describe some possibilities for a better future.

Too often the exploration and clarification of problem situations are followed, almost immediately, by the search for solutions in the secondary sense—actions that will help deal with the problem or develop the opportunity. But in many ways, outcomes are more important than actions. *What will be in place* once those actions are completed? As we saw in the last chapter, failure to specify outcomes is one of the major decision-making traps. The outcome is a Solution with a big S, while the actions leading to this outcome constitute a solution with a small s. There is great power in visualizing outcomes, just as there is a danger in formulating action strategies before getting a clear idea of desired outcomes. Stage II is about identifying or visualizing desired results, outcomes, or accomplishments. Step II-A is about envisioning possibilities. Stage III is about strategies, actions, and plans for delivering those outcomes. From another point of view, Stages II and III are about hope.

## The Psychology of Hope

Hope as part of human experience is as old as humanity. Who of us has not started sentences with "I hope . . ."? Who of us has not experienced hope or lost hope? Hope also has a long history as a religious concept. St. Paul said, "Hope that centers around things you can see is not really hope," thus highlighting the element of uncertainty. If you know that tomorrow you will receive the Oscar, you can no longer hope for it. You know it's a sure thing. Hope plays a key role in both developing and implementing possibilities for a better future. An Internet search reveals that scientific psychology is more interested in hope than one might initially believe (Erickson, Post & Page, 1975; Stotland, 1969). As mentioned earlier, Rick Snyder has written extensively about the positive and negative uses of excuses in everyday life (Snyder, 1988; Snyder, Higgins, & Stucky, 1983) and has become a kind of champion for hope (McDermott & Snyder, 1999; Snyder, McDermott, Cook, & Rapoff, 1997; Snyder, 1994a, 1994b, 1995, 1998; Snyder, Michael, & Cheavens, 1999). Indeed, he linked excuses and hope in an article entitled "Reality negotiation: From excuses to hope and beyond" (1988). He has also developed scales for measuring both dispositional hope (Snyder et al., 1991) and state hope (Snyder et al., 1996). For a full bibliography for Snyder's work on hope, go to http://www.psych.ukans. edu/faculty/rsnyder/hoperesearch.htm.

**The nature of hope.**  Snyder starts with the premise that human beings are goal directed. Hope, according to Snyder, is the process of thinking about one's goals—

*Serena is determined that she will give up smoking, drinking, and soft drugs now that she is pregnant* —of having the will, desire, or motivation to move toward these goals— *Serena is serious about her goal because she has seen the damaged children of mothers on drugs and she is also, at heart, a decent, caring person*—and of thinking about the strategies for accomplishing one's goals—*Serena knows that two or three of her friends will give her the support she needs, and she is willing to join an arduous 12-step program to achieve her goal.* Serena is hopeful. If we say that Serena has "high hopes," we mean that her goal is clear, her sense of agency (or urgency) is high, and that she is realistic in planning the pathways to her goal. Both a sense of agency and some clarity around pathways are required. Hope, of course, has emotional connotations. It is not a free-floating emotion. Rather it is the byproduct or outcome of the work of setting goals, developing a sense of agency, and devising pathways to the goal. Serena feels a mixture of positive emotions—elation, determination, satisfaction— knowing that "the will" (agency) and "the way" (pathways) have come together. Success is in sight even though she knows that there will be barriers—for instance, the ongoing lure of tobacco, wine, and soft drugs.

**The benefits of hope.** Snyder (1995) has combed the research literature to dis-cover the benefits of hope as he defines it. Here is what he has found:

> The advantages of elevated hope are many. Higher as compared with lower hope people have a greater number of goals, have more difficult goals, have success at achieving their goals, perceive their goals as challenges, have greater happiness and less distress, have superior coping skills, recover better from physical injury, and report less burnout at work, to name but a few advantages (pp. 357–358).

Counselors who do not spend a significant part of their time with clients helping them develop possibilities, clarify goals, devise strategies or pathways, and develop the sense of agency needed to bring all this to fruition are certainly shortchanging their clients. Because Stages II and III deal with possibilities, goals, commitment, pathways, and overcoming barriers, they could be named "ways of nurturing hope."

## Possible Selves

One of the characters in Gail Godwin's 1985 novel *The Finishing School* warns against getting involved with people who have "congealed into their final selves." Clients come to helpers not necessarily because they have congealed into their fi-nal selves—if this is the case, why come at all?—but because they are stuck in their current selves. Counseling is a process of helping clients get "unstuck" and develop a sense of direction. Consider the case of Ernesto. He was very young but very stuck for a variety of sociocultural and emotional reasons.

A counselor first met Ernesto in the emergency room of a large urban hospital. He was throw-ing up blood into a pan. He was a member of a street gang, and this was the third time he had been beaten up in the last year. He had been so severely beaten this time that it was likely that he would suffer permanent physical damage. Ernesto's style of life was doing him in, but it was the only one he knew. He was in need of a new way of living, a new scenario, a new way of par-ticipating in city life. This time he was hurting enough to consider the possibility of some kind of change.

Markus and Nurius (1986) use the term *possible selves* to represent "individuals' ideas of what they might become, what they would like to become, and what they are afraid of becoming" (p. 954). The counselor worked with Ernesto not by helping him explore the complex sociocultural and emotional reasons he was in this fix but principally by helping him explore his "possible selves" to discover a different purpose in life, a different direction, a different lifestyle. Step II-A is about possible selves. The notion of possible selves has captured the imagination of many helpers and of those interested in human development such as teachers (Cameron, 1999; Cross & Marcus, 1994; Hooker, Fiese, Jenkins, Morfei, & Schwagler, 1996; Strauss & Goldberg, 1999). Enter the term *possible selves* into an Internet search engine and you will find all sorts of examples of how helpers and teachers have been using this concept. In Step II-A, your job is to help clients discover their possible selves.

## SKILLS FOR IDENTIFYING POSSIBILITIES FOR A BETTER FUTURE

At its best, counseling helps clients move from problem-centered mode to discovery mode. Discovery mode involves creativity and divergent thinking. However, according to Sternberg and Lubart (1996), creativity is one of those topics in which psychology has underinvested. They present six reasons why they think this is so. Dean Simonton (2000) reviews advances in our understanding and use of creativity as part of positive psychology. However, according to Taylor, Pham, Rivkin, and Armor (1998), not just any kind of mental stimulation will do. Mental stimulation is help to the degree that it "provides a window on the future by enabling people to envision possibilities and develop plans for bringing those possibilities about. In moving oneself from a current situation toward an envisioned future one, the anticipation and management of emotions and the initiation and maintenance of problem-solving activities are fundamental tasks" (p. 429). This kind of thinking moves in the same direction as Snyder's. Not just fantasy. Not just rumination. The full problem-management and opportunity-development framework helps clients, to use Simonton's phrase, "harness the imagination."

### Creativity and Helping

One of the myths of creativity is that some people are creative and others are not. Clients, like the rest of us, can be more creative than they are. It is a question of finding ways to help them be so. Stages II and III help clients tap into their dormant creativity. A review of the requirements for creativity shows, by implication, that people in trouble often fail to use whatever creative resources they might have (see Cole & Sarnoff, 1980; Robertshaw, Mecca, & Rerick, 1978, pp. 118–120). These are the characteristics of the creative person:

- Optimism and confidence—whereas clients are often depressed and feel powerless.

- Acceptance of ambiguity and uncertainty—whereas clients may feel tortured by ambiguity and uncertainty and want to escape from them as quickly as possible.

- A wide range of interests—whereas clients may be people with a narrow range of interests or whose normal interests have been severely narrowed by anxiety and pain.
- Flexibility—whereas clients may have become rigid in their approach to themselves, others, and the social settings of life.
- Tolerance of complexity—whereas clients are often confused and looking for simplicity and simple solutions.
- Verbal fluency—whereas clients are often unable to articulate their problems, much less their goals and ways of accomplishing them.
- Curiosity—whereas clients may not have developed a searching approach to life or may have been hurt by being too venturesome.
- Drive and persistence—whereas clients may be all too ready to give up.
- Independence—whereas clients may be quite dependent or counterdependent.
- Nonconformity or reasonable risk taking—whereas clients may have a history of being very conservative and conformist, or they may get into trouble with others and with society precisely because of their particular brand of nonconformity.

A review of some of the principal obstacles or barriers to creativity (see Azar, 1995) brings further problems to the surface. Here are some of the things that can hinder innovation:

- Fear—clients are often quite fearful and anxious.
- Fixed habits—clients may have self-defeating habits or patterns of behavior that may be deeply ingrained.
- Dependence on authority—clients may come to helpers looking for the "right answers" or be quite counterdependent (the other side of the dependence coin) and fight efforts to be helped with a variety of games.
- Perfectionism—clients may come to helpers precisely because they are hounded by this problem and can accept only ideal or perfect solutions.
- Social networks—being "different" sets clients apart when they want to belong.

It is easy to say that imagination and creativity are most useful in Stages II and III, but it is another thing to help clients stimulate their own, perhaps dormant creative potential.

## Divergent Thinking

Many people habitually take a convergent-thinking approach to problem solving; that is, they look for the "one right answer." Such thinking has its uses, of course. However, many of the problem situations of life are too complex to be handled by convergent thinking. Such thinking limits the ways in which people use their own and environmental resources.

On the other hand, divergent thinking—thinking "outside the box"—assumes that there is always more than one answer. De Bono (1992) calls it "lateral thinking." In helping, that means more than one way to manage a problem or develop an opportunity. Unfortunately, divergent thinking, as helpful as it can be, is not always rewarded in our culture and sometimes is even punished. For instance, students who think divergently can be thorns in the sides of teachers. Some teachers feel comfortable only when they ask questions in such a way as to elicit the "one right answer." When students who think divergently give answers that are different from the ones expected, even though their responses might be quite useful (perhaps more useful than the expected responses), they may be ignored, corrected, or punished. Students too often learn that divergent thinking is not rewarded, at least not in school, and they may generalize their experience and end up thinking that it is simply not a useful form of behavior. Consider the following case:

> Quentin wanted to be a doctor, so he enrolled in the pre-med program at school. He did well but not well enough to get into medical school. When he received the last notice of refusal, he said to himself, "Well, that's it for me and the world of medicine. Now what will I do?" When he graduated, he took a job in his brother-in-law's business. He became a manager and did fairly well financially, but he never experienced much career satisfaction. He was glad that his marriage was good and his home life rewarding, because he derived little satisfaction from his work.

Not much divergent thinking went into handling this problem situation. No one asked Quentin what he really wanted. For Quentin, becoming a doctor was the "one right career." He didn't give serious thought to any other career related to the field of medicine, even though there are dozens and dozens of interesting and challenging jobs in the field of health care.

The case of Caroline, who also wanted to become a doctor but failed to get into medical school, is quite different from that of Quentin.

> Caroline thought to herself, "Medicine still interests me; I'd like to do something in the health field." With the help of a medical career counselor, she reviewed the possibilities. Even though she was in pre-med, she had never realized that there were so many careers in the field of medicine. She decided to take whatever courses and practicum experiences she needed to become a nurse. Then, while working in a clinic in the hills of Appalachia—an invaluable experience for her—she managed to get an MA in family-practice nursing by attending a nearby state university part time. She chose this specialty because she thought that it would enable her not only to be closely associated with delivery of a broad range of services to patients but also to have more responsibility for the delivery of these services.
>
> When Caroline graduated, she entered private practice with a doctor as a nurse practitioner in a small Midwestern town. Because the doctor divided his time among three small clinics, Caroline had a great deal of responsibility in the clinic where she practiced. She also taught a course in family-practice nursing at a nearby state school and conducted workshops in holistic approaches to preventive medical self-care. Still not satisfied, she began and finished a doctoral program in practical nursing. She taught at a state university and continued her practice. Needless to say, her persistence paid off with an extremely high degree of career satisfaction.

A successful professional career in health care always remained Caroline's aim. Using a great deal of divergent thinking and creativity, Caroline elaborated that aim into specific goals and came up with the courses of action to accomplish them. But for every success story, there are many more failures. Quentin's case is probably the norm, not Caroline's. For many, divergent thinking is either uncomfortable or too much work.

## Brainstorming: A Tool for Divergent Thinking

One excellent way of helping clients think divergently and more creatively is brainstorming. Brainstorming is a simple idea-stimulation technique for exploring the elements of complex situations. Brainstorming in Stages II and III is a tool for helping clients develop both possibilities for a better future and ways of accomplishing goals.

There are certain rules that help make this technique work: suspend judgment, produce as many ideas as possible, use one idea as a takeoff point for others, get rid of normal constraints to thinking, and produce even more ideas by clarifying items on the list. Here, then, are the rules.

**Suspend your own judgment, and help clients suspend theirs.** When brainstorming, do not let clients criticize the ideas they are generating and, of course, do not criticize them yourself. There is some evidence that this rule is especially effective when the problem situation has been clarified and defined and goals have not yet been set. In the following example, a woman whose children are grown and married is looking for ways of putting meaning into her life.

CLIENT: One possibility is that I could become a volunteer, but the very word makes me sound a bit pathetic.

HELPER: Add it to the list. Remember, we'll discuss and critique them later.

Having clients suspend judgment is one way of handling the tendency on the part of some to play a "Yes, but" game with themselves. That is, they come up with a good idea and then immediately show why it isn't really a good idea, as in the preceding example. By the same token, avoid saying such things as "I like that idea," "This one is useful," "I'm not sure about that idea," or "How would that work?" Premature approval and criticism cut down on creativity. A marriage counselor was helping a couple brainstorm possibilities for a better future. When Nina said, "We will stop bringing up past hurts," Tip, her husband, replied, "That's your major weapon when we fight. You'll never be able to give that up." The helper said, "Add it to the list. We'll look at the realism of these possibilities later on."

**Encourage clients to come up with as many possibilities as possible.** The principle is that quantity ultimately breeds quality. Some of the best ideas come along later in the brainstorming process. Cutting the process short can be self-defeating. In the following example, a man in a sex-addiction program has been brainstorming activities that might replace his preoccupation with sex.

CLIENT: Maybe that's enough. We can start putting it all together.

HELPER: It doesn't sound like you were running out of ideas.

CLIENT: I'm not. It's actually fun. It's almost liberating.

HELPER: Well, let's keep on having fun for a while.

CLIENT (pausing): Ha! I could become a monk.

Later on, the counselor, focusing on this "possibility," asked, "What would a modern-day monk who's not even a Catholic look like?" This helped the client explore the concept of sexual responsibility from a completely different perspective and to re-think the place of religion and service to others in his life. And so, within reason, the more ideas the better. Helping clients identify many possibilities for a better future increases the quality of the possibilities that are eventually chosen and turned into goals. In the end, however, do not invoke this rule for its own sake. Possibility generation is not an end in itself. Use your clinical judgment, your social intelligence, to determine when enough is enough. If a client wants to stop, often it's best to stop.

**Help clients use one idea to stimulate others.** This is called piggybacking. Without criticizing the client's productivity, encourage him or her both to develop strategies already generated and to combine different ideas to form new possibilities. In the following example, a client suffering from chronic pain is trying to come up with possibilities for a better future.

CLIENT: Well, if there is no way to get rid of all the pain, then I picture myself living a full life without pain at its center.

HELPER: Expand that a bit for me.

CLIENT: The papers are filled with stories of people who have been living with pain for years. When they're interviewed, they always look miserable. They're like me. But every once in a while there is a story about someone who has learned how to live creatively with pain. Very often they are involved in some sort of cause which takes up their energies. They don't have time to be preoccupied with pain.

A client with multiple sclerosis brought up this possibility: "I'll have a friend or two with whom I can share my frustrations as they build up." When the helper asked, "What would that look like?" the client replied, "Not just a complaining session or just a poor-me thing. It would be a normal part of a give-and-take relationship. We'd be sharing both joys and pain of our lives like other people do."

**Help clients let themselves go and develop some "wild" possibilities.** When clients seem to be "drying up" or when the possibilities being generated are quite pedestrian, you might say, "Okay, now draw a line under the items on your list and write the word 'Wild' under the line. Now let's see if you can come up with some really wild possibilities." Later, it is easier to cut suggested possibilities down to size than to expand them. The wildest possibilities often have within them at least a kernel of an idea that will work. In the following example, an older single man who is lonely is exploring possibilities for a better future.

CLIENT: I can't think of anything else. And what I've come up with isn't very exciting.

HELPER: How about getting a bit wild? You know, some crazy possibilities.

CLIENT: Well, let me think. . . . I'd start a commune and would be living in it. . . . And . . .

Clients often need permission to let themselves go, even in harmless ways. They repress good ideas because they might sound foolish. Helpers need to create an atmosphere where such apparently foolish ideas will be not only accepted but also encouraged. Help clients come up with conservative possibilities, liberal possibilities, radical possibilities, and even outrageous possibilities.

It's not always necessary to use brainstorming explicitly. As a helper, you can keep these rules in mind and then, by sharing highlights and using probes, get clients to brainstorm even though they don't know that's what they're doing. A brainstorming mentality is useful throughout the helping process.

## Future-Oriented Probes

One way of helping clients invent the future is to ask them, or get them to ask themselves, future-oriented questions related to their current unmanaged problems or undeveloped opportunities. By asking any of the following questions, helpers can encourage clients to find answers to the broader questions "What do I want?" and "What do I need?" These questions focus on outcomes—that is, on what will be in place after the clients act.

- ***What would this problem situation look like if you were managing it better?*** Ken, a college student who has been a "loner," has been talking about his general dissatisfaction with his life. In answer to this question, he said, "I'd be having fewer anxiety attacks. And I'd be spending more time with people rather than by myself."

- ***What changes in your present lifestyle would make sense?*** Cindy, who described herself as a "bored homemaker," replied, "I would not be drinking as much. I'd be getting more exercise. I would not sit around and watch the soaps all day. I'd have something meaningful to do."

- ***What would you be doing differently with the people in your life?*** Lon, a graduate student at a university near his parents' home, realized that he had not yet developed the kind of autonomy suited to his age. He mentioned these possibilities: "I would not be letting my mother make my decisions for me. I'd be sharing an apartment with one or two friends."

- ***What patterns of behavior would make life better?*** Bridget, a depressed resident in a nursing home, had this suggestion: "I'd be engaging in more of the activities offered here in the nursing home." Rick, who is suffering from lymphoma, says, "Instead of seeing myself as a victim, I'd be on the Web finding out every last thing I can about this disease and how to deal with it. I know there are new treatment options. And I'd also be getting a second or third opinion. You know, I'd be managing my lymphoma instead of just suffering from it."

- ***What current patterns of behavior would you eliminate?*** Bridget, a resident in a nursing home, adds these to her list: "I would not be putting myself down for incontinence I cannot control. I would not be complaining all the time. It gets me and everyone else down!"

- ***What would you have that you don't have now?*** Sissy, a single woman who has lived in a housing project for 11 years, said, "I'd have a place to live that's not rat infested. I'd have some friends. I wouldn't be so miserable all the time." Drew, a man tortured by perfectionism, muses, "I'd be wearing sloppy clothes, at least at times, and like it. More than that, I'd have a more realistic sense of the world and my place in it. The world is messy, it's chaotic much of the time. I'd find the beauty in the chaos."

- ***What accomplishments would you have that you don't have now?*** Ryan, a divorced man in his mid-thirties, said, "I'd have my degree in practical nursing. I'd be doing some part-time teaching. I'd be close to someone that I'd like to marry."

- ***What would this opportunity look like if you developed it?*** Enid, a woman with a great deal of talent who has been given one modest promotion in her company but who feels like a second-class citizen, had this to say: "In two years I'll be an officer of this company or have a very good job in another firm."

It is a mistake to suppose that clients will automatically gush with answers. Ask the kinds of questions just listed, or encourage them to ask themselves the questions, but then help them answer them. Create the therapeutic dialogue around possibilities for a better future. Many clients don't know how to use their innate creativity. Thinking divergently is not part of their mental lifestyle. You have to work with clients to help them produce some creative output. Some clients are reluctant to name possibilities for a better future because they sense that this will bring more responsibility. They will have to move into action mode.

## Exemplars and Models as Sources of Possibilities

Some clients can see future possibilities better when they see them embodied in others. You can help clients brainstorm possibilities for a better future by helping them identify exemplars or models. By models I don't mean superstars or people who do things perfectly. That would be self-defeating. In the next example, a marriage counselor is talking with a middle-aged, childless couple. They are bored with their marriage. When he asked them, "What would your marriage look like if it looked a little better?" he could see that they were stuck.

COUNSELOR: Maybe the question would be easier to answer if you reviewed some of your married relatives, friends, or acquaintances.

WIFE: None of them have super marriages. [Husband nods in agreement.]

COUNSELOR: No, I don't mean super marriages. I'm looking for things you could put in your marriage that would make it a little better.

WIFE: Well, Fred and Lisa are not like us. They don't always have to be doing everything together.

HUSBAND: Who says we have to be doing everything together? I thought that was your idea.

WIFE: Well, we always are together. If we weren't always together, we wouldn't be in each other's hair all the time.

COUNSELOR: All right, who else do you know who are doing things in their marriage that appeal to you. Anyone.

HUSBAND: You know Ron and Carol do some volunteer work together. Ron was saying that it gets them out of themselves. I bet they have better conversations because of it.

COUNSELOR: Now we're cooking. . . . What else? What couple do you find the most interesting?

Even though it was a somewhat torturous process, these two people were able to come up with a range of possibilities for a better marriage. The counselor had them write them down so they wouldn't lose them. At that point, the purpose was not to get the clients to commit themselves to the possibilities but to identify them.

In the following case, the client finds herself making discoveries by observing people she had not identified as models at all:

Fran, a somewhat withdrawn college junior, realizes that when it comes to interpersonal competence, she is not ready for the business world she intends to enter when she graduates. She wants to do something about her interpersonal style and a few nagging personal problems. She sees a counselor in the Office of Student Services. After a couple of discussions with him, she joins a "lifestyle" group on campus that includes some training in interpersonal skills. Even though she expands her horizons a bit from what the members of the group say about their experiences, behaviors, and feelings, she tells her counselor that she learns even more by watching her fellow group members in action. She sees behaviors that she would like to incorporate in her own style. A number of times she says to herself in the group, "Ah, there's something I never thought of." Without becoming a slavish imitator, she begins to weave some of the patterns she sees in others into her own style.

Models or exemplars can help clients name what they want more specifically. Models can be found anywhere: among the client's relatives, friends, and associates, in books, on television, in history, in movies. Counselors can help clients identify models, choose those dimensions of others that are relevant, and translate what they see into realistic possibilities for themselves.

Lockwood and Kunda (1999) have shown that, under normal circumstances, individuals can be inspired by role models so that their motivation and self-evaluations are enhanced. But not always. Bringing up role models with people who have been reviewing "best past selves" has a way of deflating people. Their best can pale in comparison with the model. Their best is none too good. This is important because in solution-focused therapies, reviewing past successes is an important part of the process. In addition, if people are asked to come up with ideas about their "best possible selves" and then are asked to review what they like about a role model, their ability to draw inspiration from the role model is impaired. In sum, using role models as sources of inspiration certainly works, but it can be tricky.

## Cases Featuring Possibilities for a Better Future

Here are a couple of cases that illustrate how helping clients develop possibilities for a better future had a substantial impact.

### The Case of Brendan: Dying Well

Brendan, a heavy drinker, had extensive and irreversible liver damage and it was clear that he was getting sicker. But he wanted to "get some things done" before he died. Brendan's action orientation helped a great deal. Over the course of a few months, a counselor helped him to name some of the things he wanted before he died or on his journey toward death. Brendan came up with the following possibilities:

- "I'd like to have some talks with someone who has a religious orientation, like a minister. I want to discuss some of the 'bigger' issues of life and death."
- "I don't want to die hopeless. I want to die with a sense of meaning."
- "I want to belong. You know, to some kind of community, people who know what I'm going through but are not sentimental about it. People not disgusted with me because of the way I've done myself in."
- "I'd like to get rid of some of my financial worries."
- "I'd like a couple of close friends with whom I could share the ups and downs of daily life. With no apologies."

## Box 15-1    Questions for Exploring Possibilities

Help clients ask themselves these kinds of questions:

- What are my most critical needs and wants?
- What are some possibilities for a better future?
- What outcomes or accomplishments would take care of my most pressing problems?
- What would my life look like if I were to develop a couple of key opportunities?
- What should my life look like a year from now?
- What should I put in place that is currently not in place?
- What are some wild possibilities for making my life better?

---

- "As long as possible, I'd like to be doing some kind of productive work, whether paid or not. I've been a flake. I want to contribute, even if in just an ordinary way."
- "I need a decent place to live, maybe with others."
- "I need decent medical attention. I'd like a doctor who has some compassion. One who could challenge me to live until I die."
- "I need to manage these bouts of anxiety and depression better."
- "I want to be get back with my family again. I want to hug my dad. I want him to hug me."
- "I'd like to make peace with one or two of my closest friends. They more or less dropped me when I got sick. But at heart, they're good guys."
- "I want to die in my home town."

Of course, Brendan didn't name all these possibilities at once. Through understanding and probes, the counselor helped name what he needed and wanted and then helped him stitch together a set of goals from these possibilities (Stage II) and ways of accomplishing them (Stage III). Box 15-1 outlines the kinds of questions you can help clients ask themselves to discover possibilities for a better future.

## The Washington Family Case

This case is more complex because it involves a family. Not only does the family as a unit have its wants and needs, but also each individual member has his or her own. Therefore, it is even more imperative to review possibilities for a better future so that competing needs can be reconciled.

Lane, the 15-year-old son of Troy and Rhonda Washington, was hospitalized with what was diagnosed as an acute schizophrenic attack. He had two older brothers, both teenagers, and two younger sisters, one 10 and one 12, all living at home. The Washingtons lived in a large city. Although both parents worked, their combined income still left them pinching pennies. They also ran into a host of problems associated with their son's hospitalization: the need to arrange ongoing help and care for Lane, financial burdens, behavioral problems among the other siblings, marital conflict, and stigma in the community ("They're a funny family with a crazy son"; "What kind of parents are they?"). To make things worse, Troy and Rhonda did not think the psychiatrist and the psychologist they met at the hospital took the time to understand their concerns. They felt that the helpers were trying to push Lane back out into the community; in their eyes, the hospital was "trying to get rid of him." Their complaint was, "They give him some pills and then give him back to you." No one explained to them that short-term hospitalization was meant to guard the civil rights of patients and avoid the negative effects of longer-term institutionalization.

When Lane was discharged, his parents were told that he might have a relapse, but they were not told what to do about it. They faced the prospect of caring for Lane in a climate of stigma without adequate information, services, or relief. Feeling abandoned, they were very angry with the mental-health establishment. They had no idea what they should do to respond to Lane's illness or to the range of family problems that had been precipitated by the episode. By chance, the Washingtons met someone who had worked for the National Alliance for the Mentally Ill (NAMI), an advocacy and education organization. This person referred them to an agency that provided support and help.

What does the future hold for such a family? With help, what kind of future can be fashioned? Social workers at the agency helped the Washingtons identify both needs and wants in seven areas (see Bernheim, 1989).

- **The home environment.** The Washingtons needed an environment in which the needs of all the family members were balanced. They didn't want their home to be an extension of the hospital. They wanted Lane taken care of, but they wanted to attend to the needs of the other children and to their own needs as well.

- **Care outside the home.** They wanted a comprehensive therapeutic program for Lane. They needed to review possible services, identify relevant services, and arrange access to those services. They needed to find a way of paying for all this.

- **Care inside the home.** They wanted all family members to know how to cope with Lane's residual symptoms. He might be withdrawn or aggressive, but they needed to know how to relate to him and help him handle behavioral problems.

- **Prevention.** Family members needed to be able to spot early warning symptoms of impending relapse. They also needed to know what to do when they saw those signs, including such things as contacting the clinic or, in the case of more severe problems, arranging for an ambulance or getting help from the police.

- **Family stress.** They needed to know how to cope with the increased stress that all this would entail. They needed forums for working out their problems. They wanted to avoid family blowups, and when blowups occurred, they wanted to manage them without damaging the social fabric of the family.

- **Stigma.** They wanted to understand and be able to cope with whatever stigma might be attached to Lane's illness. For instance, when taunted for having a "crazy brother," the children needed to know what to do and what not to do. Family members needed to know whom to tell, what to say, how to respond to inquiries, and how to deal with blame and insults.

- *Limitation of grief.* They needed to know how to manage the normal guilt, anger, frustration, fear, and grief that go with problem situations like this.

Bernheim's schema constituted a useful checklist for stimulating thinking about possibilities for a better future. The Washingtons first needed help in developing these possibilities. Then they needed help in setting priorities and establishing goals to be accomplished. This is the work of Step II-B. For positive-psychology advances in the treatment of serious mental illness, see Coursey, Alford, and Safarjan (1997).

When it comes to serious mental illness in a family, Marsh and Johnson (1997) focus not just on family burden but also on family resilience and the internal and external resources that support such resilience. This is, of course, a positive-psychology approach. They list the ways in which a helper can assist the family (p. 233):

1. Understanding and normalizing the family experience of mental illness.
2. Focusing on the strengths and competencies of their family and relatives.
3. Learning about mental illness, the mental health system, and community resources.
4. Developing skills in stress management, problem solving, and communication.
5. Resolving their feelings of grief and loss.
6. Coping with the symptoms of mental illness and its repercussions for their family.
7. Identifying and responding to the signs of impending relapse.
8. Creating a supportive family environment.
9. Developing realistic expectations for all members of the family.
10. Playing a meaningful role in their relative's treatment, rehabilitation, and recovery.
11. Maintaining a balance that meets the needs of all members of the family.

Johnson and Marsh also outline a number of intervention strategies that can help families meet these objectives:

- *Family interventions* that stress the role of the family as a support system rather than the cause of mental illness.
- *Family support and advocacy groups* such as the NAMI. These groups provide support and education, and encourage advocacy for improved services.
- *Family consultation,* which can aid in helping families determine their own goals and make informed choices regarding their use of available services.
- *Family education,* with respect to information about mental illness, caregiving, the mental-health system, community resources, and the like.
- *Family psychoeducation,* which focuses on such things as coping strategies and stress management.

In all this, you can see the outline of the solution-focused philosophy discussed in Chapter 14.

### Evaluation Questions for Step II-A

- To what degree am I an imaginative person?
- In what ways can I apply the concept of "possible selves" to myself?
- What problems do I experience as I try to help clients use their imaginations?
- Against the background of problem situations and unused opportunities, how well do I help clients focus on what they want?
- To what degree do I prize divergent thinking and creativity in myself and others?
- How effectively do I use empathic highlights, a variety of probes, and challenging to help clients brainstorm what they want?
- Besides direct questions and other probes, what kinds of strategies do I use to help clients brainstorm what they want?
- How effectively do I help clients identify models and exemplars that can help them clarify what they want?
- How easily do I move back and forth in the helping model, especially in establishing a "dialogue" between Stages I and II?
- How well do I help clients act on what they are learning?

# STEP II-B: "WHAT DO I REALLY WANT?" MOVING FROM POSSIBILITIES TO CHOICES

FROM POSSIBILITIES TO CHOICES

HELPING CLIENTS SHAPE THEIR GOALS

   Help Clients State What They Need and Want as Outcomes or Accomplishments

   Help Clients Move from Broad Aims to Clear and Specific Goals

      Good intentions

      Broad aims

      Specific goals

   Help Clients Establish Goals That Make a Difference

   Help Clients Set Goals That Are Prudent

   Help Clients Formulate Realistic Goals

      Resources: Help clients choose goals for which the resources are available

      Control: Help clients choose goals that are under their control

   Help Clients Set Goals That Can Be Sustained

   Help Clients Choose Goals That Have Some Flexibility

   Help Clients Choose Goals Consistent with Their Values

   Help Clients Establish Realistic Time Frames for the Accomplishment of Goals

NEEDS VERSUS WANTS

EMERGING GOALS

ADAPTIVE GOALS

      Satisfactory alternatives

      Coping

      Strategic self-limitation

THE "REAL-OPTIONS" APPROACH

A BIAS FOR ACTION AS A METAGOAL

EVALUATION QUESTIONS FOR STEP II-B

## From Possibilities to Choices

Once clients have developed possibilities for a better future, they need to make some choices; that is, they need to choose one or more of those possibilities and turn them into a program for constructive change. Step II-A is, in many ways, about *creativity*—getting rid of boundaries, thinking beyond one's limited horizon, moving outside the box. Step II-B is about *innovation*—turning possibilities into a practical program for change. If implemented, a goal constitutes the Solution, with a big *S*, for the client's problem or opportunity. Consider the following case:

> Bea, an African American woman, was arrested when she went on a rampage in a bank and broke several windows. She had exploded with anger because she felt that she had been denied a loan mainly because she was black and a single mother. In discussing the incident with her minister, she comes to see that she has become very prone to anger. Almost anything can get her going. She also realizes that venting her anger as she had done in the bank led to a range of negative consequences. But she is constantly "steamed up" about the "system." To complicate the picture, she tends to take her anger out on those around her, including her friends and her two children. The minister helps her look at four possible ways of dealing with her anger: venting it, repressing it, channeling it, or simply giving up and ignoring the things she gets angry at, including the injustices around her. Giving up is not in her makeup. Merely venting her anger seems to do little but make her more angry. Repressing her anger, she reasons, is just another way of giving up, and that is demeaning. And she's not very good at repressing anyway. The "channeling" option needs to be explored. In the end, Bea takes a positive-psychology approach to dealing with her frustrations. She joins a political action group involved in community organizing. She learns that she can channel her anger without giving up her values or her intensity. She also discovers that she is good at influencing others and getting things done. She begins to feel better about herself. The system doesn't seem to be such a fortress any more.

Since goals can be highly motivational, helping clients set realistic goals is one of the most important steps in the helping process.

## Helping Clients Shape Their Goals

Practical goals do not usually leap out fully formed. They need to be shaped or "designed," a term we saw in Chapter 14. Effective counselors add value by engaging clients in the kind of dialogue that will help them design, choose, craft, shape, and develop their goals. Goals are specific statements about what clients want and need.

The goals that emerge through this client-helper dialogue are more likely to be workable if they have certain characteristics. They need to be

- stated as **outcomes** rather than activities;
- **specific** enough to be verifiable and to drive action;
- **substantive** and challenging;
- both **venturesome** and **prudent**;
- **realistic** in regard to resources needed to accomplish them;
- **sustainable** over a reasonable time period;
- **flexible** without being wishy-washy;

- *congruent* with the client's *values*;
- set in a reasonable *time frame*.

Just how this package of goal characteristics will look in practice will differ from client to client. There is no one formula. From a practical point of view, these characteristics can be seen as tools that counselors can use to help clients design and shape or reshape their goals. Ineffective helpers will get lost in the details of these characteristics. Effective helpers will keep them in the backs of their minds and, in a second-nature manner, turn them into helpful "sculpting" probes at the right time. These characteristics, then, take on life through the following flexible principles.

## Help Clients State What They Need and Want as Outcomes or Accomplishments

The goal of counseling, as emphasized again and again, is neither discussing nor planning nor engaging in activities. Helping is about Solutions with a big S. "I want to start doing some exercise" is an activity rather than an outcome. "Within six months I will be running three miles in less than 30 minutes at least four times a week" is an outcome, a pattern of behavior that will be in place by a certain time. If a client says, "My goal is to get some training in interpersonal communication skills," then she is stating her goal as a set of activities—a solution with a small *s*— rather than as an accomplishment. But if she says that she wants to become a better listener as a wife and mother, then she is stating her goal as an accomplishment, even though "better listener" needs further clarification. Goals stated as outcomes provide direction for clients.

You can help clients describe what they need and want by using this "past-participle approach"—drinking stopped, number of marital fights decreased, anger habitually controlled. Stating goals as outcomes or accomplishments is not just a question of language. Helping clients state goals as accomplishments rather than activities helps them avoid directionless and imprudent action. If a woman with breast cancer says that she thinks she should join a self-help group, she should be helped to see what she wants to get out of such a group. Joining a group and participating in it are activities. She wants support. She wants to feel supported. Goals, at their best, are expressions of what clients need and want. Clients who know what they want are more likely to work not just harder but also smarter.

Consider the case of Chester, a former Marine suffering from post-traumatic stress disorder:

Chester was involved in the Kosovo peacekeeping effort. During a patrol, he and three of his buddies shot and killed four civilians who were out to kill their neighbors. Afterward he began acting in strange ways, wandering around at times in a daze. He was given a medical discharge and sent home. Although he seemed to recover, he lived an aimless life. He went to college but dropped out during the first semester. He became rather reclusive but never really engaged in odd behavior. Rather he was sinking into the landscape. He moved in and out of a number of low-paying jobs. He also became less careful about his person. He said to a counselor, "You know, I used to be very careful about the way I dressed. Kind of proud of myself in the Marine tradition. Don't get me wrong; I'm not a bum and don't smell or anything, but I'm not myself." The whole direction of Chester's life was wrong; he was headed for serious trouble. He was bothered by thoughts about the war and had taken to sleeping whenever he felt like it, day or night, "just to make it all stop."

Ed, Chester's counselor, had a good relationship with Chester. He helped Chester tell his story and challenged some of his self-defeating thinking. He went on to help Chester focus on what he wanted from life. They moved back and forth between Stage I and Stage II, between problems and possibilities for a better future. Eventually, Chester began talking about his real needs and wants—that is, what he needed to accomplish to "get back to his old self." Here is an excerpt from their dialogue:

CHESTER:  I've got to stop hiding in my hole. I'm going to get out and see people more. I'm going to stop feeling so damn sorry for myself. Who wants to be with a nothing!

COUNSELOR:  What will Chester's life look like a year or two from now?

CHESTER:  One thing for sure, he will be seeing women again. He might not be married, but he will probably have a special girlfriend. And she will see him as an ordinary guy.

Here Chester talks about changes as patterns of behavior that will be in place. He is painting a picture of what he wants to be. He is designing some goals. The counselor's probe reinforces this outcome approach.

## Help Clients Move from Broad Aims to Clear and Specific Goals

Specific rather than general goals tend to drive behavior. Therefore, broad goals need to be translated into more specific goals and tailored to the needs and abilities of each client. Skilled helpers use probes to help clients move from the general to the specific.

Chester said that he wanted to become "more disciplined." His counselor helped him make that more specific.

COUNSELOR:  What areas do you want to focus on?

CHESTER:  Well, if I'm going to put more order in my life, I need to look at the times I sleep. I've been going to bed whenever I feel like it and getting up whenever I feel like it. It was the only way I could get rid of those thoughts and the anxiety. But I'm not nearly as anxious as I used to be. Things are calming down.

COUNSELOR:  So more disciplined means a more regular sleep schedule because there's no particular reason now for not having one.

CHESTER:  Yeah, sleeping whenever I want is just a bad habit. And I can't get things done if I'm asleep.

Chester goes on to translate "more disciplined" into other problem-managing needs and wants related to school, work, and care of his person. Greater discipline, once translated into specific patterns of behavior, will have a decidedly positive impact on his life.

Counselors often add value by helping clients move from good intentions and vague desires to broad aims and then on to quite specific goals.

**Good intentions.** "I need to do something about this" is a statement of intent. However, even though good intentions are a good start, they need to be translated into aims and goals. In the following example, the client, Jon, has been discussing his relationship with his wife and children. The counselor has been helping him see that his "commitment to work" is perceived negatively by his family. Jon is open to challenge and is a fast learner.

JON: Boy, this session has been an eye-opener for me. I've really been blind. My wife and kids don't see my investment—rather, my overinvestment—in work as something I'm doing for them. I've been fooling myself, telling myself that I'm working hard to get them the good things in life. In fact, I'm spending most of my time at work because I like it. My work is mainly for me. It's time for me to realign some of my priorities.

The last statement is a good intention, an indication on Jon's part that he wants to do something about a problem now that he sees it more clearly. It may be that Jon will now go out and put a different pattern of behavior in place without further help from the counselor. Or he may benefit from some help in realigning his priorities.

**Broad aims.** A broad aim is more than a good intention. It has content; that is, it identifies the area in which the client wants to work and makes some general statement about that area. Let's return to the example of Jon and his overinvestment in work:

JON: I don't think I'm spending so much time at work in order to run away from family life. But family life is deteriorating because I'm just not around enough. I must spend more time with my wife and kids. Actually, it's not just a case of must. I want to.

Jon moves from a declaration of intent to an aim or a broad goal, spending more time at home. But he still has not created a picture of what that would look like.

**Specific goals.** To help Jon move toward greater specificity, the counselor uses such probes as "Tell me what 'spending more time at home' will look like."

JON: I'm going to consistently spend three out of four weekends a month at home. During the week, I'll work no more than two evenings.

COUNSELOR: So you'll be at home a lot more. Tell me what you'll be doing with all this time.

Notice how much more specific Jon's statement is than "I'm going to spend more time with my family." He sets a goal as a specific pattern of behavior he wants to put in place. But his goal as stated deals with quantity, not quality. The counselor's probe is really a challenge. It's not just the amount of time Jon is going to spend with his family but also the kinds of things he will be doing—quality time, some call it. But a client trying to come to grips with work-life balance once said to me, "My family, especially my kids, don't make the distinction between quantity and quality. For them quantity is quality. Or there's no quality without a chunk of quantity." This warrants further discussion because maybe the family wants a relaxed rather than an intense Jon at home.

This example brings up the difference between *instrumental* goals and *higher-order* or *ultimate* goals. Jon's ultimate goal is "a good family life." Such a goal, once spelled out, will differ from family to family and from culture to culture. Think of your own definition. When Jon says that one of his goals is spending more time at home, he is talking about an instrumental goal. Unless he's there, he can't do things with his wife and kids. Although just "being there" is a goal because it is a pattern of behavior *in place*, it is certainly not Jon's ultimate goal. But Jon is not worried about the ultimate goal. When he is there, they have a rich family life together; that's not the problem. However, instrumental goals are *strategies* for achieving higher-order goals, so it's important to make sure that the client has clarity about the higher-order goal. If Jon were spending a lot of time at the office because he didn't like being with his wife and kids or because there was a great deal of conflict at home, then his higher-order goal would be something like "experiencing the stimulation of an exciting workplace" (if home life was dull) or "peace of mind" (if home life was full of conflict). When you are helping clients design and shape instrumental goals, make sure they can answer the instrumental-for-what question.

Helping clients move from good intentions to more and more specific goals is a shaping process. Consider the example of a couple whose marriage has degenerated into constant bickering, especially about finances.

- *Good intention*: "We want to straighten out our marriage."
- *Broad aim*: "We want to handle our decisions about finances in a much more constructive way."
- *Specific goal*: "We try to solve our problems about family finances by fighting and arguing. We'd like to reduce the number of fights we have and begin making mutual decisions about money. We yell instead of talking things out. We need to set up a month-by-month budget. Otherwise, we'll be arguing about money we don't even have. We'll have a trial budget ready the next time we meet with you."

Having sound household finances is a fine goal—a goal in itself. Reducing unproductive conflict is also a fine goal. In this case, however, installing a sound, fair, and flexible household budget system is also instrumental to establishing peace at home. Declarations of intent, broad goals, and specific goals can all drive constructive behavior, but specific goals have the best chance. Is it possible to get clients to be too specific about their goals? Yes, if they get lost in the planning details and crafting the goal becomes more important than the goal itself.

If the goal is clear enough, the client will be able to determine progress toward the goal. For many clients, being able to measure progress is an important incentive. If goals are stated too broadly, it is difficult to determine both progress and accomplishment. "I want to have a better relationship with my wife" is a very broad goal, difficult to verify. "I want to socialize more, you know, with couples we both enjoy" comes closer, but "socialize more" needs more clarity.

It is not always necessary to count things to determine whether a goal has been reached, though sometimes counting is helpful. Helping is about living more fully, not about accounting activities. At a minimum, however, desired outcomes need to be capable of being verified in some way. For instance, a couple might say something like "Our relationship is better, not because we've stop squabbling. In fact, we've discovered that we like to squabble. But life is better because the meanness has gone out of our squabbling. We accept each other more. We listen more carefully, we talk about more personal concerns, we are more relaxed, and we make more mutual decisions about issues that affect us both." This couple does not need a scientific experiment to verify that they have improved their relationship.

## Help Clients Establish Goals That Make a Difference

Outcomes and accomplishments are meaningless if they do not have the required impact on the client's life. The goals clients choose should have substance to them—that is, some significant contribution toward managing the original problem situation or developing some opportunity.

Vitorio ran the family business. His son, Anthony, worked in sales. After spending a few years learning the business and getting an MBA part time at a local university, Anthony wanted more responsibility and authority. His father never thought that he was "ready." They began arguing quite a bit, and their relationship suffered from it. Finally, a friend of the family persuaded them to spend time with a consultant-counselor who worked with small family businesses. He spent relatively little time listening to their problems. After all, he had seen this same problem over and over again—the reluctance and conservatism of the father, the pushiness of the son.

Vitorio wanted the business to stay on a tried-and-true course. Anthony wanted to be the company's marketer, to move it into new territory. After a number of discussions with the consultant-counselor, they settled on this scenario: A "marketing department" headed by Anthony would be created. He could divide his time between sales and marketing as he saw fit, provided that he maintained the current level of sales. Vitorio agreed not to interfere. They would meet once a month with the consultant-counselor to discuss problems and progress. Vitorio insisted that the consultant's fee come from increased sales. After some initial turmoil, the bickering decreased dramatically. Anthony easily found new customers, although they demanded modifications in the product line, which Vitorio reluctantly approved. Both sales and margins increased to the point that another person was needed in sales.

Not all issues in family businesses are handled as easily. In fact, a few years later, Anthony left the business and founded his own. But the goal package they worked out—the deal they cut—made quite a difference both in the father-son relationship and in the business.

Second, goals have substance to the degree that they help clients stretch themselves. As Locke and Latham (1984) note: "Extensive research . . . has established that, within reasonable limits, the . . . more challenging the goal, the better the resulting performance. . . . People try harder to attain the hard goal. They exert more effort. . . . In short, people become motivated in proportion to the level of challenge with which they are faced. . . . Even goals that cannot be fully reached will lead to high effort levels, provided that partial success can be achieved and is rewarded" (pp. 21–26). Consider the following case:

A young woman was a quadriplegic as a result of an auto accident. In the beginning, she was full of self-loathing: "The accident was all my fault; I was just stupid." She was close to despair. Over time, however, with the help of a counselor, she came to see herself, not as a victim of her own "stupidity" but as someone who could bring hope to young people with life-changing afflictions. In her spare time, she visited young patients in hospitals and rehabilitation centers, got some to join self-help groups, and generally helped people like herself to manage an impossible situation in a more humane way. One day, she said to her counselor, "The best thing I ever did was to stop being a victim and become a fellow traveler with people like myself. The last two years, though bitter at times, have been the best years of my life." She had set her goal quite high—becoming an outgoing helper instead of remaining a self-centered victim—but it proved to be quite realistic.

Of course, when it comes to goals, challenging should not mean impossible. There seems to be a curvilinear relationship between goal difficulty and goal performance. If the goal is too easy, people see it as trivial and ignore it. If the goal is too difficult, it is not accepted. However, this difficulty-performance ratio differs from person to person. What is small for some is big for others (see the later section titled "Adaptive Goals.")

## Help Clients Set Goals That Are Prudent

Although the helping model described in this book encourages a bias toward action on the part of clients, action needs to be both directional and wise. Discussing and setting goals should contribute to both direction and wisdom. The following case begins poorly but ends well:

> Harry was a sophomore in college who was admitted to a state mental hospital because of some bizarre behavior at the university. He was one of the disc jockeys for the university radio station. He came to the notice of college officials one day when he put on an attention-getting performance that included rather lengthy dramatizations of grandiose religious themes. In the hospital, it was soon discovered that this quite pleasant, likable young man was actually a loner. Everyone who knew him at the university thought that he had many friends, but in fact he did not. The campus was large, and his lack of friends went unnoticed.
>
> Harry was soon released from the hospital but returned weekly for therapy. At one point, he talked about his relationships with women. Once it became clear to him that his meetings with women were perfunctory and almost always took place in groups—he had imagined that he had a rather full social life with women—Harry launched a full program of getting involved with the opposite sex. His efforts ended in disaster, however, because Harry had some basic sexual and communication problems. He also had serious doubts about his own worth and therefore found it difficult to make a gift of himself to others. He ended up in the hospital again.
>
> The counselor helped Harry get over his sense of failure by emphasizing what Harry could learn from the "disaster." With the therapist's help, Harry returned to the problem-clarification and new-perspectives part of the helping process and then established more realistic short-term goals regarding getting back "into community." The direction was the same—establishing a realistic social life—but the goals were now more prudent because they were "bite-size." Harry attended socials at a local church where a church volunteer provided support and guidance.

Harry's leaping from problem clarification to action without taking time to discuss possibilities and set reasonable goals was part of the problem rather than part of the solution. His lack of success in establishing solid relationships with women actually helped him see his problem with women more clearly. There are two kinds of prudence: playing it safe is one, doing the wise thing is the other. Problem management and opportunity development should be venturesome. They are about making wise choices rather than playing it safe.

## Help Clients Formulate Realistic Goals

Setting goals that demand clients stretch can help clients energize themselves. They rise to the challenge. On the other hand, goals set too high can do more harm than good. Locke and Latham (1984) put it succinctly:

> Nothing breeds success like success. Conversely, nothing causes feelings of despair like perpetual failure. A primary purpose of goal setting is to increase the motivation level of the individual. But goal setting can have precisely the opposite effect if it produces a yardstick that constantly makes the individual feel inadequate (p. 39).

A goal is realistic if the client has access to the resources needed to accomplish it and the goal is under the client's control, not hampered by external circumstances.

**Resources: Help clients choose goals for which the resources are available.** It does little good to help clients develop specific, substantive, and verifiable goals if the resources needed for their accomplishment are not available. Consider the case of Rory, who, because of a merger and extensive restructuring, has had to take a demotion. He now wants to leave the company and become a consultant.

INSUFFICIENT RESOURCES: Rory does not have the assertiveness, marketing savvy, industry expertise, and interpersonal style needed to become an effective consultant. Even if he did, he does not have the financial resources needed to tide him over while he develops a business.

SUFFICIENT RESOURCES: Challenged by the outplacement counselor, Rory changes his focus. Graphic design is an avocation of his. He is not good enough to take a technical position in the company's design department, but he does apply for a supervisory role in that department. He is good with people, very good at scheduling and planning, and knows enough about graphic design to discuss issues meaningfully with the members of the department.

Rory combines his managerial skills with his interest in graphic design to move in a more realistic direction. The move is challenging, but it can have a substantial impact on his work life. For instance, the opportunity to hone his graphic design skills will open up further career possibilities.

**Control: Help clients choose goals that are under their control.** Sometimes clients defeat their own purposes by setting goals that are not under their control. For instance, it is common for people to believe that their problems would be solved if only other people would not act the way they do. In most cases, however, we do not have any direct control over the ways others act. Consider the following example:

Tony, a 16-year-old boy, felt that he was the victim of his parents' inability to relate to each other. Each tried to use him in the struggle, and at times he felt like a Ping-Pong ball. A counselor helped him see that he could probably do little to control his parents' behavior but that he might be able to do quite a bit to control his *reactions* to his parents' attempts to use him. For instance, when his parents started to fight, he could simply leave instead of trying to "help." If either tried to enlist him as an ally, he could say that he had no way of knowing who was right. Tony also worked at creating a good social life outside the home. That helped him weather the tensions he experienced when at home.

Tony needed a new way of managing his interactions with his parents to minimize their attempts to use him as a pawn in their own interpersonal game.

Goals are not under clients' control if they are blocked by external forces that they cannot influence. "To live in a free country" may be an unrealistic goal for a person living in a totalitarian state because one cannot change internal politics, nor can one change emigration laws in one's own country or immigration laws in other countries. "To live as freely as possible in a totalitarian state," however, might well be an aim that could be translated into realistic goals.

## Help Clients Set Goals That Can Be Sustained

Clients need to commit themselves to goals that have staying power. One separated couple said that they wanted to get back together again. They did so only to get divorced again within six months. Their goal of getting back together again was achievable but not sustainable. Perhaps they should have asked themselves, "What do we need to do not only to get back together but also to stay together? What would our marriage have to look like to become and remain workable?" In discretionary-change situations, the issue of sustainability needs to be visited early on.

Many Alcoholics Anonymous-like programs work because of their "one day at a time" approach. The goal of being, let us say, drug free has to be sustained only over a single day. The next day is a new era. In a previous example, Vitorio and Anthony's arrangement had enough staying power to produce good results in the short term. It also allowed them to reset their relationship and to improve the business. The goal was not designed to produce a lasting business arrangement because, in the end, Anthony's aspirations were bigger than the family business.

## Help Clients Choose Goals That Have Some Flexibility

In many cases, goals have to be adapted to changing realities. Therefore, there might be some trade-offs between goal specificity and goal flexibility in uncertain situations. Napoleon noted this when he said, "He will not go far who knows from the first where he is going." Sometimes making goals too specific or too rigid does not allow clients to take advantage of emerging opportunities.

> Even though he liked the work and even the company he worked for, Jessie felt like a second-class citizen. He thought that his supervisor gave him most of the dirty work and that there was an undercurrent of prejudice against Hispanics in his department. Jessie wanted to quit and get another job, one that would pay the same relatively good wages he was now earning. A counselor helped Jessie challenge his choice. Even though the economy was booming, the industry in which Jessie was working was in recession. There were few jobs available for workers with Jessie's skills. The counselor helped Jessie choose an interim goal that was more flexible and more directly related to coping with his present situation. The interim goal was to use his time preparing himself for a better job outside his current industry. In six months to a year, he could be better prepared for a career in the "new economy." Jessie began volunteering for special assignments that helped him learn some new skills and took some crash courses dealing with computers and the Internet. He felt good about what he was learning and more easily ignored the prejudice.

Counseling is a living, organic process. Just as organisms adapt to their changing environments, the choices clients make need to be adapted to their changing circumstances.

## Help Clients Choose Goals Consistent with Their Values

Although helping is a process of social influence, it remains ethical only if it respects, within reason, the values of the client. Values are criteria we use to make decisions. Helpers can challenge clients to reexamine their values, but they should not encourage clients to perform actions that are not in keeping with their values.

> The son of Vincente and Consuela Garza is in a coma in the hospital after an automobile accident. He needs a life-support system to remain alive. His parents are experiencing a great deal of uncertainty, pain, and anxiety. They have been told that there is practically no chance that their son will ever come out of the coma. One possibility is to terminate the life-support system. The counselor should not urge them to terminate the life-support system if that is counter to their values. But she can help them explore and clarify the values involved. In this case, the counselor suggests that they discuss their decision with their clergyman. In doing so, they find out that the termination of the life-support system would not be against the tenets of their religion. Now they are free to explore other values that relate to their decision.

Some problems involve a client's trying to pursue contradictory goals or values. Chester, the ex-Marine, wants to get an education, but he also wants to make a decent living as soon as possible. Going to school full time would put him in debt, but failing to get a college education would lessen his chances of securing the kind of job he wants. The counselor helps him identify and use his values to consider some trade-offs. Chester chooses to work part time and go to school part time. He chooses a job in an office instead of a job in construction. Even though the latter pays better, it would be much more exhausting and would leave him with little energy for school.

## Help Clients Establish Realistic Time Frames for the Accomplishment of Goals

Goals that are to be accomplished "sometime or other" probably won't be accomplished at all. Therefore, helping clients put some time frames in their goals can add value. Greenberg (1986) talks about immediate, intermediate, and final outcomes. Here's what they look like when applied to Janette's problem situation. She suffers in a variety of ways because she lets others take advantage of her. She needs to become more assertive. She needs to stand up for her own rights.

- *Immediate outcomes* are changes in attitudes and behaviors evident in the helping sessions themselves. For Janette, the helping sessions constitute a safe forum for her to become more assertive. In her dialogues with her counselor, she learns and practices the skills of being more assertive.

- *Intermediate outcomes* are changes in attitudes and behaviors that lead to further change. It takes Janette a while to transfer her assertiveness skills both to the workplace and to her social life. She chooses relatively safe situations to practice being more assertive. For instance, she stands up to her mother more often.

- *Final outcomes* refer to the completion of the overall program for constructive change through which problems are managed and opportunities developed. It takes more than two years for Janette to become assertive in a consistent day-to-day way.

The next example deals with a young man who has been caught shoplifting. Here, too, there are immediate, intermediate, and final outcomes.

> Jensen, a 22-year-old on probation for shoplifting, was seeing a counselor as part of a court-mandated program. An immediate need in his case was overcoming his resistance to his court-appointed counselor and developing a working alliance with her. Because of the counselor's skill and her unapologetic caring attitude that had some toughness in it, he quickly came to see her as "on his side." Their relationship became a platform for establishing further goals. An intermediate outcome was attitudinal in nature. Brainwashed by what he saw on television, Jensen thought that America owed him some of its affluence and that personal effort had little to do with it. The counselor helped him see that his entitlement attitude was unrealistic and that hard work played a key role in most payoffs. There were two significant final outcomes in Jensen's case. First, he made it through the probation period free of any further shoplifting attempts. Second, he acquired and kept a job that helped him pay his debt to the retailer.

Taussig (1987) talks about the usefulness of setting and executing minigoals early in the helping process. Consider the case of Gaston:

> Gaston, a 16-year-old school dropout and loner, was arrested for arson. Though he lived in the inner city and came from a single-parent household, it was difficult to discover just why he had

turned to arson. He had torched a few structures that seemed relatively safe to burn. No one was injured. Was his behavior a cry for help? Social rage expressed in vandalism? Just a way of getting some kicks? The social worker assigned to the case found these questions too speculative to be of much help. Instead of looking for the root causes of Gaston's malaise, she tried to help him set some simple goals that appealed to him and that could be accomplished relatively quickly. One goal was social support. The counselor helped Gaston join a social club at a local youth center. A second goal was having a role model. Gaston struck up a friendship with one of the more active members of the center, a dropout who had gotten a high school equivalency degree. He also received some special attention from one of the adult monitors of the center. This was the first time he had experienced the presence of a strong adult male in his life. A third goal was broadening his view of the world. A group of college students who did volunteer work in both the black and the white communities invited Gaston and a couple of the other boys to help them in a housing facility for the elderly located in a white neighborhood. This was the first time he had been engaged in any kind of work outside the black community. The experience helped him push back the walls a bit. He saw white people with real needs. The accomplishment of these minigoals helped Gaston become a bit more realistic about the world around him. He enjoyed the camaraderie of the volunteer group and began experiencing himself in a new, more constructive way.

It is not suggested here that goal setting is a façile answer to intractable social problems. But the achievement of sequenced minigoals can go a long way toward making a dent in intractable problems.

There is no such thing as a set time frame for every client. Some goals need to be accomplished now, some soon; others are short-term goals; still others are long term. Consider the case of a priest who had been unjustly accused of child molestation.

- A *"now" goal*: some immediate relief from debilitating anxiety attacks and keeping his equilibrium during the investigation and court procedings.

- A *"soon" goal*: obtaining the right kind of legal aid.

- A *short-term goal*: winning the court case.

- A *long-term goal*: reestablishing his credibility in the community and learning how to live with those who would continue to suspect him.

There is no particular formula for helping all clients choose the right mix of goals at the right time in the right sequence. Although helping is based on problem-management principles, it remains an art.

It is not always necessary, then, to make sure that each goal in a client's program for constructive change has all the characteristics outlined in this chapter. For some clients, identifying broad goals is enough to kick start the entire problem-management and opportunity-development process. They shape the goals themselves. For others, some help in formulating more specific goals is called for. The principle is clear: Help clients develop goals that have some sort of agency—if not urgency—built in. In one case, this may mean helping a client deal with clarity; in another, with substance; in still another, with realism, values, or time frame. Box 16-1 outlines some questions that you can help clients ask themselves to choose goals from among possibilities.

## NEEDS VERSUS WANTS

In some cases, what clients want and what they need coincide. The lonely person wants a better social life and needs some kind of community to be more engaged in

## Box 16-1    Questions for Shaping Goals

- Is the goal stated in outcome or results language?
- Is the goal specific enough to drive behavior? How will I know when I have accomplished it?
- If I accomplish this goal, will it make a difference? Will it really help manage the problems and opportunities I have identified?
- Does this goal have "bite" while remaining prudent?
- Is it doable?
- Can I sustain this goal over the long haul?
- Does this goal have some flexibility?
- Is this goal in keeping with my values?
- Have I set a realistic time frame for the accomplishment of the goal?

life. In other cases, clients might not want what they need. The alcoholic may need a life of total abstention but wants to drink moderately. Brainstorming possibilities for a better future should focus on the package of needs and wants that makes sense for this particular client. Consider the case of Irv:

Irv, a 41-year-old entrepreneur, collapsed one day at work. He had not had a physical in years. He was shocked to learn that he had both a mild heart condition and multiple sclerosis. His future was uncertain. The father of one of his wife's friends had multiple sclerosis but had lived and worked well into his seventies. But no one knew what the course of the disease would be. Irv had made his living by developing and then selling small businesses. He loved his work and wanted to continue to do it. But what he needed was a less physically demanding work schedule. Working 60 to 70 hours per week was no longer in the cards. Furthermore, he had always plowed the money he received from selling one business into starting up another. But now he needed to think of the future financial well-being of his wife and three children. Up to this point, his philosophy had been that the future would take care of itself. It was very wrenching for him to move from a lifestyle he wanted to one he needed.

Irv was a voluntary client who had to look at needs instead of wants. Involuntary clients often need to be challenged to look beyond their wants to their needs. One woman who voluntarily led a homeless life was attacked and severely beaten on the street. But she still wanted the freedom that came with her lifestyle. When a court-appointed counselor challenged her to consider the kinds of freedom she wanted, she admitted that freedom from responsibility was at the core. "I want to do what I want to do when I want to do it." It was her choice to live the way she did. The counselor helped her explore the consequences of her choices and tried to help her look at other options. How could she be "free" and not at risk? Was there some kind of trade-off between what she wanted and what she needed? In the end, of course, the decision was hers.

In the following case, the client, dogged by depression, was ultimately able to integrate what he wanted with what he needed:

Milos had come to the United States as a political refugee. The last few months in his native land had been terrifying. He had been jailed and beaten. He got out just before another crackdown. Once the initial euphoria of having escaped had subsided, he spent months feeling confused and disorganized. He tried to live as he had in his own country, but the North American culture was too invasive. He thought he should feel grateful, and yet he felt hostile. After two years of misery, he began seeing a counselor. He had resisted getting help because "back home" he had been "his own man."

In discussing these issues with a counselor, it gradually dawned on him that he *wanted* to reestablish links with his native land but that he *needed* to integrate himself into the life of his host country. He saw that the accomplishment of both these broad aims would be very freeing. He began finding out how other immigrants who had been here longer than he had accomplished this goal. He spent time in the immigrant community, which differed from the refugee community. In the immigrant community, there was a long history of keeping links to the homeland culture alive. But the immigrants had also adapted to their adopted country in practical ways that made sense to them. The friends he made became role models for him. The more active he became in the immigrant community, the more his depression lifted.

In this case goals responded to a mixture of needs and wants. If Milos had focused only on one or the other, he would have remained unhappy.

## EMERGING GOALS

It is not always a question of *designing* and *setting* goals in an explicit way. Rather, goals can naturally emerge through the client-helper dialogue. Often when clients talk about problems and unused opportunities, possible goals and action strategies bubble up. Clients, once they are helped to clarify a problem situation through a combination of probing, empathic highlights, and challenge, begin to see more clearly what they want and what they have to do to manage the problem. Indeed, some clients must first act in some way before they find out just what they want to do. Once goals begin to emerge, counselors can help clients clarify them and find ways of implementing them. However, "emerge" should not mean that clients wait around until "something comes up." Nor should it mean that clients try many different solutions in the hope that one of them will work. These kinds of "emergence" tend to be self-defeating.

Though goals do often emerge, explicit goal setting is not to be underrated. Taussig (1987) showed that clients respond positively to goal setting even when goals are set very early in the counseling process. A client-centered, "no one right formula" approach seems to be best. Although all clients need focus and direction in managing problems and developing opportunities, what focus and direction will look like will differ from client to client.

## ADAPTIVE GOALS

Collins and Porras (1994) coined the term "big, hairy, audacious goals" (BHAGs) for "superstretch" goals. The term, however, fits better into the hype of business than the practicalities of helping. It is true that some clients are looking for big goals. They believe, and perhaps rightly so, that without big goals their lives will not be substantially different. But even for clients who choose goals that can be called "big" in one way or another, a bit-by-bit approach to achieving these goals is needed. It is usually better to take a big goal and divide it into smaller pieces lest the big goal on its own seem too daunting. The meaning of the phrase "within reasonable limits" will differ from client to client.

**Satisfactory alternatives.**  While difficult or "stretch" goals are often the most motivational, this is not true in every case. Some clients choose to make very substantive changes in their lives, but others take a more modest approach. Wheeler and Janis (1980) caution against the search for the "absolute best" goal all the time: "Sometimes it is more reasonable to choose a satisfactory alternative than to continue searching for the absolute best. The time, energy, and expense of finding the best possible choice may outweigh the improvement in the choice" (p. 98). Consider the following case:

> Joyce, a buyer for a large retail chain, is nearing middle age. She has centered most of her non-working life on her aging mother. Joyce had even turned down promotions because the new positions would have demanded more travel and longer hours. Her mother had been pampered by her now-deceased husband and her three children and allowed to have her way all her life. She now played the role of the tyrannical old woman who constantly feels neglected and who can never be satisfied. Though Joyce knew that she could live much more independently without abandoning her mother, she found it very difficult to move in that direction. Guilt stood in the way of any change in her relationship with her mother. She even said that being a virtual slave to her mother's whims was not as bad as the guilt she experienced when she stood up to her mother or "neglected" her.
>
> The counselor helped Joyce experiment with a few new ways of dealing with her mother. For instance, Joyce went on a two-week trip with friends even though her mother objected, saying that it was ill timed. Although the experiments were successful in that no harm was done to Joyce's mother and Joyce did not experience excessive guilt, counseling did not help her restructure her relationship with her mother in any substantial way. The experiments, however, did give her a sense of greater freedom. For instance, she felt freer to say no to this or that demand of her mother. This provided enough slack, it seems, to make Joyce's life more livable.

In this case, counseling helped the client fashion a life that was "a little bit better," though not as good as the counselor thought it could be. When asked, "What do you want?" Joyce had in effect replied, "I want a bit more slack and freedom, but I do not want to abandon my mother." Joyce's "new" lifestyle did not differ dramatically from the old. But perhaps it was enough for her. It was a case of choosing a satisfactory alternative rather than the best.

Leahey and Wallace (1988) offer the following example of another client in adaptive mode:

> "For the last five years, I've thought of myself as a person with low self-esteem and have read self-help books, gone to therapists, and put things off until I felt I had good self-esteem. I just need to get on with my life, and I can do that with excellent self-esteem or poor self-esteem. Treatment isn't really necessary. Being a person with enough self-esteem to handle situations is good enough for me." (p. 216)

The following client, putting a more positive spin on the problem situation itself, takes a more adaptive route:

> "I would say that I am completely cured. . . . I can still pinpoint these conditions which I had thought to be symptoms. . . . These worries and anxieties make me prepare thoroughly for the daily work I have to do. They prevent

me from being careless. They are expressions of the desire to grow and to develop." (Weisz, Rothbaum, & Blackburn, 1984, p. 964)

In some cases, clients will be satisfied with "surface" solutions such as the elimination of symptoms. For instance, a couple is satisfied with reducing and managing the petty annoyances they both experience in their relationship. Although the very structure of the relationship may be problematic because some fundamental inequalities or inequities are built into the relationship, this couple doesn't want to do much about restructuring the relationship to avoid the annoyances they experience.

Some helpers, reviewing these examples, would be disappointed. Others would see them as legitimate examples of adapting to rather than changing reality. However, all these clients did act to achieve some kind of goal, however minimal. They did *something* about the way they thought and behaved. And they felt that their lives were better because of it.

**Coping.** Choosing an adaptive rather than a stretch goal has been associated with *coping* (Coyne & Racioppo, 2000; Folkman & Moskowitz, 2000; Lazarus, 2000; Snyder, 1999). All human beings cope rather than conquer at times. In fact, in human affairs as a whole, coping probably outstrips conquering. And sometimes people have no other choice. It's cope or succumb. For some, coping has a bad reputation because it seems to be associated with mediocrity. But in many difficult situations, helping clients cope is one of the best things helpers can do. Coping often has an enormous upside. A young mother with three children has just lost her husband. When someone asks, "How's she doing?" the response is, "She's coping quite well." She's not letting her grief get the better of her. She is taking care of the children and helping them deal with their sense of loss. She's moving along on all the tasks that a death in a family entails. At this stage, what could be more positive than that?

Folkman and Moskowitz (2000), from a positive-psychology point of view, see positive affect as playing an important role in coping. And so they ask how positive affect is generated and sustained in the face of chronic stress. They suggest three ways:

• *Positive reappraisal:* Help clients reframe situations to see them in a positive light. For instance, Victor, recovering from multiple injuries received when he fell off his bicycle, sees the entire rehabilitation process as "one big daunting glob." Taken as a whole, it looks undoable. However, the rehabilitation counselor first helps Victor picture himself once more engaging in the ordinary task of everyday life, even riding a bicycle. That is, she helps him separate the very desirable end state from the arduous set of activities that will get him there. Victor does not ever have to cope with the "big glob." He needs to cope with each day. Victor is rebuilding his body. Every day he is doing something to forge a link in the recovery chain. Each week he is helped to see that there is something he can now do that he was not able to do the previous week. Victor has low moments. Of course. But he also has moments of positive affect that keep him going.

• *Problem-focused coping:* Help clients deal with problems one at a time as they come up. For instance, Agnes is caring for her husband who has multiple sclerosis. There is a certain unpredictability and uncontrollability associated with her husband's disease. However, she does not have to cope with his multiple sclerosis. Rather, each day or each week or each stage brings its own set of problems. Her counselor can help her "pursue realistic, attainable goals by focusing on specific proximal tasks or problems related to caregiving" (Folkman and Moskowitz, p. 650). Agnes is heartened by the very fact that she faces and deals with each problem as it arises. The sense of mastery and control she experiences is accompanied by positive affect. Even in the face of great stress, she is buoyed up enough to move on to the next step or stage with grace.

• *Infusing ordinary events with positive meaning:* In one study, Folkman and Moskowitz asked the participants, all caregivers for people with AIDS, to describe something they did or something they experienced that made them feel good and helped them get through the day. More than 99% of the caregivers interviewed talked about some such event. The point is that, even at times of great stress, people note and remember positive events. The events were not "big deals." Rather they were "ordinary events," such as having dinner with a friend or seeing some flowers in a hospital room or receiving a compliment from someone. But these events together with the positive affect they produced helped them get through the day.

Lazarus (2000) adds a note of caution to all of this. He notes that so-called positively valenced emotions such as love and hope are often mixed with negative feelings and are therefore experienced as distressing. It is painful for caregivers to see those they love in pain. And so-called negatively valenced emotions such as anger are not unequivocally negative. Anger can be experienced as positive or is often mixed with positive feelings. While counselors can help clients under great stress do things that will increase the kind of positive affect that makes their lives more livable, there are limits. In other words, Lazarus is cautioning us to use but be careful with positive-psychology approaches.

**Strategic self-limitation.** Robert Leahy (1999) relates the kinds of reluctance and resistance reviewed in Chapter 9 to goal setting under the rubric of "strategic self-limitation." Reluctant and resistant behaviors serve the purpose of setting limits on change. All change carries some risk and uncertainty, and these can be distressing in themselves. Putting up barriers to change limits both risk and uncertainty. It is the client's way of saying, "Enough is enough. I don't want to engage in a change program that will lead to further effort, stress, failure, and regret." The strategies such clients use are the ordinary ones: attacking the therapist, failing to do homework assignments, being emotionally volatile, getting mired down in a "this won't work" mentality, and so forth. Helpers, even though they can point out to clients the ways they are engaging in what Leahy calls "self-handicapping," don't choose goals for clients. There is a huge difference between best possible goals and goals that are possible for this client in this set of circumstances.

The main point, however, is that helping clients cope with the adversities of life does not mean that you are shortchanging them. When you are helping them adapt rather than conquer, you are not failing. Neither are they. When it comes to outcomes, there is no one universal rule of success.

## THE "REAL-OPTIONS" APPROACH

How can you help clients set goals if the future is uncertain—as it always is to one degree or another? One way is through the "real-options" approach. Borrowed from business settings (Trigeorgis, 1999), it has applications to personal life. The trick is flexibility. If the future is uncertain, it pays to have a broad range of options open. There is no use investing a great deal of time and energy designing a goal that will have to be changed because the client's world changes. The economics are poor. Therefore, help clients choose one or more backup goals to take care of such eventualities. If a client comes up with three viable possibilities, he or she may pursue one while holding the other two in reserve. In this way, the client has not only direction but also a contingency plan. If the world changes, the client can choose the best goals—that is, the one that best fits the circumstances at the time. Consider this example:

> Linda is a young woman working for a computer firm in Mexico. Born and raised in Iraq (Linda is not her real name), she has made a tortuous journey through South and Central America as an illegal immigrant. Her journey included prostitution and a range of harrowing, even life-threatening experiences. The upside of all this is that she has learned to live by her wits. After returning from an illegal trip to the United States, she has one goal: to live there permanently. She takes counsel with a friend of hers, a lawyer in Mexico, telling him of her plan to live as an illegal in the United States. Both intelligent and socially savvy, she feels that she can pull it off.
>
> Her lawyer friend, knowing that her ultimate goal is to live permanently in the United States, helps her review a range of instrumental goals—goals in themselves but also steps toward helping her achieve her ultimate goal. They dialogue about possibilities. Options other than living by her wits as an illegal immigrant include political refugee status, obtaining a green card, marrying a U.S. citizen, marrying a foreigner who is most likely to get a green card, and being included in the quota of immigrants allowed permanent resident status because they have essential skills such as those needed in booming technology industries. A set of strategies would be needed to pursue each of these. Linda's future is certainly filled with risk and uncertainty. She now has to choose an instrumental goal that she thinks offers the best possibility for success. But now she also has fallback options.

While clients can identify and develop further goal options as the risky and uncertain world changes, doing so upfront has advantages. We have already seen how the real-options approach applies to clients fashioning a career in medicine. If the ultimate goal is to have a satisfying career in health care, the first goal might be to become a doctor. However, if there is risk and uncertainty in the pursuit of this goal, then a cluster of other career possibilities in health care can be reviewed and chosen as standby positions. The economics are better.

So choosing need not be a once-and-forever decision. The client thinks, "I'll go this route until another of my real options seems better." Furthermore, having real options helps the client kill an option that is no longer working. Or an option that is not working at this time can be put on the back burner. The real-options approach provides freedom and flexibility. This approach keeps clients from falling into the status quo decision-making trap outlined earlier.

## A BIAS FOR ACTION AS A METAGOAL

Although clients set goals that are directly related to their problem situations, there are also metagoals or superordinate goals that would make them more effective in pursuing the goals they set and in leading fuller lives. The overall goal of helping clients become more effective in problem management and opportunity development was mentioned in Chapter 1. Another metagoal is to help clients become more effective "agents" in life—doers rather than mere reactors, preventers rather than fixers, initiators rather than followers.

> Lawrence was liked by his superiors for two reasons. First, he was competent—he got things done. Second, he did whatever they wanted him to do. They moved him from job to job when it suited them. He never complained. However, as he matured and began to think more of his future, he realized that there was a great deal of truth in the adage "If you're not in charge of your own career, no one is." After a session with a career counselor, he outlined the kind of career he wanted and presented it to his superiors. He pointed out to them how this would serve both the company's interests and his own. At first they were taken aback by Lawrence's assertiveness, but then they agreed. Later, when they seemed to be sidetracking him, he stood up for his rights. Assertiveness was his bias for action.

The doer is more likely to pursue stretch goals rather than adaptive goals in managing problems. The doer is also more likely to move beyond problem management to opportunity development.

### Evaluation Questions for Step II-B

- To what degree am I helping clients choose specific goals from among a number of possibilities?
- How well do I challenge clients to translate good intentions into broad goals and broad goals into specific, actionable goals?
- To what extent do I help clients shape goals that have the characteristics outlined in Box 16-1?
- How effectively do I help clients establish goals that take into consideration both needs and wants?
- To what degree do I help clients become aware of goals that are naturally emerging from the helping process?
- How well do I help clients identify real-option goals when the future is both risky and uncertain?
- How effectively do I help clients choose the right mix of adaptive and stretch goals?
- How well do I help clients explore the consequences of the goals they are setting?
- How do I help clients make a bias toward action one of their metagoals?

# Step II-C: "What Am I Willing to Pay for What I Want?" Commitment

**Helping Clients Commit Themselves to a Better Future**
Help Clients Set Goals That Are Worth More Than They Cost
Help Clients Set Appealing Goals
Help Clients Own the Goals They Set
Help Clients Deal with Competing Agendas

**Great Expectations: Client Self-Efficacy**
The Nature of Self-Efficacy
Help Clients Develop Self-Efficacy
Skills
Corrective feedback
Positive feedback
Using success as a reinforcer
Models
Providing encouragement
Reducing fear and anxiety

**Stage II and Action**

**The Shadow Side of Goal Setting**

**Evaluation Questions for Step II-C**

As mentioned earlier, Step II-C is not really a step in the true sense of the term but a dimension of the goal-setting process. Clients may formulate goals, but that does not mean that they are willing to pay for them. Once clients state what they want and set goals, the battle is joined, as it were. It is as if clients' "old selves" or old lifestyles begin vying for resources with their potential "new selves" or new lifestyles. On a more positive note, history is full of examples of people whose strength of will to accomplish some goal has enabled them to do seemingly impossible things.

> A woman with two sons in their twenties was dying of cancer. The doctors thought she could go at any time. However, one day she told the doctor that she wanted to live to see her older son get married in six months. The doctor talked vaguely about "trusting in God" and "playing the cards she had been dealt." Against all odds, the woman lived to see her son get married. Her doctor was at the wedding. During the reception, he went up to her and said, "Well, you got what you wanted. Despite the way things are going, you must be deeply satisfied." She looked at him wryly and said, "But Doctor, my second son will get married someday."

Although the job of counselors is not to encourage clients to heroic efforts, counselors should not undersell clients, either.

In Step II-C, which is usually intermingled with the other two steps of Stage II, counselors help their clients pose and answer such questions as these:

- Why should I pursue this goal?
- Is it worth it?
- Is this where I want to invest my limited resources of time, money, and energy?
- What competes for my attention?
- What are the incentives for pursuing this agenda?
- How strong are competing agendas?

Again, there is no formula. Some clients, once they establish goals, race to accomplish them. At the other end of the spectrum are clients who, once they decide on goals, stop dead in the water. Furthermore, the same client might speed toward the accomplishment of one goal and crawl toward another. Or start out fast and then slow to a crawl. The job of the counselor is to help clients face up to their commitments.

## HELPING CLIENTS COMMIT THEMSELVES TO A BETTER FUTURE

There is a difference between initial commitment to a goal and an ongoing commitment to a strategy or plan to accomplish the goal. The proof of initial commitment lies in goal-accomplishing action. For instance, one client who chose as a goal a less abrasive interpersonal style began to engage in an "examination of conscience" each evening to review what his interactions with people had been like that day. In doing so, he discovered, somewhat painfully, that in some of his interactions he actually moved beyond abrasiveness to contempt. That forced him back to a deeper analysis of the problem situation and the blind spots associated with it. Being dismissive of people he did not like or who were "not important" had become ingrained in his interpersonal lifestyle.

There are several things you can do to help clients in their initial commitment to goals and the kind of action that is a sign of that commitment: You can help clients set goals that are "cost-effective," help them make goals appealing, help them enhance their sense of ownership, and help them deal with competing agendas.

## Help Clients Set Goals That Are Worth More Than They Cost

Here we revisit the economics of helping. Cost-effectiveness could have been included in the characteristics of workable goals outlined in the previous chapter, but it is considered here instead because of its close relationship to commitment. Some goals that can be accomplished carry too high a cost in relation to their payoff. It may sound overly technical to ask whether any given goal is "cost-effective," but the principle remains important. Skilled counselors help clients budget rather than squander their resources—work, time, emotional energy.

> Eunice discovered that she had a terminal illness. In talking with several doctors, she found out that she would be able to prolong her life a bit through a combination of surgery, radiation treatment, and chemotherapy. However, no one suggested that these would lead to a cure. She also found out what each form of treatment and each combination would cost, not so much in monetary terms, but in added anxiety and pain. Ultimately, she decided against all three, since no combination of them promised much for the quality of the life that was being prolonged. Instead, with the help of a doctor who was an expert in hospice care, she developed a scenario that would ease both her anxiety and her physical pain as much as possible.

It goes without saying that another patient might have made a different decision. Costs and payoffs are relative. Some clients might value an extra month of life no matter what the cost.

Since it is often impossible to determine the cost-benefit ratio of any particular goal, counselors can add value by helping clients understand the consequences of choosing a particular goal. For instance, a client who sets her sights on a routine job with minimally adequate pay might find that this outcome takes care of some of her immediate needs but proves to be a poor choice in the long run. Helping clients foresee the consequences of their choices may not be easy. Another woman with cancer felt that she was no longer able to cope with the sickness and depression that came with her chemotherapy treatments. She decided abruptly one day to end the treatment, saying that she didn't care what happened. No one helped her explore the consequences of her decision. Eventually, when her health deteriorated, she had second thoughts about the treatments, saying, "There are still a number of things I must do before I die." But it was too late. Some reasonable challenge on the part of a helper might have helped her make a better decision.

The balance-sheet methodology outlined in Chapter 19 is a tool you can use selectively to help clients weigh costs against benefits in choosing both goals and the programs to implement goals. The balance sheet, as used in Chapter 19, also helps clients choose best-fit strategies for accomplishing their goals.

## Help Clients Set Appealing Goals

Just because goals will help in managing a problem situation or developing an opportunity and are cost-effective does not mean that they will automatically appeal to the client. Setting appealing goals is common sense, but it is not always easy to do. For instance, for many if not most addicts, a drug-free life is not immediately appealing, to say the least.

> A counselor tries to help Chester work through his resistance to giving up prescription drugs. While he listens and is empathic, the counselor also challenges the way Chester has come to think about drugs and his dependency on them. One day the counselor says something about "giving up the crutch and walking straight." In a flash Chester sees himself not as a drug addict but as a "cripple." A friend of his had lost a leg in a land-mine explosion in Kosovo. He remembered how his friend had longed for the day when he could be fitted with a prosthesis and throw his crutches away. The image of "throwing away the crutch" and "walking straight" proved to be very appealing to Chester.

An incentive is a promise of a reward. As such, incentives can contribute to developing a climate of hope around problem management and opportunity development. A goal is appealing if there are incentives for pursuing it. Counselors need to help clients in their search for incentives throughout the helping process. Ordinarily, negative goals—giving up something that is harmful—need to be translated into positive goals—getting something that is helpful. It was much easier for Chester to commit himself to returning to school than to giving up prescription drugs, because school represented something he was getting. Images of himself with a degree and of holding some kind of professional job were solid incentives. The picture of him "throwing away the crutch" proved to be an important incentive in cutting down on drug use.

## Help Clients Own the Goals They Set

In Chapters 10–12 we discussed how important it is for clients to "own" the problems and unused opportunities they talk about. It is also important for them to own the goals they set. It is essential that the goals clients choose be the clients' rather than the helpers' or someone else's goals. Various kinds of probes can be used to help clients discover what they want to do to manage some dimension of their problem situations more effectively. For instance, Carl Rogers, in a film of a counseling session (Rogers, Perls, & Ellis, 1965), is asked by a woman what she should do about her relationship with her daughter. He says to her, "I think you've been telling me all along what you want to do." She knew what she wanted the relationship to look like, but she was asking for his approval. If he had given it, the goal would have become, to some degree, his goal instead of hers. At another time, Rogers asks, "What is it that you want me to tell you to do?" This question puts the responsibility for goal setting where it belongs—on the shoulders of the client.

> Cynthia was dealing with a lawyer because of an impending divorce. Discussions about what would happen to the children had taken place, but no decision had been reached. One day she came in and said that she had decided on mutual custody. She wanted to work out such details as which residence, hers or her husband's, would be the children's principal one and so forth. The lawyer asked her how she had reached her decision. She said that she had been talking to her husband's parents (she was still on good terms with them) and that they had suggested this arrangement. The lawyer challenged Cynthia to take a closer look at her decision. "Let's start from zero," he said, "and you tell me what kind of living arrangements *you* want and why." He did not think that it was wise to help her carry out a decision that was not her own.

Choosing goals suggested by others enables clients to blame others if they fail to reach the goals. Also, if they simply follow other people's advice, they often fail to explore the down-the-road consequences.

Commitment to goals can take different forms: compliance, buy-in, and ownership. The least useful is mere compliance. "Well, I guess I'll have to change some of my habits if I want to keep my marriage afloat" does not augur well for sustaining changes in behavior. But it may be better than nothing. Buy-in is a level up from compliance: "Yes, these changes are essential if we are to have a marriage that makes sense for both of us. We say we want to preserve our marriage, but now we have to prove it to ourselves." This client has moved beyond mere compliance. But like mere compliance, sometimes buy-in alone does not provide enough staying power because it depends too much on reason. "This is logical" is far different from "This is what I really want!" Ownership is a higher form of commitment. It means that the client can say, "This goal is not someone else's, it's not just a good idea; it is mine, it is what I want to do." Consider the following case:

> A counselor worked with a manager whose superiors had intimated that he would not be moving much further in his career unless he changed his style in dealing with the members of his team and other key people with whom he worked within the organization. At first the manager resisted setting any goals. "What they want me to do is a lot of hogwash. It won't do anything to make the business better," was his initial response. One day, when asked whether accomplishing what "they" wanted him to do would cost him that much, he pondered a few moments and then said, "No, not really." That got him started. He moved beyond resistance.
>
> With a bit of help from the counselor, he identified a few areas of his managerial style that could well be "polished up." Within a few months he got much more into the swing of things. Given the favorable responses to his changed behavior that he had gotten from the people who reported to him, he was able to say, "Well, I now see that this makes sense. But I'm doing it because it has a positive effect on the people in the department. It's the right thing to do." Buy-in had arrived. A year later, he moved up another notch. He became much more proactive in finding ways to improve his style. He delegated more, gave people feedback, asked for feedback, held a couple of managerial retreats, joined a human-resource task force, and routinely rewarded his direct reports for their successes. Now he began to say such things as "This is actually fun." Ownership had arrived. The people in his department began to see him as one of the best executives in the company. This process took over two years.

The manager did not have a personality transformation. He did not change his opinion of some of his superiors and was right in pointing out that they didn't follow their own rules. But he did change his behavior because he gradually discovered meaningful incentives to do so.

The use of contracts to structure the helping process itself was discussed in Chapter 3. Self-contracts—that is, contracts that clients make with themselves— can also help clients commit themselves to new courses of action. Although contracts are promises clients make to themselves to behave in certain ways and to attain certain goals, they are also ways of making goals more focused. It is not only the expressed or implied promise that helps but also the explicitness of the commitment. Consider the following example in which one of Dora's sons disappears without a trace.

> About a month after one of Dora's two young sons disappeared, she began to grow listless and depressed. She was separated from her husband at the time the boy disappeared. By the time she saw a counselor a few months later, a pattern of depressed behavior was quite pro-

nounced. Although her conversations with the counselor helped ease her feelings of guilt—for instance, she stopped engaging in self-blaming rituals—she remained listless. She shunned relatives and friends, kept to herself at work, and even distanced herself emotionally from her other son. She resisted developing images of a better future, because the only better future she would allow herself to imagine was one in which her son had returned.

   Some strong challenging from Dora's sister-in-law, who visited her from time to time, helped jar her loose from her preoccupation with her own misery. "You're trying to solve one hurt, the loss of Bobby, by hurting Timmy and hurting yourself. I can't imagine in a thousand years that this is what Bobby would want!" her sister-in-law screamed at her one night. Afterward Dora and the counselor discussed a "recommitment" to Timmy, to herself, to the extended family, and to their home. Through a series of contracts, she began to reintroduce patterns of behavior that had been characteristic of her before the tragedy. For instance, she contracted to opening her life up to relatives and friends once more, to creating a much more positive atmosphere at home, to encouraging Timmy to have his friends over, and so forth. Contracts worked for Dora because, as she said to the counselor, "I'm a person of my word."

When Dora first began implementing these goals, she felt she was just going through the motions. However, what she was really doing was acting herself into a new mode of thinking. Contracts helped Dora in both her initial commitment to a goal and her movement to action. In counseling, contracts are not legal documents but human instruments to be used if they are helpful. They often provide both the structure and the incentives some clients need.

## Help Clients Deal with Competing Agendas

Clients often set goals and formulate programs for constructive change without taking into account competing agendas—other things in their lives that soak up time and energy, such as job, family, and leisure pursuits. The world is filled with distractions. For instance, one manager wanted to begin developing computer and Internet-related skills, but the daily push of business and a divorce set up competing agendas and sapped his resources. Not one of the goals of his self-development agenda was accomplished. Programs for constructive change often involve a rearrangement of priorities. If a client is to be a full partner in the reinvention of his marriage, he cannot spend as much time "with the boys." Or the underemployed blue-collar worker might have to put aside some parts of her social life if she wants a more fulfilling job. She eventually discovers a compromise. A friend introduces her to the job search possibilities on the Internet. She discovers that she can work full time to support herself, do a better job looking for new employment on the Internet than by using traditional methods, and still have some time for a reasonable social life.

   This is not to suggest that all competing agendas are frivolous. Sometimes clients have to choose between right and right. The woman who wants to expand her horizons by getting involved in social settings outside the home still has to figure out how to handle the tasks at home. This is a question of balance, not frivolity. The single parent who wants a promotion at work needs to balance her new responsibilities with involvement with her children. A counselor who had worked with a two-career couple as they made a decision to have a child helped them think of competing agendas once the pregnancy started. A year after the baby was born, they saw the counselor again for a couple of sessions to work on some issues that had come up. However, they started the session by saying, "Are we glad that you talked about competing agendas when we were struggling with the decision to become

**Box 17-1    Questions on Client Commitment**

You can help clients ask themselves these kinds of questions as they struggle with committing themselves to a program of constructive change:

- What is my state of readiness for change in this area at this time?
- How badly do I want what I say I want?
- How hard am I willing to work?
- To what degree am I choosing this goal freely?
- How highly do I rate the personal appeal of this goal?
- How do I know I have the courage to work on this?
- What's pushing me to choose this goal?
- What incentives do I have for pursuing this change agenda?
- What rewards can I expect if I work on this agenda?
- If this goal is in any way being imposed by others, what am I doing to make it my own?
- What difficulties am I experiencing in committing myself to this goal?
- In what way is it possible that my commitment is not a true commitment?
- What can I do to get rid of the disincentives and overcome the obstacles?
- What can I do to increase my commitment?
- In what ways can the goal be reformulated to make it more appealing?
- To what degree is the timing for pursuing this goal poor?
- What do I have to do to stay committed?
- What resources can help me?

parents! After the baby was born, we went back time and time again to review what we said about managing competing and conflicting priorities. It helped stabilize us for the last two years."

Even self-contracts have a shadow side. There is no such thing as a perfect contract. Most people don't think through the consequences of all the provisions of a contract, whether it be marriage, employment, or self-contracts designed to enhance a client's commitment to goals. And even people of goodwill unknowingly add covert codicils to contracts they make with themselves and others: "I'll pursue this goal—until it begins to hurt," or "I won't be abusive—unless she pushes me to the wall." The codicils are buried deep in the decision-making process and only gradually make their way to the surface.

Box 17-1 indicates the kinds of questions you can help clients ask themselves about their commitment to their change agendas.

## GREAT EXPECTATIONS: CLIENT SELF-EFFICACY

The role of expectations in life is being explored more broadly, more deeply, and more practically (Kirsch, 1999). Clients need to find the motivation to seize their goals and run with them. The more they find their motivation within themselves the better. "Self-regulation" is the ideal. Helping clients choose goals, commit to them, and develop a sense of agency and assertiveness (Galassi & Bruch, 1992) are part of the self-regulation picture. Expectations, whether "great" or not, are also part of the self-regulation picture. Here we look at client expectations through the lens of "self-efficacy" (Bandura, 1986, 1989, 1991, 1995, 1997; Cervone, 2000; Cervone & Scott, 1995; Lightsey, 1996; Locke & Latham, 1990; Maddux, 1995; Schwarzer, 1992). Self-efficacy is an extremely useful concept when it comes to constructive change. It is impossible to do justice to it here. What follows will, hopefully, pique your interest and help you relate self-efficacy to helping. You can feast on the vast self-efficacy literature later.

### The Nature of Self-Efficacy

As Bandura (1995) notes, "Perceived self-efficacy refers to beliefs in one's capabilities to organize and execute the courses of action required to manage prospective situations. Efficacy beliefs influence how people think, feel, motivate themselves, and act" (p. 2). People's expectations of themselves and can-do beliefs have a great deal to do with their willingness to put forth effort to cope with difficulties, the amount of effort they will expend, and their persistence in the face of obstacles. Clients with higher self-efficacy will make bolder choices, moving from adaptation to stretch goals. Clients tend to take action if two conditions are fulfilled:

- *Outcome expectations:* Clients tend to act if they see that their actions will most likely lead to certain desirable results or accomplishments: "I will end up with a better relationship with Sophie."

- *Self-efficacy beliefs:* People tend to act if they were reasonably sure that they have the wherewithal—for instance, working knowledge, skill, time, stamina, guts, and other resources—to successfully engage in the kind of behavior that would lead to the desired outcomes. "I have the ability deal with the conflicts Sophie and I have. I can do this. I'm going to do this."

Now let's see these two factors operating together in a few examples. Yolanda, who has had a stroke, not only believes that participation in a rather painful and demanding physical rehabilitation program will literally help her get on her feet again (an outcome expectation) but also that she has what it takes to inch her way through the program (a self-efficacy belief). She therefore enters the program with a very positive attitude and makes good progress. Yves, on the other hand, is not convinced that an aggressive drug rehabilitation program will lead to a more fulfilling life (a negative outcome expectation), even though he knows he could "get through" the program (a self-efficacy belief). So he says no to the therapist. Even though the therapist has "promised" him a "drug free" life, Yves keeps saying to himself, "Drug free for what?" He sees being drug free as an instrumental goal. But he has not yet come up with an attractive ultimate goal. Xavier is convinced that a series of radiation and chemotherapy treatments would help him (a positive outcome expectation), but he does not feel that he has the stamina and courage to go through with them (a negative self-efficacy expectation). He, too, refuses the treatment.

Outcome expectations and self-efficacy beliefs are factors not just in helping but in everyday life. Do an Internet search on that term and you will find a rich literature covering all facets of life—for instance, applications to education (Lopez, Lent, Brown, and Gore, 1997; Multon, Brown, & Lent, 1991; Smith & Nadya, 1999; Zimmerman, 1996), health care (O'Leary, 1985; Schwarzer & Fuchs, 1995), physical rehabilitation (Altmaier, Russell, Kao, Lehmann, & Weinstein, 1993), and work (Donnay & Borgen, 1999).

## Help Clients Develop Self-Efficacy

People's sense of self-efficacy can be strengthened in a variety of ways (see Mager, 1992). Self-efficacy is not a paradigm that applies only to the weak. Take the case of Nick, a very strong manager who wanted to change his abrasive supervisory style but was doubtful that he could do so. "After all these years, I am what I am," he would say. It would have been silly to merely tell him, "Nick, you can do it; just believe in yourself." It was necessary to help him do a number of things to help strengthen his sense of self-efficacy in supervision.

**Skills.** *Make sure that clients have the skills they need to perform desired tasks.* Self-efficacy is based on ability and the conviction that the ability can be used to get a task done. Nick first read about and then attended some training sessions on building such "soft" skills as listening, responding with empathic highlights, giving feedback that is soft on the person yet hard on the problem, and constructive challenging. In truth, he had many of these skills, but they lay dormant. These short training experiences put him back in touch with some things he could do but didn't do. A caution: Merely acquiring skills does not by itself increase clients' self-efficacy. The way they acquire them must give them a sense of their competence. "I now have these skills and I am positive that I can use them to get this task done."

**Corrective feedback.** *Provide feedback that is based on deficiencies in performance, not on deficiencies in the client's personality.* Corrective feedback can help clients develop a sense of self-efficacy by clearing away barriers to the use of resources. Since I attended many meetings with Nick, I routinely described the ups and downs of his performance. I'd say such things as this:

> "Nick, in yesterday's meeting, you listened to and responded to everyone's ideas. Let me make a suggestion. You don't have to respond, as you did, in a positive way to every suggestion. Crap is still crap. Do some sorting as you listen and respond. Show why good ideas are good and why lousy ideas are bad. Then, whether the ideas are good or bad, everyone learns something."

When corrective feedback sounds like a personality attack, the client's sense of self-efficacy declines. The feedback helped boost Nick's self-efficacy belief because it pointed out that he could be decent and listen well and still use his excellent critical abilities. People would leave the room enlightened, not angry. When you give feedback to a client, you would do well to ask yourself, "In what ways will this feedback help increase the client's sense of self-efficacy?"

**Positive feedback.** *Provide positive feedback and make it as specific as corrective feedback.* Positive feedback strengthens clients' self-efficacy by emphasizing their strengths and reinforcing what they do well. This is especially true when feedback is specific. Too often negative feedback is very detailed, while positive feedback is perfunctory: "Nice job." This and other throwaway phrases probably sound like clichés. Here's one bit of feedback I gave Nick:

> "Yesterday, you interrupted Jeff, who was engaging in another one of his monologues. You summarized his main ideas. Then, with a few questions, you showed him why only part of his plan was viable. The others were glad you took Jeff on. He learned something. And you saved a lot of time."

The formula for giving specific positive feedback goes something like this. "Here's what you did. Here's the positive outcome it had. And here's the wider upbeat impact." Helping Nick see the value of this pattern of behavior helped him engage in it more frequently and increased his sense of self-efficacy: "I can combine the hard stuff and the soft stuff." Clients need to interpret feedback as information they need to accomplish tasks.

**Using success as a reinforcer.** *Challenge clients to engage in actions that produce positive results.* Even small successes can increase a client's sense of self-efficacy. Success is reinforcing. Often success in a small endeavor will give a client the courage to try something more difficult: "I can do even more." Nick began delegating a few minor tasks to some of his direct reports. They handled their assignments very well. When I commented, "They seem to be doing pretty well," Nick replied, "I think that I can safely begin to put more on their plates. They like it, and I like seeing them succeed." Successful delegation increased Nick's sense of supervisory self-efficacy. He could say to himself more assuredly, "I can delegate without worrying whether it's going to get done or not." Make sure, however, that the link between success and increased self-confidence is forged. A series of successes on its own does not necessarily increase the strength of a client's self-efficacy beliefs. Success has to be linked to a sense of increased competence.

**Models.** *Help clients increase their own sense of self-efficacy by learning from others.* I asked Nick to name the best manager in the division. He mentioned a name. "What's he like?" I asked. Of course, Nick talked about how competent this guy was, how effective he was in getting results, and how tough he was. Tongue in cheek, I remarked, "But I suppose that he's not very good with people." Nick exploded. "Of course he's fair. He's as good at all of this soft stuff as anyone else." He went on to name ways in which the guy was "good with people." Then suddenly he stopped, looked at me, and smiled. "Caught me, didn't you?" Learning makes clients more competent and increases their self-efficacy. Learning from models is, as we have seen, a bit tricky. Nick had too much pride to think that he could learn very much from others.

**Providing encouragement.** *Support clients' self-efficacy beliefs without being patronizing.* We took a brief look at encouragement in Chapter 11. However, if your support is to increase clients' sense of self-efficacy, your support must be real, and what you support in them must be real. Encouragement and support must be tailored to each client and in each instance. A supportive remark to one client might sound patronizing to another. Had I patronized Nick—"Give it a try, Nick, I know that you can do it"—I would have been dead. My encouragement was, let's say, more subtle and indirect.

**Reducing fear and anxiety.** *Help clients overcome their fears.* Fear blocks clients' sense of self-efficacy. If clients fear that they will fail, they will be reluctant to act. Therefore, procedures that reduce fear and anxiety help heighten their sense of self-efficacy. Deep down, Nick was fearful of two things regarding changing his supervisory style: messing up the business and making a fool of himself. As he tentatively changed some of his supervisory practices, business results held steady. He even noticed that two of his team members seemed to become more productive. Helping him allay his fear of making a fool of himself by being too soft was a bit trickier. His behavior outside the office came to the rescue. Although he was often an ogre in the office, Nick was very upbeat when we visited teams out in the field. He was as good at "rallying the troops" as anyone I had ever seen. And he was real. Discussions about his two different styles helped him get rid of fears that he would make a fool of himself with his direct reports by engaging them instead of driving them.

## STAGE II AND ACTION

The work of Step II-A—developing possibilities for a better future—is just what some clients need. It frees them from thinking solely about problem situations and unused resources and enables them to begin fashioning a better future. Once they identify some of their wants and needs and consider a few possible goals, they move into action.

> Francine is depressed because her aging and debilitated father has been picking on her even though she has put off marriage to take care of him. Some of the things he says to her are quite hurtful. The situation has begun to affect her productivity at work. A counselor suggests to her that the hurtful things her father says to her are not her father but his illness speaking. This gives her a whole new perspective and frees her to think about other possibilities. She spends a bit of time brainstorming answers to the counselor's question—"What do you want for both yourself and your father?" She says things like, "I'd like both of us to go through this with our dignity intact" and "I'd like to be living the kind of life he would want me to have if his mind weren't so clouded." Once she brainstorms some possibilities for a better arrangement with her father, she needs little further help. Her usual resourcefulness returns. She gets on with life.

For other clients, Step II-B is the trigger for action. Shaping goals helps them see the future in a very different way. Once they have a clear idea of just what they want or need, they go for it.

> Nero, a man in his early twenties, had a car accident while driving under the influence of alcohol—an accident that took his wife's life. Strangely, Nero is filled with self-pity rather than remorse. The counselor, at her wits' end, confronts him about remaining wrapped up in himself. She says, "Who's the most decent person you know?" After fudging around a bit, he names Saul, an uncle. "Describe his lifestyle to me," she urges. "What makes him so decent?" With some prodding, Nero describes the lifestyle of this decent man. Then she says, "Do the description again, but instead of saying 'Saul' say 'Nero.'" Nero sweats, but the session has an enormous impact on him. The picture of the contrast between his uncle's lifestyle and his own

haunts him for days after. But he begins to stop feeling so sorry for himself. He visits his wife's parents and begs their forgiveness. He begins to see that there are other people in the world besides Nero.

For still other clients, the search for incentives for commitment is the trigger for action. Once they see what's in it "for me"—a kind of upbeat and productive self-ishness, if you will—they move into action.

Callahan is seeing a consultant because he is very distressed. He owns and runs a small business. A few of his employees have gotten together and filed a workplace discrimination suit against him. The "troublemakers" he calls them, meaning a few women, a couple of Hispanics, and three African Americans. The consultant finds out that Callahan believes that they are "decent workers." The fact is that they are more than decent. Callahan tells the counselor that he is paying them "scale." The fact is that he is underpaying them. Callahan also says that he doesn't expect his supervisors to "bend over backwards to become their friends." The fact is that some supervisors—all but two are white males—are sometimes abusive.

The counselor convinces Callahan to attend an excellent program on diversity "before some court orders you to." A couple of weeks after returning from the program, he has a session with the consultant. He says that he had never even once considered the advantages of diversity in the workplace. All the term had meant to him was "a bunch of politicians looking for votes." Now that he saw the business reasons for diversity, he knew there were a few things he could do, but he still needed the consultant's help and guidance. "I don't want to look like a soft jerk." Callahan's newly acquired "human touch" is far from being soft. He remains a rather rough-and-tough business guy.

Callahan didn't change his stripes overnight, but finding a package of incentives certainly helped him move toward much-needed action. Who knows, the whole situation might have even made a dent in his deeply ingrained prejudices.

## THE SHADOW SIDE OF GOAL SETTING

Despite the advantages of goal setting outlined in the last three chapters, some helpers and clients seem to conspire to avoid goal setting as an explicit process. It is puzzling to see counselors helping clients explore problem situations and unused opportunities but stopping short of asking them what they want and helping them set goals. As Bandura (1990) put it, "Despite this unprecedented level of empirical support [for the advantages of goal setting], goal theory has not been accorded the prominence it deserves in mainstream psychology" (p. xii). Years ago, the same concern was expressed differently. A U.S. developmental psychologist was talking to a Russian developmental psychologist. The Russian said, "It seems to me that American researchers are constantly seeking to explain how a child came to be what he is. We in the USSR are striving to discover how he can become what he not yet is" (see Bronfenbrenner, 1977, p. 528). One of the main reasons that counselors do not help clients develop realistic life-enhancing goals is that they are not trained to do so.

There are other reasons. First, some clients see goal setting as very rational, perhaps too rational. Their lives are so messy, and goal setting seems so sterile. Both helpers and clients object to this overly rational approach. There is a dilemma. On the one hand, many clients need or would benefit from a rigorous application of the problem-management process, including goal setting. On the other hand, they resist its rationality and discipline. They find it alien. Second, goal setting means that clients have to move out of the relatively safe harbor of discussing problem situations and of exploring the possible roots of those problems in the past and move into the uncharted waters of the future. This can be uncomfortable for clients and helpers alike. Third, clients who set goals and commit themselves to them move beyond the victim-of-my-problems game. Victimhood and self-responsibility make poor bedfellows.

A fourth reason that clients might resist developing goals is that goal setting involves clients' placing demands on themselves, making decisions, committing themselves, and moving to action. If I say, "This is what I want," then at least logically, I must also say, "Here is what I am going to do to get it. I know the price and I'm willing to pay it." Since this demands work and pain, clients will not always be grateful for this kind of "help." Fifth, goals, though liberating in many respects, also hem clients in. If a woman chooses a career, she might not be able to have the kind of marriage she would like. If a man commits himself to one woman, he can no longer play the field.

There is some truth in the ironic statement "There is only one thing worse than not getting what you want, and that's getting what you want." The responsibilities accompanying getting what you want—a drug-free life, a renewed marriage, custody of the children, a promotion, the peace and quiet of retirement, freedom from an abusing husband—often open up a new set of problems. Even good solutions create new problems. It is one thing for parents to decide to give their children more freedom; it is another thing for them to watch them use that freedom. Finally, there is a phenomenon called post-decisional depression. Once choices are made, clients begin to have second thoughts that often keep them from acting on their decisions.

As to action, some clients move into action too quickly. The focus on the future liberates them from the past, and the first few possibilities are very attractive. They fail to get the kind of focus and direction provided by Step II-B. So they go off half-cocked. Failing to weigh alternatives and shape goals often means that they have to do the process all over again.

Effective helpers know what lurks in the shadows of goal setting both for themselves and for their clients and are prepared to manage their own part of it and help clients manage theirs. Helpers must receive training in the entire problem-management process and be able to share a picture of the entire process with every client. Then goal setting, described in the client's language, will be a natural part of the process. Artful helpers weave goal setting, under whatever name, into the flow of helping. They do so by moving easily back and forth among the stages and steps of the helping process even in brief therapy.

# Evaluation Questions for Step II-C

- What do I need to do to help clients commit themselves to a better future?
- What do I do to make sure that the goals that clients set are really their goals and not mine or those of a third party?
- How effectively do I help clients examine the benefits of goals they are choosing as measured against the costs?
- In what ways do I help clients focus on the appealing dimensions of the goals being set?
- How effectively do I perceive and deal with the misgivings clients have about the goals they are formulating?
- To what degree do I help clients enter into self-contracts with respect to the accomplishment of goals?
- What am I doing to help clients identify, explore, and manage competing agendas?
- What do I do to help clients move to initial goal-accomplishing action?
- What do I do to help clients acquire and increase their sense of self-efficacy?

# STAGE III: HELPING CLIENTS DEVELOP STRATEGIES TO ACCOMPLISH THEIR GOALS

Stage III is an important part of solution-focused helping. Once clients are helped to establish goals, they often still need help planning the actions or strategies that will enable them to accomplish their goals. These actions or strategies are solutions with a small *s*. Chapter 18 provides an introduction to Stage III and discusses strategies in terms of possibilities. Just as Step II-A focused on goals in terms of possibilities for a better future, Step III-A talks about the many different paths to any given goal. It is about options. Chapter 19, Step III-B, is about helping clients choose the strategies or paths are best fit the needs, style, and resources of the client. It is about choosing the right options. Finally, Chapter 20, Step III-C, is about taking chosen goals and strategies and turning them into a problem-managing or opportunity-developing plan. Planning, whether formal or informal, often plays a key role in problem management and opportunity development.

# STEP III-A: "HOW MANY WAYS ARE THERE TO GET WHAT I NEED AND WANT?" ACTION STRATEGIES

INTRODUCTION TO STAGE III

    Step III-A: Strategies

    Step III-B: Best-fit strategies

    Step III-C: Plans

MANY DIFFERENT PATHS TO GOALS

   Help Clients Brainstorm Strategies for Accomplishing Goals

   Develop Frameworks for Stimulating Clients' Thinking About Strategies

    Individuals

    Models and exemplars

    Communities

    Places

    Things

    Organizations

    Programs

"WHAT SUPPORT DO I NEED TO WORK FOR WHAT I WANT?"

"WHAT WORKING KNOWLEDGE AND SKILLS WILL HELP ME GET WHAT I NEED AND WANT?"

LINKING STRATEGIES TO ACTION

EVALUATION QUESTIONS FOR STEP III-A

# INTRODUCTION TO STAGE III

Planning, in its broadest sense, includes all the steps of Stages II and III; that is, it deals with solutions with a big *S* and a small *s*. In a narrower sense, planning deals with identifying, choosing, and organizing the strategies needed to accomplish goals. Whereas Stage II is about outcomes—goals or accomplishments "powerfully imagined"—Stage III is about the activities or the work needed to produce those outcomes.

Clients, when helped to explore what is going wrong in their lives, often ask, "Well, what should I do about it?" That is, they focus on the actions they need to take to "solve" things. But, as we shall see, action, though essential, is valuable only to the degree that it leads to problem-managing and opportunity-developing outcomes. Accomplishments or outcomes, also essential, are valuable only to the degree that they have a constructive impact on the life of the client. The distinction between action, outcomes, and impact is seen in the following example:

> Lacy, a 40-year-old single woman, is making a great deal of progress in controlling her drinking through her involvement with an AA program. She engages in certain activities—for instance, she attends AA meetings, follows the 12 steps, stays away from situations that would tempt her to drink, and calls fellow AA members when she feels depressed or when the temptation to drink is pushing her hard. The outcome is that she has stayed sober for over seven months. She feels that this is quite an accomplishment. The impact of all this is very rewarding. She feels better about herself, and she has had both the energy and the enthusiasm to do things that she has not done in years—developing a circle of friends, getting interested in church activities, and doing a bit of travel.
>
> But Lacy is also struggling with a troubled relationship with a man. In fact, her drinking was, in part, an ineffective way of avoiding the problems in the relationship. She knows that she no longer wants to tolerate the psychological abuse she has been getting from her male friend, but she's afraid of the vacuum she will create by cutting off the relationship. She is, therefore, trying to determine what she wants, almost fearing that ending the relationship might turn out to be the best option.
>
> She has engaged in a number of activities in attempting to manage the relationship. For instance, she has become much more assertive with her friend. She now cuts off contact whenever her companion becomes abusive. And she no longer lets him make all the decisions about what they are going to do together. But the relationship remains troubled. Even though she is doing many things, there is no satisfactory outcome. She has not yet determined what the outcome should be; that is, she has not determined what kind of relationship she would like and if it is possible to have such a relationship with this man. Nor has she determined to end the relationship.
>
> Finally, after one seriously abusive episode, she tells him that she is ending the relationship. She does what she has to do to sever all ties with him (action), and the outcome is that the relationship ends and stays ended. The impact is that she feels liberated but lonely. The helping process needs to be recycled to help her with this new problem.

Stage III has three steps, in our usual definition of *step*. They are all aimed at action on the part of the client.

**Step III-A: Strategies.**  Help clients develop possible strategies for accomplishing their goals. "What kind of actions will help me get what I need and want?"

**Step III-B: Best-fit strategies.**  Help clients choose strategies tailored to their preferences and resources. "What actions are best for me?"

**The Skilled-Helper Model**

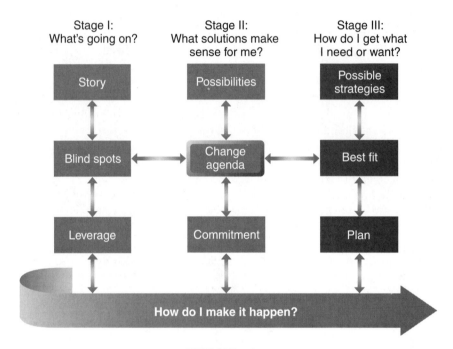

FIGURE 18-1
The Helping Model—Stage III

**Step III-C: Plans.** Help clients formulate actionable plans. "What should my campaign for constructive change look like? What do I need to do first? second? third?"

Stage III, highlighted in Figure 18-1, adds the final pieces to a client's planning a program for constructive change. Stage III deals with the "game plan." However, these three "steps" constitute *planning* for action and should not be confused with action itself. Without action, a program for constructive change is nothing more than a wish list. The implementation of plans is discussed in Part Six.

Strategy is the art of identifying and choosing realistic courses of action for achieving goals and doing so under adverse conditions, such as war. The problem situations in which clients are immersed constitute adverse conditions; often clients are at war with themselves and the world around them. Helping clients develop strategies to achieve goals can be the most thoughtful, humane, and fruitful way of being with them. This step in the counseling process is another that helpers sometimes avoid because it is too "technological." They do their clients a disservice. Clients with goals but no clear idea of how to accomplish them are still at sea.

Strategies are actions that help clients accomplish their goals. Step III-A, developing a range of possible strategies to accomplish goals, is a powerful exercise. Clients who feel hemmed in by their problems and unsure of the viability of their goals are liberated through this process. Clients who see clear pathways to their goals have a greater sense of self-efficacy: "I can do this."

# MANY DIFFERENT PATHS TO GOALS

Once again it is a question of helping clients stimulate their imaginations and engage in divergent thinking. Most clients do not instinctively seek different routes to goals and then choose the ones that make most sense.

## Help Clients Brainstorm Strategies for Accomplishing Goals

Brainstorming, discussed in Chapter 15, plays an important part in strategy development. The more routes to the achievement of a goal, the better. Consider the case of Karen, who has come to realize that heavy drinking is ruining her life. Her goal is to stop drinking. She feels that it simply would not be enough to cut down; she has to stop. At first, she thought the way forward was simple enough: Whereas before she drank, now she wouldn't. Because of the novelty of not drinking, she was successful for a few days; then she fell off the wagon. This happened a number of times until she finally realized that she could use some help. Stopping drinking, at least for her, is not as simple as it first seemed.

A counselor at a city alcohol and drug treatment center helps her explore a number of techniques that could be used in an alcohol-management program. Together they come up with the following possibilities:

- Just stop cold turkey and get on with life.
- Join Alcoholics Anonymous.
- Move someplace declared "dry" by local government.
- Take a drug that causes nausea if followed by alcohol.
- Replace drinking with other rewarding behaviors.
- Join some self-help group other than Alcoholics Anonymous.
- Get rid of all liquor in the house.
- Take the "pledge" not to drink; to make it more binding, take it in front of a minister.
- Join a residential hospital detoxification program.
- Avoid friends who drink heavily.
- Change other social patterns; for instance, find places other than bars and cocktail lounges to socialize.
- Try hypnosis to reduce the urge to drink.
- Use behavior modification techniques to develop an aversion for alcohol; for instance, pair painful but safe electric shocks with drinking or even thoughts about drinking.
- Change self-defeating patterns of self-talk, such as "I have to have a drink" or "One drink won't hurt me."
- Become a volunteer to help others stop drinking.
- Read books and view films on the dangers of alcohol.
- Stay in counseling as a way of getting support and challenge for stopping.

- Share intentions to stop drinking with family and close friends.
- Spend a week with an acquaintance who does a great deal of work in the city with alcoholics, and go with him on his rounds.
- Walk around skid row meditatively.
- Have a discussion with members of the family about the impact drinking has on them.
- Eat foods, such as sweets, that can help reduce the craving for alcohol.
- Get a hobby or an avocation that demands time and energy.
- Substitute a range of self-enhancing activities, such as exercise or surfing the Web, for drinking.

This list contains many more items than Karen would have thought of had she not been stimulated by the counselor to take a census of possible strategies. One of the reasons that clients are clients is that they are not very creative in looking for ways of getting what they want. Once goals are established, getting them accomplished is not just a matter of hard work. It is also a matter of imagination.

If a client is having a difficult time coming up with strategies, the helper can "prime the pump" by offering a few suggestions. Driscoll (1984) put it well.

> Alternatives are best sought cooperatively, by inviting our clients to puzzle through with us what is or is not a more practical way to do things. But we must be willing to introduce the more practical alternatives ourselves, for clients are often unable to do so on their own. Clients who could see for themselves the more effective alternatives would be well on their way to using them. That clients do not act more expediently already is in itself a good indication that they do not know how to do so. (p. 167)

Although the helper may need to suggest alternatives, he or she can do so in such a way that the principal responsibility for evaluating and choosing possible strategies stays with the client. For instance, there is the "prompt and fade" technique. The counselor can say, "Here are some possibilities. . . . Let's review them and see whether any of them make sense to you or suggest further possibilities." Or "Here are some of the things that people with this kind of problem situation have tried. . . . How do they sound to you?" The "fade" part of this technique keeps it from being advice giving. It remains clear that the client must think over these strategies, choose the right ones, and commit to them.

Elton, a graduate student in counseling psychology, is plagued with perfectionism. Although he is an excellent student, he worries about getting things right. After he writes a paper or practices counseling, he agonizes over what he could have done better. This kind of behavior puts him on edge when he practices counseling with his fellow trainees. They tell him that his "edge" makes them uncomfortable and interferes with the flow of the helping process. One student says to him, "You make me feel as if I'm not doing the right things as a client."

Elton realizes that "less is more"—that becoming less preoccupied with the details of helping—will make him a more effective helper. His goal is to become more relaxed in the helping sessions, free his mind of the "imperatives" to be perfect, and learn from mistakes rather than expending an excessive amount of effort trying to avoid them. He and his supervisor talk about ways he can free himself of these inhibiting imperatives.

SUPERVISOR: What kinds of things can you do to become more relaxed?

ELTON: I need to focus my attention on the client and the client's goals instead of being preoccupied with myself.

SUPERVISOR: So a basic shift in your orientation right from the beginning will help.

ELTON: Right. . . . And this means getting rid of a few inhibiting beliefs.

SUPERVISOR: Such as . . .

ELTON: That technical perfection in the helping model is more important than the relationship with the client. I get lost in the details of the model and have forgotten that I'm a human being with another human being.

SUPERVISOR: So "rehumanizing" the helping process in your own mind will help. . . . Any other internal behaviors need changing?

ELTON: Another belief is that I have to be the best in the class. That's my history, at least in academic subjects. Being as effective as I can be in helping a client has nothing to do with competing with my fellow students. Competing is a distraction. I know it's in my bones. It might have been all right in high school, but . . .

SUPERVISOR: Okay, so the academic-game mentality doesn't work here . . .

ELTON (interrupting): That's precisely it. Even the practicing we do with one another is real life, not a game. You know that a lot of us talk about real issues when we practice.

SUPERVISOR: You've been talking about getting your attitudes right and the impact that can have on helping sessions. Are there any external behaviors that might also help?

ELTON (pauses): I'm hesitating because it strikes me how I'm in my head too much, always figuring me out. . . . On a much more practical basis, I like what Jerry and Philomena do. Before each session with their "clients" in their practice sessions, they spend 5 or 10 minutes reviewing just where the client is in the overall helping process and determining what they might do in the next session to add value and move things forward. That puts the focus where it belongs, on the client.

SUPERVISOR: So a mini-prep for each session can help you get out of your world and into the client's.

ELTON: Also, in debriefing the training videos we make each week, I now see that I always start by looking at my behavior instead of what's happening with the client. . . . Oh, there's another thing I can do. I can share just what we've been discussing here with my training partner. She can help me refocus myself.

SUPERVISOR: I'm not sure whether you bring up the perfectionism issues when you're the "client" in the practice sessions or in the weekly lifestyle group meetings.

ELTON (hesitating): Well, not really. I'm just coming to realize how pervasive it is in my life. . . . To tell you the truth I think I haven't brought it up because I'd rather have my fellow trainees see me as competent, not perfectionistic. . . . Well, the cat is out of the bag with you, so I guess it makes sense to put it on my lifestyle group agenda.

This dialogue, which includes empathy, probes, and challenges on the part of the supervisor, produces a number of strategies that Elton can use to develop a more client-focused mentality. He ends by saying that all these can be reinforced through his interactions with his training partner.

## Develop Frameworks for Stimulating Clients' Thinking About Strategies

How can helpers find the right probes to help clients develop a range of strategies? Simple frameworks can help. Consider the following case:

Jackson has terminal cancer. He has been in and out of the hospital several times over the past few months, and he knows that he probably will not live more than a year. He would like the year to be as full as possible, and yet he wants to be realistic. He hates being in the hospital, especially a large hospital, where it is so easy to be anonymous. One of his goals is to die outside the hospital. He would like to die as benignly as possible and retain possession of his faculties as long as possible. How is he to achieve these goals?

You can use probes and prompts to help clients discover possible strategies by assisting them in investigating resources in their lives, including people, models, communities, places, things, organizations, programs, and personal resources.

**Individuals.** What individuals might help clients achieve their goals? Jackson gets the name of a local doctor who specializes in the treatment of chronic cancer-related pain. The doctor teaches people how to use a variety of techniques to manage pain. Jackson says that perhaps his wife and daughter can learn how to give simple injections to help him control the pain. A friend of his has mentioned that his father got excellent hospice care and died at home. Also, he thinks that talking every once in a while with a friend whose wife died of cancer, a man he respects and trusts, will help him find the courage he needs.

**Models and exemplars.** Are there people presently doing what clients want to do? One of Jackson's fellow workers died of cancer at home. Jackson visited him there a couple of times. That's what gave him the idea of dying at home, or at least outside the hospital. He noticed that his friend never allowed himself to engage in poor-me talk. He refused to see dying as anything but part of living. This touched Jackson deeply at the time, and now reflecting on that experience may help him develop the same kind of upbeat attitude.

**Communities.** What communities of people are there through which clients might identify strategies for implementing their goals? Even though Jackson has not been a regular churchgoer, he does know that the parish within which he resides has some resources for helping those in need. A brief investigation reveals that the parish has developed a relatively sophisticated approach to providing various services for the sick. He also does an Internet search and discovers that there are of number of self-help groups for people like him.

**Places.** Are there particular places that might help? Jackson immediately thinks of Lourdes, the shrine to which Catholic believers flock with all sorts of human problems. He doesn't expect miracles, but he feels that he might experience life more deeply there. It's a bit wild, but why not a pilgrimage? He still has the time and money to do it. He also finds a high-tech place—an Internet chat room for cancer patients *and* their caregivers. This helps him get out of himself and, at times, become a helper instead of a client.

**Things.** What things exist that can help clients achieve their goals? Jackson has read about the use of combinations of drugs to help stave off pain and the side effects of chemotherapy. He has heard that certain kinds of electric stimulation can ward off chronic pain. He explores all these possibilities with his doctor and even arranges for second opinions.

**Box 18-1    Questions on Developing Strategies**

- Now that I know what I want, what do I need to do?
- Now that I know my destination, what are the different routes for getting there?
- What actions will get me to where I want to go?
- Now that I know the gaps between what I have and what I want and need, what do I need to do to bridge those gaps?
- How many ways are there to accomplish my goals?
- How do I get started?
- What can I do right away?
- What do I need to do later?

**Organizations.** What groups or institutions are available to help clients? Jackson runs across an organization that helps young cancer patients get their wishes. He volunteers. In his role as helper, he finds he receives as much help, motivation, and solace as he gives.

**Programs.** Do any ready-made programs exist to help clients in this position? He learns that a new hospice in his part of town has three programs. One helps people who are terminally ill stay in the community as long as they can. A second makes provision for part-time residents. The third is a residential program for those who can spend little or no time in the community. The goals of these programs are practically the same as Jackson's.

Box 18-1 outlines some questions that you can help clients ask themselves to develop strategies for accomplishing goals.

## "WHAT SUPPORT DO I NEED TO WORK FOR WHAT I WANT?"

Step III-A can also be seen as helping clients get the resources, both internal and environmental, they need to pursue goals. Many clients do not know how to mobilize needed resources. One of the most important resources is social support. A great deal is said in the literature about the kind of support helpers should provide their clients (Alford & Beck, 1997; Arkowitz, 1997; Castonguay, 1997; Yalom & Bugental, 1997). In a sense, this entire book is about that kind of support. But if clients are to pursue goals "out there" in their real lives, they also need social support. Unfortunately, as Robert Putnam (2000) shows with a great deal of evidence, such support is not always easy to find. His central thesis is that in North American society, the supply of "social capital"—both informal social connectedness and formal civic engagement—has fallen dangerously low. Putnam reports that we belong to fewer organizations that hold meetings, know our neighbors less, meet with friends less frequently, and even socialize with our families less often. This is the environment in which clients must do the work of constructive change.

However, social support is a key element in change (see Basic Behavioral Science Task Force of the National Advisory Mental Health Council, 1996).

> Social support has . . . been examined as a predictor of the course of mental illness. In about 75% of studies with clinically depressed patients, social-support factors increased the initial success of treatment and helped patients maintain their treatment gains. Similarly, studies of people with schizophrenia or alcoholism revealed that higher levels of social support are correlated with fewer relapses, less frequent hospitalizations, and success and maintenance of treatment gains. (p. 628)

In a study on weight loss and maintaining the loss (Wing & Jeffery, 1999), clients who enlisted the help of friends were much more successful than clients who took the solo path. This is called "social facilitation" and is quite different from dependence. Social facilitation, a positive-psychology approach, is energizing, while dependence is often depressing. Therefore, a culture of social isolation does not bode well for clients. Of course, all of this reinforces what we already know through common sense. Which of us has not been helped through difficult times by family and friends?

When it comes to social support, there are two categories of clients. First, there are those who lead an impoverished social life. The objective with this group is to help them find social resources, to get back into community in some productive way. But as Putnam (2000) points out, even when clients, at least on paper, have a social system, they may not use it very effectively. This second group provides counselors with a different challenge, that is, helping clients tap into those human resources in a way that helps them manage problem situations more effectively.

Indeed, the Basic Behavorial Science Task Force (1996) study previously quoted showed that people who are highly distressed and therefore most in need of social support may be the least likely to receive it because their expressions of distress drive away potential supporters. Which of us, at one time or another, has not avoided a distressed friend or colleague? Therefore, distressed clients can be helped to learn how to modulate their expressions of distress. Who wants to help whiners? On the other hand, potential supporters can learn how to deal with distressed friends and colleagues, even when the latter let themselves become whiners.

The Task Force study suggests two general strategies for fostering social support: helping clients mobilize or increase support from existing social networks and "grafting" new ties onto impoverished social networks. Both of these come into play in the following case:

> Casey, a bachelor whose job involved frequent travel literally around the world, fell ill. He had many friends, but they were spread around the world. Because he was neither married nor in a marriage-like relationship, he had no primary caregiver in his life. He received excellent medical care, but his psyche fared poorly.
>
> Once out of the hospital, he recuperated slowly, mainly because he was not getting the social support he needed. In desperation he had a few sessions with a counselor, sessions that proved to be quite helpful. The counselor challenged him to "ask for help" from his local friends.

He had underplayed his illness with them because he didn't want to be a "burden." He discovered that his friends were more than ready to help. But since their time was limited, he, with some hesitancy, "grafted" onto his rather sparse hometown social network some very caring people from the local church. He was fearful that he would be deluged with piety, but instead he found people like himself. Moreover, they were, in the main, socially intelligent. They knew how much or how little care to give. In fact, most of the time their care was simple friendship. Finally, he hired a couple of students from a local university to do word processing and run errands for him from time to time. They also provided some social support.

As the Task Force authors note, it's important not only that people be available to provide support but also that those needing support perceive that it is available. This may mean, as in Casey's case, working with the client's attitudes and openness to receive support.

Eventually, all clients have to make it without the help of a counselor. Therefore, effective helpers right from the beginning try to help them explore the social-support dimensions of problem situations. At the action arrow stage, questions like the following are appropriate: Who might help you do this? Who's going to challenge you when you want to give up? With whom can you share these kinds of concerns? Who's going to give you a pat on the back when you accomplish your goal?

Although social support is often key, it is not the only resource clients need to pursue their goals. Effective helpers build some kind of resource census into the helping process.

## "WHAT WORKING KNOWLEDGE AND SKILLS WILL HELP ME GET WHAT I NEED AND WANT?"

It often happens that people get into trouble or fail to get out of it because they lack the needed life skills or coping skills to deal with problem situations. If this is the case, then helping clients find ways of learning the life skills they need to cope more effectively is an important broad strategy. Indeed, the use of skills training as part of therapy—what years ago Carkhuff (1971) called "training as treatment"—might be essential for some clients. Challenging clients to engage in activities for which they don't have the skills is compounding rather than solving their problems. What kinds of working knowledge and skills does this client need to get where he or she wants to go? Consider the following case:

Jerzy and Zelda fell in love. They married and enjoyed a relatively trouble-free honeymoon period of about two years. Eventually, however, the problems that inevitably arise from living together in such intimacy asserted themselves. They found, for instance, that they counted too heavily on positive feelings for each other and now, in their absence, could not "communicate" about finances, sex, and values. They lacked certain critical interpersonal communication skills. Furthermore, they lacked understanding of each other's developmental needs. Jerzy had little working knowledge of the developmental demands of a 20-year-old woman; Zelda had little working knowledge of the kinds of cultural blueprints that were operative in the lifestyle of her 29-year-old husband. The relationship began to deteriorate. Since they had few problem-solving skills, they didn't know how to handle their situation.

Jerzy and Zelda needed skills. This is hardly surprising. Lack of requisite interpersonal communication and other life skills is often at the heart of relationship breakdowns. One marriage counselor I know works with groups of four couples. Training in communication skills is part of the process. He separates men from women and trains them in tuning in, active listening, and sharing empathic highlights. For skills practice, he begins by pairing a woman with a woman and a man with a man. Next he pairs a man and a woman, but not spouses, for skills practice. Finally, spouses are paired, taught a simple version of the problem-management process outlined in this book, and then helped to use the skills they have learned to engage in problem solving with each other. In sum, he equips them with two sets of life skills: interpersonal communication and problem solving.

The literature is filled with programs designed to equip clients with the working knowledge and skills they need to manage problems and lead fuller lives. Some of them focus on specific problems. For instance, Deffenbacher and his associates (Deffenbacher, Thwaites, Wallace, & Oetting, 1994; Deffenbacher, Oetting, Huff, & Thwaites, 1995) have devised and evaluated programs for general anger reduction. Although programs such as these need to be tailored to individual clients, they are often gold mines of strategies for accomplishing goals. Tailoring such generic programs to clients will be discussed in the next chapter.

## LINKING STRATEGIES TO ACTION

Although all the steps of the helping process can and should stimulate action on the part of the client, this is especially true of Step III-A, which deals with possible actions. Many clients, once they begin to see what they can do to get what they want, begin acting immediately. They don't need a formal plan. Here are a couple of examples of clients who, once they were helped to identify strategies for implementing their goals, acted on them.

> Jeff had been in the army for about ten months. He found himself both overworked and, perhaps not paradoxically, bored. He had a couple of sessions with one of the educational counselors on the base. During these sessions, Jeff began to see quite clearly that not having a high school diploma was working against him. The counselor mentioned that he could finish high school while in the army. Jeff realized that this possibility had been pointed out to him during the orientation talks, but he hadn't paid any attention to it. He had joined the army because he wasn't interested in school and, being unskilled, couldn't find a job. Now he decided that he would get a high school diploma as soon as possible.
>
> Jeff obtained the authorization needed from his company commander to go to school. He found out what courses he needed and enrolled in time for the next school session. It didn't take him long to finish. Once he received his high school diploma, he felt better about himself and found that opportunities for more interesting jobs opened up for him in the army. Achieving his goal of getting a high school diploma helped him manage the problem situation.

Jeff was one of those fortunate ones who, with a little help, quickly set a goal (the "what") and identified and implemented the strategies (the "how") to accomplish it. Notice, too, that his goal of getting a diploma was also a means to other goals: feeling good about himself and getting better job opportunities in the army.

Grace's road to problem management was quite different from Jeff's. She needed much more help.

As long as she could remember, Grace had been a fearful person. She was especially afraid of being rejected and of being a failure. As a result, she had an impoverished social life. She had held a series of jobs that were safe but boring. She became so depressed that she made a half-hearted attempt at suicide, probably more an expression of anguish and a cry for help than a serious attempt to get rid of her problems by getting rid of herself.

During her stay in the hospital, Grace had a few therapy sessions with one of the staff psychiatrists. The psychiatrist was supportive and helped her handle both the guilt she felt because of the suicide attempt and the depression that had led to the attempt. Just talking to someone about things she usually kept to herself seemed to help. She began to see her depression as a case of "learned helplessness." She saw quite clearly how she had let her choices be dictated by her fears. She also began to realize that she had a number of underused resources. For instance, she was intelligent and, though not good-looking, attractive in other ways. She had a fairly good sense of humor, though she seldom gave herself the opportunity to use it. She was also sensitive to others and basically caring.

After Grace was discharged from the hospital, she returned for a few outpatient sessions. She got to the point where she wanted to do something about her general fearfulness and her passivity, especially the passivity in her social life. A psychiatric social worker taught her relaxation and thought-control techniques that helped her reduce her anxiety. As she became less anxious, she was in a better position to do something about establishing some social relationships. With the social worker's help, she set goals of acquiring a couple of friends and becoming a member of some social group. However, she was at a loss as to how to proceed. She thought that friendship and a fuller social life were things that should happen "naturally." She soon came to realize that many people had to work at acquiring a more satisfying social life, that for some people there was nothing automatic about it at all.

The social worker helped Grace identify various kinds of social groups that she might join. She was then helped to see which of these would best meet her needs without placing too much stress on her. She finally chose to join an arts and crafts group at a local YMCA. The group gave her an opportunity to begin developing some of her talents and to meet people without having to face demands for intimate social contact. It also gave her an opportunity to take a look at other, more socially oriented programs sponsored by the Y. In the arts and crafts program, she met a couple of people she liked and who seemed to like her. She began having coffee with them once in a while and then an occasional dinner.

Grace still needed support and encouragement from her helper, but she was gradually becoming less anxious and feeling less isolated. Once in a while, she would let her anxiety get the better of her. She would skip a meeting at the Y and then lie about having attended. However, as she began to let herself trust her helper more, she revealed this self-defeating game. The social worker helped her develop coping strategies for those times when her anxiety seemed to be higher.

Grace's problems were more severe than Jeff's, and she did not have as many immediate resources. Therefore, she needed both more time and more attention to develop goals and strategies.

### Evaluation Questions for Step III-A

How effectively do I do the following?

- Use probes, prompts, and challenges to help clients identify possible strategies
- Help clients engage in divergent thinking with respect to strategies
- Help clients brainstorm as many ways as possible to accomplish their goals
- Use some kind of framework in helping clients be more creative in identifying strategies
- Help clients identify and begin to acquire the resources they need to accomplish their goals
- Help clients identify and develop the skills they need to accomplish their goals
- Help clients see the action implications of the strategies they identify

# STEP III-B: "WHAT STRATEGIES ARE BEST FOR ME?" BEST-FIT STRATEGIES

"WHAT'S BEST FOR ME?" THE CASE OF BUD

HELPING CLIENTS CHOOSE BEST-FIT STRATEGIES
    Specific strategies
    Robust strategies
    Realistic strategies
    Strategies in keeping with clients' values

STRATEGY SAMPLING

A BALANCE-SHEET METHOD FOR CHOOSING STRATEGIES
  A Sample Balance Sheet
  Realism in Using the Balance Sheet

LINKING STEP III-B TO ACTION

THE SHADOW SIDE OF SELECTING STRATEGIES
    Wishful thinking
    Playing it safe
    Avoiding the worst outcome
    Striking a balance

EVALUATION QUESTIONS FOR STEP III-B

## "What's Best for Me?" The Case of Bud

In the last two steps of Stage III, clients are in decision-making mode once again. After brainstorming strategies for accomplishing goals, they need to choose strategies ("packages") that best fit their situations and resources and turn them into some kind of plan for constructive change. Whether these steps are done with the kind of formality outlined here is not the point. Counselors, understanding the "technology" of planning, can add value by helping clients find ways of accomplishing goals (getting what they need and want) in a systematic, flexible, personalized, and cost-effective way. Step III-B involves ways of helping clients choose the strategies that are best for them. Step III-C deals with turning those strategies into some kind of step-by-step plan.

Some clients, once they are helped to develop a range of strategies to implement goals, move forward on their own; that is, they choose the best strategies, put together action plans, and implement them. Others, however, need help in choosing strategies that best fit their situations, and so we add Step III-B to the helping process. It is useless to have clients brainstorm if they don't know what to do with all the action strategies they generate.

Consider the case of Bud, a man who was helped to discover two best-fit strategies for achieving emotional stability in his life. With these, he achieved outcomes that surpassed anyone's wildest expectations.

One morning, Bud, then 18 years old, woke up unable to speak or move. He was taken to a hospital, where catatonic schizophrenia was diagnosed. After repeated admissions to hospitals, where he underwent both drug and electroconvulsive therapy (ECT), his diagnosis was changed to paranoid schizophrenia. He was considered incurable.

A quick overview of Bud's earlier years suggests that much of his emotional distress was caused by unmanaged life problems and the lack of human support. He was separated from his mother for four years when he was young. They were reunited in a city new to both of them, and there he suffered a great deal of harassment at school because of his "ethnic" looks and accent. There was simply too much stress and change in his life. He protected himself by withdrawing. He was flooded with feelings of loss, fear, rage, and abandonment. Even small changes became intolerable. His catatonic attack occurred in the autumn on the day of the change from daylight saving to standard time. It was the last straw.

In the hospital, Bud became convinced that he and many of his fellow patients could do something about their illnesses. They did not have to be victims of themselves or of the institutions designed to help them. Reflecting on his hospital stays and the drug and ECT treatments, he later said he found his "help" so debilitating that it was no wonder that he got crazier. Somehow Bud, using his own inner resources, managed to get out of the hospital. Eventually, he got a job, found a partner, and got married.

One day, after a series of problems with his family and at work, Bud felt himself becoming agitated and thought he was choking to death. His doctor sent him to the hospital "for more treatment." There Bud had the good fortune to meet Sandra, a psychiatric social worker who was convinced that many of the hospital's patients were there because of lack of support before, during, and after their bouts of illness. She helped him see his need for social support, especially at times of stress. In the inpatient counseling groups that she ran, Sandra also discovered that Bud had a knack for helping others. Bud's broad goal was still emotional stability, and he wanted to do whatever was necessary to achieve it. Finding human support and helping others cope with their problems—instrumental goals—were his best strategies for achieving the stability he wanted.

> Outside, Bud started a self-help group for ex-patients like himself. In the group, he was a full-fledged participant. Sandra, the social worker, also coached Bud's wife on how to provide support for him at times of stress. As to helping others, Bud not only founded a self-help group but also turned it into a network of self-help groups for ex-patients.

This is an amazing example of a client who focused on one broad goal—emotional stability; translated it into a number of immediate, practical goals; discovered two broad strategies—finding ongoing emotional support and helping others—for accomplishing those goals; translated the strategies into practical applications; and by doing all that, found the emotional stability he was looking for.

## HELPING CLIENTS CHOOSE BEST-FIT STRATEGIES

The criteria for choosing goal-accomplishing strategies are somewhat like the criteria for choosing goals outlined in Step II-B. These criteria are reviewed briefly here through a number of examples. Strategies to achieve goals should be, like goals themselves, specific, robust, prudent, realistic, sustainable, flexible, cost-effective, and in keeping with the client's values. Let's take a look at a few of these criteria as they apply to choosing strategies.

**Specific strategies.** Strategies for achieving goals should be specific enough to drive behavior. In the preceding example, Bud's two broad strategies for achieving emotional stability—tapping into human support and helping others—were translated into quite specific strategies: keeping in touch with Sandra, getting help from his wife, participating in a self-help group, starting a self-help group, and founding and running a self-help organization. Contrast Bud's case with Stacy's.

> Stacy was admitted to a mental hospital because she had been exhibiting bizarre behavior in her neighborhood. She dressed in a slovenly way and went around admonishing the residents of the community for their "sins." Her condition was diagnosed as schizophrenia, simple type. She had been living alone for about five years, since the death of her husband. It seems that she had become more and more alienated from herself and others. In the hospital, medication helped control some of her symptoms. She stopped admonishing others and took reasonable care of herself, but she was still quite withdrawn. She was assigned to "milieu" therapy, a euphemism meaning that she was helped to follow the more or less benign routine of the hospital—a bit of work, a bit of exercise, some programmed opportunities for socializing. She remained withdrawn and usually seemed moderately depressed. No therapeutic goals had been set, and the nonspecific program to which she was assigned was totally inadequate.

So-called milieu therapy did nothing for Stacy because in no way was it specific to her needs. It was a general program that was only marginally better than drug-focused standard care. Bud's strategies, on the other hand, proved to be powerful. They not only helped him gain stability but also gave him a new perspective on life.

**Robust strategies.** Strategies are robust to the degree that they challenge clients to use their resources and, when implemented, actually achieve goals. Not only was Stacy's program too general, but it also lacked bite. Bud's strategies, on the other hand, were substantive, especially the strategy of starting and running a self-help organization. What could be done for Stacy?

> A newly hired psychiatrist, who had been influenced by Corrigan's (1995) notion of "champions of psychiatric rehabilitation," saw immediately that Stacy needed more than either standard

psychiatric or milieu-centered care. He involved her in a new comprehensive social-learning program, which included cognitive restructuring, social-skills training, and behavioral-change interventions based on incentives, shaping, modeling, and rewards. Stacy responded very well to the new, rather intensive program. She was discharged within six months and, with the help of an outpatient extension of the program, remained in the community.

For Stacy, this program proved to be not only robust but also specific, prudent, realistic, sustainable, flexible, cost-effective, and in keeping with her values. It was cost-effective in two ways. First, it was the best use of Stacy's time, energy, and psychological resources. Second, it helped her and others like her to get back into the community and stay there. It was in keeping with her values because, even though some staff members at the hospital had concluded that all she wanted was "to be left alone," Stacy did value human companionship and freedom. She did better in a community setting.

**Realistic strategies.** If clients choose strategies that are beyond their resources, they are doing themselves in. Strategies are realistic when they can be carried out with the resources the client has, are under the client's control, and are unencumbered by obstacles. Bud's strategies would have appeared unrealistic to most clients and helpers. But this highlights an important point. Just as we should help clients set stretch goals whenever possible, so we should not underestimate what clients are capable of doing. In the following case, Desmond moves from unrealistic to realistic strategies for getting what he wants:

Desmond was in a halfway house after leaving a state mental hospital. From time to time, he still had bouts of depression that would incapacitate him for a few days. He wanted to get a job because he thought that a job would help him feel better about himself, become more independent, and manage his depression better. He answered job advertisements in a rather random way and was constantly turned down after being interviewed. He simply did not yet have the kinds of resources needed to put himself in a favorable light in job interviews. Moreover, he was not yet ready for a regular, full-time job.

On his own, Desmond does not do well in choosing strategies to achieve even modest goals. But here's what happened next:

A local university received funds to provide outreach services to halfway houses in the metropolitan area. The university program included finding companies that were willing, on a win-win basis, to work with halfway-house residents. A counselor from the program helped Desmond get in contact with companies that had specific programs to help people with psychiatric problems. He found two that he thought would fit his needs. Some of their best workers had a variety of disabilities, including psychiatric problems. After a few interviews, Desmond got a job in one of these companies that fitted his situation and capabilities. The entire work culture was designed to provide the kind of support he needed.

There is, of course, a difference between realism and allowing clients to sell themselves short. Robust strategies that make clients stretch for a valued goal can be most rewarding. Bud's case is an exceptional example of that.

**Strategies in keeping with clients' values.** Make sure that the strategies that clients choose are consistent with their values. Let's return to the case of the priest who had been unjustly accused of child molestation.

In preparing for the court case, the priest and his lawyer had a number of discussions. The lawyer wanted to do everything possible to destroy the credibility of the accusers. He had dug

**Box 19-1    Questions on Best-Fit Strategies**

- Which strategies will be most useful in helping me get what I need and want?
- Which strategies are best for this situation?
- Which strategies best fit my resources?
- Which strategies will be most economic in the use of my resources?
- Which strategies are most powerful?
- Which strategies best fit my preferred way of acting?
- Which strategies best fit my values?
- Which strategies will have the fewest unwanted consequences?

into their past and dredged up some dirt. The priest objected to these tactics. "If I let you do this," he said, "I descend to their level. I can't do that." The priest discussed this with his counselor, his superiors, and another lawyer. He stuck to his guns. They prepared a strong case without out the sleaze.

After the trial was over and he was acquitted, the priest said that his discussion about the lawyer's preferred tactics was one of the most difficult issues he had to face. Something in him said that since he was innocent, any means to prove his innocence was allowed. Something else told him that this was not right. The counselor helped him clarify and challenge his values but made no attempt to impose either his own or the lawyer's values on his client.

Box 19-1 outlines the kinds of questions you can help clients answer as they choose best-fit strategies.

## STRATEGY SAMPLING

Some clients find it easier to choose strategies if they first sample some of the possibilities. Consider this case:

Two business partners were in conflict over ownership of the firm's assets. Their goals were to see justice done, to preserve the business, and, if possible, to preserve their relationship. A colleague helped them sample some possibilities. Under her guidance, they discussed with a lawyer the process and consequences of bringing their dispute to the courts, they had a meeting with a consultant-counselor who specialized in these kinds of disputes, and they visited an arbitration firm.

In this case, the sampling procedure had the added effect of giving them time to let their emotions simmer down. They agreed to go the consultant-counselor route.

Karen, the woman who, with the help of her counselor, brainstormed a wide range of strategies for disengaging from alcohol, decided to sample some of the possibilities.

Surprised by the number of program possibilities there were to achieve the goal of getting liquor out of her life, Karen decided to sample some of them. She went to an open meeting of Alcoholics Anonymous, attended a meeting of a women's lifestyle-issues group, visited the hospital that had the residential treatment program, and joined up for a two-week trial physical fitness program at a YMCA. She engaged in these activities frantically. She tried them out and then discussed them with her counselor. Her search for the programs that were best for her did occupy her energies and strengthened her resolve to do something about her alcoholism.

Of course, some clients could use strategy sampling as a way of putting off action. That was certainly not the case with Bud. His attending the meeting of a self-help group after leaving the hospital was a form of strategy sampling. Although he was impressed by the group, he thought that he could start a group limited to ex-patients that would focus more directly on the kinds of issues he and other ex-patients were facing.

## A Balance-Sheet Method for Choosing Strategies

Some form of balance sheet can be used to help clients make decisions in general. The methodology could be used for any key decision related to the helping process—to get help in the first place, to work on one problem rather than another, or to choose this rather than that goal. Balance sheets deal with the acceptability and unacceptability of both benefits and costs. A balance-sheet approach, applied to choosing strategies for achieving goals, poses questions such as the following:

- What are the benefits of choosing this strategy for myself? for significant others?
- To what degree are these benefits acceptable to me? to significant others?
- In what ways are these benefits unacceptable to me? to significant others?
- What are the costs of choosing this strategy for myself? for significant others?
- To what degree are these costs acceptable to me? to significant others?
- In what ways are these costs unacceptable to me? to significant others?

Let's return to Karen. She used the balance-sheet method to assess the viability not of a goal but of strategies to achieve a goal. Karen's goal was to stop drinking. One possible strategy for accomplishing that goal was to spend a month as an inpatient at an alcoholic treatment center. This possibility appealed to her. However, since choosing this strategy would be a serious decision, the counselor, Joan, helped Karen use a balance sheet to weigh possible costs and benefits. After filling it out, Karen and Joan discussed Karen's findings. She chose to consider the pluses and minuses for herself and for her husband and children.

### A Sample Balance Sheet

#### Benefits of Choosing the Residential Program

- *For me.*  It would help me because it would be a dramatic sign that I want to do something to change my life. It's a clean break, as it were. It would also give me time just for myself. I'd get away from all my commitments to family, relatives, friends, and work. I see it as an opportunity to do some planning. I'd have to figure out how I would act as a sober person.

- *For significant others.* I'm thinking mainly of my family here. It would give them a breather, a month without an alcoholic wife and mother around the house. I'm not saying that to put myself down. I think it would give them time to reassess family life and make some decisions about any changes they'd like to make. I think something dramatic like my going away would give them hope. They've had very little reason to hope for the last five years.

### Acceptability of benefits

- *For me.* I feel torn here. But looking at it just from the viewpoint of acceptability, I feel kind enough toward myself to give myself a month's time off. Also, something in me longs for a new start in life. And it's not just time off. The program is a demanding one.

- *For significant others.* I think that my family would have no problems in letting me take a month off. I'm sure that they'd see it as a positive step from which all of us would benefit.

### Unacceptability of benefits

- *For me.* Going away for a month seems such a luxury, so self-indulgent. Also, even though taking such a dramatic step would give me an opportunity to change my current lifestyle, it would also place demands on me. My fear is that I would do fine while in the program but that I would come out and fall on my face. I guess I'm saying it would give me another chance at life, but I have misgivings about having another chance. I need some help here.

- *For significant others.* The kids are young enough to readjust to a new me. But I'm not sure how my husband would take this "benefit." He has more or less worked out a lifestyle that copes with my being drunk a lot. Though I have never left him and he has never left me, still I wonder whether he wants me back sober. Maybe this belongs under the "cost" part of this exercise. I need some help here. And, of course, I need to talk to my husband about all this. I also notice that some of my misgivings relate not to a residential program as such but to a return to a lifestyle free of alcohol. Doing this exercise helped me see that more clearly.

### Costs of Choosing the Residential Program

- *For me.* Well, there's the money. I don't mean the money just for the program, but I would be losing four weeks' wages. But I've lost a lot of wages through drinking. The major cost seems to be the commitment I have to make about a lifestyle change. And I know the residential program won't be all fun. I don't know exactly what they do there, but some of it must be demanding. Probably a lot of it.

- *For significant others.* It's a private program, and it's going to cost the family a lot of money. The services I have been providing at home will be missing for a month. It could be that I'll learn things about myself that will make it harder to live with me—though living with a drunken spouse and mother is no joke.

What if I come back more demanding of them—I mean, in good ways? I need to talk this through more thoroughly.

Acceptability of costs

- *For me.* I have no problem at all with the money or with whatever the residential program demands of me physically or psychologically. I'm willing to pay. What about the costs of the demands the program will place on me for substantial lifestyle changes? Well, in principle I'm willing to pay what that costs. But I'm not sure what these are. I need some help here.

- *For significant others.* They will have to make financial sacrifices, but I have no reason to think that they would be unwilling. Still, I can't be making decisions for them. I see much more clearly the need to have a counseling session with my husband and children present. I think they're also willing to have a "new" person around the house, even if it means making adjustments and changing their lifestyle a bit. I want to check this out with them, but I think it would be helpful to do this with the counselor. I think they will be willing to come.

Unacceptability of costs

- *For me.* Although I'm ready to change my lifestyle, I hate to think that I will have to accept some dumb, dull life. I think I've been drinking, at least in part, to get away from dullness; I've been living in a fantasy world, a play world a lot of the time. A stupid way of doing it, perhaps, but it's true. I have to do some life planning of some sort. I need some help here.

- *For significant others.* It strikes me that my family might have problems with a sober me if it means that I will strike out in new directions. I wonder if they want the traditional homebody wife and mother. I don't think I could stand that. All this should come out in the meeting with the counselor.

Karen concludes, "All in all, it seems like the residential program is a good idea. There is something much more substantial about it than an outpatient program. But that's also what scares me."

Karen's use of the balance sheet helps her make an initial program choice, but it also enables her to discover issues that she has not yet worked out completely. By using the balance sheet, she returns to the counselor with work to do. This highlights the usefulness of exercises and other forms of structure that help clients take more responsibility for what happens both in the helping sessions and outside.

## Realism in Using the Balance Sheet

Now let's look at a more practical and flexible approach to using the balance sheet. It is not to be used with every client to work out the pros and cons of every course of action. Tailor the balance sheet to the needs of the client. Choose the parts of the balance sheet that will add the most value with *this* client pursuing *this* goal or set of goals. In fact, one of the best uses of the balance sheet is not to use it directly at all. Keep it in the back of your mind whenever clients are making decisions. Use it as a filter to listen to clients. Then turn relevant parts of it into probes to help clients focus on issues they may be overlooking. "How will this decision affect the significant people in your life?" is a probe that originates in the balance sheet. "Is there any downside to that strategy?" might help a client who is being a bit too optimistic. There's no formula.

## LINKING STEP III-B TO ACTION

Some clients are filled with great ideas for getting things done but never seem to do anything. They lack the discipline to evaluate their ideas, choose the best ones, and turn them into action. Often this kind of work seems too tedious to them, even though it is precisely what they need. Consider the following case:

> Clint came away from the doctor feeling depressed. He was told that he was in the high-risk category for heart disease and that he needed to change his lifestyle. He was cynical, a man very quick to anger, a man who did not readily trust others. Venting his suspicions and hostility did not make these feelings go away; it only intensified them. Therefore, one critical lifestyle change was to change this pattern and develop the ability to trust others. He developed three broad goals: reducing mistrust of others' motives; reducing the frequency and intensity of such emotions as rage, anger, and irritation; and learning how to treat others with consideration. Clint read through the strategies suggested to help people pursue these broad goals (see Williams, 1989). They included
>
> * keeping a hostility log to discover the patterns of cynicism and irritation in one's life;
> * finding someone to talk to about the problem, someone to trust;
> * "thought stopping"—catching oneself in the act of indulging in hostile thoughts or in thoughts that lead to hostile feelings;
> * talking sense to oneself when tempted to put others down;
> * developing empathic thought patterns—that is, walking in the other person's shoes;
> * learning to laugh at one's own silliness;
> * using a variety of relaxation techniques, especially to counter negative thoughts;
> * finding ways of practicing trust;
> * developing active listening skills;
> * substituting assertive for aggressive behavior;
> * getting perspective, seeing each day as one's last;
> * practicing forgiving others without being patronizing or condescending.
>
> Clint prided himself on his rationality (though his "rationality" was one of the things that got him into trouble). So, as he read down the list, he chose strategies that could form an "experiment," as he put it. He decided to talk to a counselor (for the sake of objectivity), keep a hostility log (data gathering), and use the tactics of thought stopping and talking sense to himself whenever he felt that he was letting others get under his skin. The counselor noted to himself that none of these necessarily involved changing Clint's attitudes toward others. However, he did not challenge Clint at this point. His best bet was that through "strategy sampling" Clint would learn more about his problem, that he would find that it went deeper than he thought. Clint set himself to his experiment with vigor.

Clint chose strategies that fit his values. The problem was that the values themselves needed reviewing. But Clint did act, and action gave him the opportunity to learn.

## THE SHADOW SIDE OF SELECTING STRATEGIES

The shadow side of decision making, discussed in Chapter 14, is certainly at work in clients' choosing strategies to implement goals. Goslin (1985) puts it well:

> In defining a problem, people dislike thinking about unpleasant eventuali-
> ties, have difficulty in assigning . . . values to alternative courses of action,
> have a tendency toward premature closure, overlook or undervalue long-
> range consequences, and are unduly influenced by the first formulation of
> the problem. In evaluating the consequences of alternatives, they attach ex-
> tra weight to those risks that can be known with certainty. They are more
> subject to manipulation . . . when their own values are poorly thought
> through. . . . A major problem . . . for . . . individuals is knowing when to
> search for additional information relevant to decisions. (pp. 7, 9)

In choosing courses of action, clients often fail to evaluate the risks involved and
to determine whether the risk is balanced by the probability of success. Gelatt,
Varenhorst, and Carey (1972) suggest four ways in which clients may try to deal
with the factors of risk and probability: wishful thinking, playing it safe, avoiding
the worst outcome, and achieving some kind of balance. The first three are often
pursued without reflection and therefore lie in the "shadows."

**Wishful thinking.** In this case, clients choose a course of action that might (they
hope) lead to the accomplishment of a goal regardless of risk, cost, or probability.
For instance, Jenny wants her ex-husband to increase the amount of support he is
paying for the children. She tries to accomplish this by constantly nagging him and
trying to make him feel guilty. She doesn't consider the risk (he might get angry and
stop giving her anything), the cost (she spends a great deal of time and emotional
energy arguing with him), or the probability of success (he does not react favorably
to nagging). Wishful-thinking clients operate blindly, engaging in courses of action
without taking into account their usefulness. At its worst, this is a reckless approach.
Clients who "work hard" and still "get nowhere" may be engaged in wishful thinking, persevering in using means they prefer but that are of doubtful efficacy. Effective helpers find ways of challenging wishful thinking: "Jenny, let's review what
you've been doing to get Tom to pay up and how successful you've been."

**Playing it safe.** In this case, clients choose only safe courses of action, ones that
have little risk and a high degree of probability of producing at least limited success. For instance, Liam, a manager in his early forties, is very dissatisfied with the
way his boss treats him at work. His ideas are ignored, the delegation he is supposed
to have is preempted, and his boss does not respond to his attempts to discuss career development. His goals center around his career. He wants to let his boss know
about his dissatisfaction, and he wants to learn what his boss thinks about him and
his career possibilities. These are instrumental goals, of course, since his overall
goal is to carve out a career path. However, he fails to bring these issues up when
his boss is "out of sorts." On the other hand, when things are going well, Liam
doesn't want to "upset the applecart." He drops hints about his dissatisfaction, even
joking about them at times. He tells others in hopes that word will filter back to his
boss. During formal appraisal sessions, he allows himself to be intimidated by his
boss. However, in his own mind, he is doing whatever could be expected of a "reasonable" man. He does not know how safe he is playing it. The helper says, "Liam,
you're playing pretty safe with your boss. And, while it's true that you haven't upset
him, you're still in the dark about your career prospects."

**Avoiding the worst outcome.** Often clients choose means that are likely to help them avoid the worst possible result. They try to minimize the maximum danger, often without identifying what that danger is. Crissy, dissatisfied with her marriage, sets a goal to be "more assertive." However, even though she has never said this either to herself or to her counselor, the maximum danger for her is losing her partner. Therefore, her "assertiveness" is her usual pattern of compliance, with some frills. For instance, every once in a while, she tells her husband that she is going out with friends and will not be around for supper. He, without her knowing it, actually enjoys these breaks. At some level of her being, she realizes that her absences are not putting him under any pressure. She continues to be assertive in this way. But she never sits down with her husband to review where they stand with each other. That might be the beginning of the end. Early in one session, the counselor says, "What if some good friend were to say to you, 'Bill has you just where he wants you.' How would you react?" Crissy is startled, but she comes away from the session much more realistic.

**Striking a balance.** Ideally, clients choose strategies for achieving goals that balance risks against the probability of success. This "combination" approach is the most difficult to apply because it involves the right kind of analysis of problem situations and opportunities, choosing goals with the right edge, being clear about one's values, ranking a variety of strategies according to these values, and estimating how effective any given course of action might be. Even more to the point, it demands challenging the blind spots that might distort these activities. Since some clients have neither the skill nor the will for this combination approach, it is essential that their counselors help them engage in the kind of dialogue that will help them face up to this impasse.

### Evaluation Questions for Step III-B

How well am I doing the following as I try to help clients choose goal-accomplishing strategies that are best for them?

- Helping clients choose strategies that are clear and specific, that best fit their capabilities, that are linked to goals, that have power, and that are suited to clients' styles and values
- Helping clients engage in and benefit from strategy sampling
- Helping clients in selected cases use the balance sheet as a way of choosing strategies by outlining the principal benefits and costs for self, others, and relevant social settings
- Helping clients manage the shadow side of selecting courses of action—that is, wishful thinking, playing it too safe, focusing on avoiding the worst possible outcome rather than on getting what they want, and wasting time by trying to spell out a perfectly balanced set of strategies
- Helping clients use the act of choosing strategies to stimulate problem-managing action

# Step III-C: "What Kind of Plan Will Help Me Get What I Need and Want?" Helping Clients Make Plans

No Plan of Action: The Case of Frank

How Plans Add Value to Clients' Change Programs

    Plans help clients develop needed discipline

    Plans keep clients from being overwhelmed

    Formulating plans helps clients search for more useful ways of accomplishing goals—that is, even better strategies

    Plans provide opportunities to evaluate the realism and adequacy of goals

    Plans make clients aware of the resources they will need to implement their strategies

    Formulating plans helps clients uncover unanticipated obstacles to the accomplishment of goals

Shaping the Plan: Three Cases

    The case of Wanda

    The case of Harriet: The economics of planning

    The case of Frank revisited

Humanizing the Technology of Constructive Change

    Build a Planning Mentality into the Helping Process Right from the Start

    Adapt the Constructive-Change Process to the Client's Style

    Devise a Plan for the Client and Then Work with the Client to Revise It as Needed

Tailoring Ready-Made Programs to Clients' Needs

    A prevention program for pedophilia

    A program for helping people on welfare become successful at work

    General well-being programs: Exercise

Evaluation Questions for Step III-C

After identifying and choosing strategies to accomplish goals, clients need to organize these strategies into plans. This is the work of Step III-C. In this step, counselors help clients come up with plans, sequences of actions—"What should I do first, second, and third?"—that will get them what they want, their goals.

## NO PLAN OF ACTION: THE CASE OF FRANK

The lack of a plan—that is, a clear step-by-step process to accomplish a goal—keeps some clients mired in their problem situations. Consider the case of Frank, a vice president of a large West Coast corporation.

> Frank was a go-getter. He was very astute about business and had risen quickly through the ranks. Vince, the president of the company, was in the process of working out his own retirement plans. From a business point of view, Frank was the heir apparent. But there was a glitch. Vince was far more than a good manager; he was a leader. He had a vision of what the company should look like five to ten years down the line. Early on, he saw the power of the Internet and used it wisely to give the business a competitive edge.
>
> Though tough, Vince related well to people. People constituted the human capital of the company. He knew that products *and* people kept customers happy. He also took to heart the results of a millennium survey of some 2 million employees in the United States. One of the sentences in the summary of the survey results haunted him: "People join companies but leave supervisors." In the "war for talent," he couldn't afford supervisors who alienated their team members.
>
> Frank was quite different. He was a "hands-on" manager, meaning, in his case, that he was slow to delegate tasks to others, however competent they might be. He kept second-guessing others when he did delegate, reversed their decisions in a way that made them feel put down, listened poorly, and took a fairly short-term view of the business: "What were last week's figures like?" He was not a leader but an "operations" man. His direct reports called him a micromanager.
>
> One day, Vince sat down with Frank and told him that he was considering him as his successor down the line but that he had some concerns. "Frank, if it were just a question of business acumen, you could take over today. But my job, at least in my mind, demands a leader." Vince went on to explain what he meant by a leader and to point out the things in Frank's style that had to change.
>
> So Frank did something that he never thought he would do. He began seeing a coach. Roseanne had been an executive with another company in the same industry but had opted to become a coach for family reasons. Frank chose her because he trusted her business acumen. That's what meant most to him. They worked together for over a year, often over lunch and in hurried meetings early in the morning or late in the evening. And, indeed, he valued their dialogues about the business.
>
> Frank's ultimate aim was to become president. If getting the job meant that he had to try to become the kind of leader his boss had outlined, so be it. Since he was very bright, he came up with some inventive strategies for moving in that direction. But he could never be pinned down to an overall program with specific milestones by which he could evaluate his progress. Roseanne pushed him, but Frank was always "too busy" or would say that a formal program was "too stifling." That was odd, since formal planning was one of his strengths in the business world.
>
> Frank remained as astute as ever in his business dealings. But he merely dabbled in the strategies meant to help him become the kind of leader Vince wanted him to be. Frank had the opportunity of not just correcting some mistakes but of developing and expanding his managerial style. But he blew it. At the end of two years, Vince appointed someone else president of the company.

Frank never got his act together. He never put together the kind of change program needed to become the kind of leader Vince wanted as president. Why? Frank had two significant blind spots that the coach did not help him overcome. First, he never really took Vince's notion of leadership seriously. So he wasn't really ready for a change program. He thought the president's job was his, that business acumen alone would win out in the end. Second, he thought he could change his management style at the margins, when more substantial changes were called for.

Roseanne never challenged Frank as he kept "trying things" that never led anywhere. Maybe things would have been different if she had said something like this: "Come on, Frank, you know you don't really buy Vince's notion of leadership. But you can't just give lip service to it. Vince will see right through it. We're just messing around. You don't want a program because you don't believe in the goal. Let's do something or call these meetings off." In a way, she was a co-conspirator because she, too, relished their business discussions. When Frank didn't get the job, he left the company, leaving Roseanne to ponder her success as an executive but her failure as a coach.

## How Plans Add Value to Clients' Change Programs

Some clients, once they know what they want and some of the things they have to do to get what they want, get their act together, develop a plan, and move forward. Other clients need help. Since some clients (and some helpers) fail to appreciate the power of a plan, it is useful to start by reviewing the advantages of planning.

Not all plans are formal. "Little plans," whether called such or not, are formulated and executed throughout the helping process. Tess, an alcoholic who wants to stop drinking, feels the need for some support. She contacts Lou, a friend who has shaken a drug habit, tells him of her plight, and enlists his help. He readily agrees. Objective accomplished. This "little plan" is part of her overall change program. Change programs are filled with setting "little objectives" and developing and executing "little plans" to achieve them.

Formal planning usually focuses on the sequence of "big steps" clients must take to get what they need or want. Clients are helped to answer the question, "What do I need to do first, second, and third?" The most formal version of planning takes strategies for accomplishing goals, divides them into workable steps, puts the steps in order, and assigns a timetable for the accomplishment of each step.

Formal planning, provided that it is adapted to the needs of individual clients, has a number of advantages.

**Plans help clients develop needed discipline.** Many clients get into trouble in the first place because they lack discipline. Planning places reasonable demands on clients to develop discipline. Desmond, the halfway-house resident discussed in the last chapter, needed discipline and benefited greatly from a formal job-seeking program. Indeed, ready-made programs such as the 12-step program of Alcoholics Anonymous are in themselves plans that demand or at least encourage self-discipline.

**Plans keep clients from being overwhelmed.** Plans help clients see goals as doable, keeping the steps toward the accomplishment of goals "bite-size." Amazing things can be accomplished by taking bite-size steps toward substantial goals. Bud, the ex-psychiatric patient who ended up creating a network of self-help groups for ex-patients, started with the bite-size step of participating in one of those groups himself. He did not become a self-help entrepreneur overnight. It was a step-by-step process.

**Formulating plans helps clients search for more useful ways of accomplishing goals—that is, even better strategies.** Sy Johnson was an alcoholic. When Mr. Johnson's wife and children, working with a counselor, began to formulate a plan for coping with their reactions to his alcoholism, they realized that the strategies they had been trying were hit-or-miss. With the help of an Al-Anon self-help group, they went back to the drawing board. Mr. Johnson's drinking had introduced a great deal of disorder into the family. Planning would help them restore order.

**Plans provide opportunities to evaluate the realism and adequacy of goals.** This aspect of planning is an example of the "dialogue" that should take place among the stages of the helping process. When Walter, a middle manager who had many problems in the workplace, began tracing out a plan to cope with the loss of his job and with a lawsuit filed against him by his former employer, he realized that his initial goals—getting his job back and filing and winning a countersuit—were unrealistic. His revised goals included getting his former employer to withdraw the suit and getting into better shape to search for a job by participating in a self-help group of managers who had lost their jobs.

**Plans make clients aware of the resources they will need to implement their strategies.** When Dora was helped by a counselor to formulate a plan to pull her life together after the disappearance of her younger son, she realized that she lacked the social support needed to carry out the plan. She had retreated from friends and even relatives, but now she knew she had to get back into community. Normalizing life demanded ongoing social involvement and support. A goal of finding the support needed to get back into community was added to her constructive-change program.

**Formulating plans helps clients uncover unanticipated obstacles to the accomplishment of goals.** Ernesto, a U.S. soldier who had accidentally killed an innocent bystander during his stint in Kosovo, was seeing a counselor because of the difficulty he was having returning to civilian life. Only when he began pulling together and trying out plans for normalizing his social life did he realize how ashamed he was of what had happened to him in the military. He felt so flawed because of what had happened that it was almost impossible to involve himself intimately with others. Helping him deal with his shame became one of the most important parts of the healing process.

Formulating plans will not solve all our clients' problems, but it is one way of making time an ally instead of an enemy. Many clients engage in aimless activity in their efforts to cope with problem situations. Plans help clients make the best use of their time. Finally, planning itself has a hefty shadow side. For a good review of the shadow side of planning, see Dorner (1996, pp. 153–183).

## SHAPING THE PLAN: THREE CASES

Plans need "shape" to drive action. A formal plan identifies the activities or actions needed to accomplish a goal or a subgoal, puts those activities into a logical but flexible order, and sets a time frame for the accomplishment of each key step. Therefore, there are three simple questions:

- What are the concrete things that need to be done to accomplish the goal or the subgoal?
- In what sequence should these be done? What should be done first, second, third, and so on?
- What is the time frame? What should be done today, tomorrow, next month?

If clients choose goals that are complex or difficult, it is useful to help them establish subgoals as a way of moving step-by-step toward the ultimate goal. For instance, once Bud decided to start an organization of self-help groups composed of ex-patients from mental hospitals, there were a number of subgoals he needed to accomplish before the organization would become a reality. His first step was to set up a test group. This instrumental goal provided the experience needed for further planning. A later step was to establish some kind of charter for the organization. "Charter in place" was one of the subgoals leading to his main goal.

In general, the simpler the plan the better. However, simplicity is not an end in itself. The question is not whether a plan or program is complicated but whether it is well shaped and designed to produce results. If complicated plans are broken down into subgoals and the strategies or activities needed to accomplish them, they are as capable of being achieved as simpler ones, assuming the time frame is realistic. In schematic form, shaping looks like this:

Subprogram 1 (a set of activities) leads to subgoal 1 (usually an instrumental goal).

Subprogram 2 leads to subgoal 2.

Subprogram n (the last in the sequence) leads to the accomplishment of the ultimate goal.

**The case of Wanda.**  Consider Wanda, a client who set a number of goals to manage a complex problem situation. One of her goals was finding a job. The plan leading to this goal had a number of steps, each of which led to the accomplishment of a subgoal. The following subgoals were part of Wanda's job-finding program. They are stated as accomplishments (the outcome or results approach).

Subgoal 1: Resumé written.

Subgoal 2: Kind of job wanted determined.

Subgoal 3: Job possibilities canvassed.

Subgoal 4: Best job prospects identified.

Subgoal 5: Job interviews arranged.

Subgoal 6: Job interviews completed.

Subgoal 7: Offers evaluated.

The accomplishment of these subgoals leads to the accomplishment of the overall goal of Wanda's plan—that is, getting the kind of job she wants.

Wanda also had to set up a step-by-step process or program to accomplish each of these subgoals. For instance, the process for accomplishing the subgoal "job possibilities canvassed" included such things as doing an Internet search on one or more of the many job search sites, reading the "Help Wanted" sections of the local newspapers, contacting friends or acquaintances who could provide leads, visiting employment agencies, reading the bulletin boards at school, and talking with someone in the job placement office. Sometimes the sequencing of activities is important, sometimes not. In Wanda's case, it's important for her to have her resumé completed before she begins to canvass job possibilities, but when it comes to using different methods for identifying job possibilities, the sequence does not make any difference.

**The case of Harriet: The economics of planning.** Harriet, an undergraduate student at a small state college, wants to become a counselor. Although the college offers no formal program in counseling psychology, with the help of an advisor, she identifies several undergraduate courses that would provide some of the foundation for a degree in counseling. One is called Social Problem-Solving Skills; a second is Effective Interpersonal Communication Skills; a third is Developmental Psychology: The Developmental Tasks of Late Adolescence and Early Adulthood. Harriet takes the courses as they come up. The first course she can enroll in is Social Problem-Solving Skills. The good news is that it includes a great deal of practice in the skills. The bad news is that it assumes competence in interpersonal communication skills. Too late she realizes that she is taking the courses out of optimal sequence. She would have gotten much more from the course had she taken the communication skills course first.

Harriet also volunteers for the dormitory peer-helper program run by the Center for Student Services. The center's counselors are very careful in choosing people for the program, but they don't offer much training. It is a learn-as-you-go approach. Harriet realizes that the developmental psychology course would have helped her enormously in this program. It would have helped her understand both herself and her peers better. She finally realizes that she needs a better plan. In the next semester, she drops out of the peer-counselor program. She sits down with one of the center's psychologists, reviews the schools offerings with him, decides which courses will help her most, and determines the proper sequencing of these courses. The psychologist also suggests a couple of courses she could take in a local community college. Harriet's opportunity-development program would have been much more efficient had it been better shaped in the first place.

**The case of Frank revisited.** Let's see what planning might have done for Frank, the vice president who needed leadership skills. In this fantasy, Frank, like Scrooge, gets a second chance.

What does Frank need to do? To become a leader, Frank decides to reset his managerial style with his subordinates by involving them more in decision making.

### Box 20-1    Questions on Planning

Here are some questions you can help clients ask themselves to come up with a viable plan for constructive change:

- Which sequence of actions will get me to my goal?
- Which actions are most critical?
- How important is the order in which these actions take place?
- What is the best time frame for each action?
- Which step of the program needs substeps?
- How can I build informality and flexibility into my plan?
- How do I gather the resources, including social support, needed to implement the plan?

He wants to listen more, set work objectives through dialogue, ask subordinates for suggestions, and delegate more. He knows he should coach his direct reports in keeping with their individual needs, give them feedback on the quality of their work, recognize their contributions, and reward them for achieving results beyond their objectives.

In what sequence should Frank do these things? Frank decides that the first thing he will do is call in each subordinate and ask, "What do you need from me to get your job done? How can I add value to your work? And what management style on my part would help you most?" Their dialogue around these issues will help him tailor his supervisory interventions to the needs of each team member. The second step is also clear. The planning cycle for the business year is about to begin, and each team member needs to know what his or her objectives are. It is a perfect time to begin setting objectives through dialogue rather than simply assigning them. So Frank sends a memo to each of his direct reports, asking them to review the company's strategy and business plan and the strategy and plan for each of their functions, and to write down what they think their key managerial objectives for the coming year should be. He asks them to include stretch goals.

What is Frank's time frame? Frank calls in each of his subordinates immediately to discuss what they need from him. He completes his objective-setting sessions with them within three weeks. He puts off further action on delegation until he gets a better reading on their performance. This is a rough idea of what a plan for Frank might have looked like and how it might have improved his chaotic and abortive effort to change his managerial style—on the condition, of course, that he was convinced that a different approach to management and supervision made personal and business sense.

Box 20-1 lists questions you can use to help clients think systematically about crafting a plan to get what they need and want.

# HUMANIZING THE TECHNOLOGY
# OF CONSTRUCTIVE CHANGE

Some years ago, I lent a friend of mine an excellent, though somewhat detailed, book on self-development. About two weeks later, he came back, threw the book on my desk, and said, "Who would go through all of that!" I retorted, "Anyone really interested in self-development." That was the righteous, not the realistic, response. Planning in the real world seldom looks like planning in textbooks. Textbooks do provide useful frameworks, principles, and processes, but they are seldom used. Most people are too impatient to do the kind of planning outlined in the previous section. One reason for the dismal track record of discretionary change mentioned earlier is that even when clients do set realistic goals, they lack the discipline to develop reasonable plans. The detailed work of planning is too burdensome.

Therefore, Stages II and III of the helping process together with their six steps need a human face. If helpers skip the goal-setting and planning steps clients need, they shortchange them. On the other hand, if they are pedantic, mechanistic, or awkward in their attempts to help clients engage in these steps—if they fail to give these processes a human face—helpers run the risk of alienating the people they are trying to serve. Clients might well say, "I'm getting a lot of boring garbage from him." Here, then, are some principles to guide the constructive-change process from Step II-A through Step III-C.

## Build a Planning Mentality into the
## Helping Process Right from the Start

A constructive-change mind-set should permeate the helping process from the very beginning. This is part of the hologram metaphor—the whole model should be found in each of its parts—mentioned in Chapter 2. Helpers need to see clients as self-healing agents capable of changing their lives, not just as individuals mired in problem situations. Even while listening to a client's story, the helper needs to begin thinking of how the situation can be remedied and through probes find out what approaches to change the client is thinking about—no matter how tentative these ideas might be. As mentioned earlier, helping clients act in their real world at the start of the helping process helps them develop some kind of initial planning mentality. If helping is to be solution-focused, thinking about strategies and plans must be introduced early. When a client tells of some problem, the helper can ask early on, "What have you done so far to try to cope with the problem?"

> Cora, a battered spouse, did not want to leave her husband because of the kids. Right from the beginning, the helper saw Cora's problem situation from the point of view of the whole helping process. While she listened to Cora's story, without distorting it, she saw possible goals and strategies. Within the helping sessions, the counselor helped Cora learn a great deal about how battered women typically respond to their plight and how dysfunctional some of those responses are. Cora also learned how to stop blaming herself for the violence and to overcome her fears of developing more active coping strategies. At home, she confronted her husband and stopped submitting to the violence in a vain attempt to avoid further abuse. She also joined a local self-help group for battered women. There she found social support and learned how to invoke both police protection and recourse to the courts. Further sessions with the counselor helped her gradually change her identity from battered woman to survivor and, eventually, to doer. She moved from simply facing problems to developing opportunities.

Constructive-change scenarios like this must be in the helper's mind from the start, not as preset programs to be imposed on clients but as part of a constructive-change mentality.

## Adapt the Constructive-Change Process to the Client's Style

Setting goals, devising strategies, and making and implementing plans can be done formally or informally. There is a continuum. Some clients actually like the detailed work of devising plans; it fits their style.

> Gitta sought counseling as she entered the "empty nest" period of her life. Although there were no specific problems, she saw too much emptiness as she looked into the future. The counselor helped her see this period of life as a normal experience rather than a psychological problem. It was a developmental opportunity and challenge (see Raup & Myers, 1989). It was an opportunity to reset her life. After spending a bit of time discussing some of the maladaptive responses to this transitional phase of life, they embarked on a review of possible scenarios. Gitta loved brainstorming, getting into the details of the scenarios, weighing choices, setting strategies, and making formal plans. She had been running her household this way for years. So the process was familiar even though the content was new.

Here's another case:

> Connor, in rebuilding his life after a serious automobile accident, very deliberately planned both a rehabilitation program and a career change. Keeping to a schedule of carefully planned actions not only helped him keep his spirits up but also helped him accomplish a succession of goals. These small triumphs buoyed his spirits and moved him, however slowly, along the rehabilitation path.

Both Gitta and Connor readily embraced the positive-psychology approach embedded in constructive-change programs. They thrived on both the work and the discipline to develop plans and execute them. Many, if not most, people, however, are not like Gitta and Connor. The distribution is skewed toward the "I hate all this detail and won't do it" end of the continuum.

Kirschenbaum (1985) challenges the notion that planning should always provide an exact blueprint for specific actions, their sequencing, and the time frame. There are three questions:

- How specific do the activities have to be?
- How rigid does the order have to be?
- How soon does each activity have to be carried out?

Kirschenbaum suggests that, at least in some cases, being less specific and rigid about actions, sequencing, and deadlines can "encourage people to pursue their goals by continually and flexibly choosing their activities" (p. 492). That is, flexibility in planning can help clients become more self-reliant and proactive. Rigid planning strategies can lead to frequent failure to achieve short-term goals.

Consider the case of Yousef, a single parent with a mentally retarded son. He was challenged one day by a colleague at work. "You've let your son become a ball and chain, and that's not good for you or him!" his friend said. Yousef smarted from the remark, but eventually—and reluctantly—he sought counseling. He never discussed any kind of extensive change program with his helper, but with some stimulation from her, he began doing little things differently at home. When he came home from work especially tired and frustrated, he had a friend in the apartment building stop by. This helped him to refrain from taking his frustrations out on his son. Then, instead of staying cooped up over the weekend, Yousef found simple things to do that eased tensions, such as going to the zoo and to the art museum with a woman friend and his son. He discovered that his son enjoyed these pastimes immensely despite his limitations. In short, he discovered little ways to blend caring for his son with a better social life. His counselor had a constructive-change mentality right from the beginning but did not try to engage Yousef in overly formal planning activities.

On the other hand, a slipshod approach to planning—"I will have to pull myself together one of these days"—is also self-defeating. We need only look at our own experiences to see that such an approach is fatal. Overall, counselors should help clients embrace the kind of rigor in planning that makes sense for them in their situations. There are no formulas; there are only client needs and common sense. Some things need to be done now, some later. Some clients need more slack than others. Sometimes it helps to spell out the actions that need to be done in quite specific terms; at other times it is necessary only to help clients outline them in broad terms and leave the rest to their own sound judgment. If therapy is to be brief, help clients start doing things that lead to their goals. Then, in a later session, help them review what they have been doing, drop what is not working, continue what is working, add more effective strategies, and put more organization in their programs. If you have a limited number of sessions with a client, you can't engage in extensive goal setting and planning. "What can I do that will add the most value?" is the ongoing challenge in brief therapy.

## Devise a Plan for the Client and Then Work with the Client to Revise It as Needed

The more experienced helpers become, the more they learn about the elements of program development and the more they come to know what kinds of programs work for different clients. They build up a stockpile of useful programs and know how to stitch pieces of different programs together to create new programs. And they can use their knowledge and experience to fashion plans for clients who lack the skills or the temperament to pull together plans for themselves. Of course, their objective is not to foster dependence but to help clients grow in self-determination. For instance, a helper can first offer a plan as a sketch or in outline form rather than as a detailed program. Then the helper can work with the client to fill out the sketch and adapt it to the client's needs and style. Consider the following case:

> Katrina, a woman who dropped out of high school but managed to get a high school equivalency diploma, was overweight and reclusive. Over the years, she had restricted her activities because of her weight. Sporadic attempts at dieting had left her even heavier. Because she was chronically depressed and had little imagination, she was not able to come up with any kind of coherent plan. Once her counselor understood the dimensions of Katrina's problem situation, she pulled together an outline of a change program that included such things as blame reduction, the redefinition of beauty, decreasing self-imposed social restrictions, and cognitive restructuring activities aimed at lessening depression (see Robinson & Bacon, 1996). She also gathered information from health-care sources about obesity and suggestions for dealing with

it. She presented these to Katrina in a simple format, adding detail only for the sake of clarity. She added further detail as Katrina got involved in the planning process and in making choices.

Although this counselor pulled together elements of a range of already existing programs, counselors are, of course, free to make up their own programs based on their expertise and experience. The point is to give clients something to work with, something to get involved in. The elaboration of the plan emerges through dialogue with the client and in the kind of detail the client can handle.

The ultimate test of the effectiveness of plans lies in the problem-managing and opportunity-developing action clients engage in to get what they need and want. There is no such thing as a good plan in and of itself. Results, not planning or hard work, are the final arbiter. The next and final chapter deals with turning planning into accomplishments.

## TAILORING READY-MADE PROGRAMS TO CLIENTS' NEEDS

There are many ready-made programs for clients with particular problems. They are often tried-and-true constructive-change programs. The 12-step approach of Alcoholics Anonymous is one of the most well known. It has been adapted to other forms of substance abuse and addiction. Systematic desensitization, a behavioral approach, has been used to treat clients with PTSD, or post-traumatic stress disorder (Frueh, de Arellano, & Turner, 1997). This program includes sessions in muscle relaxation, the development of a fear hierarchy, and, finally, weekly sessions in the systematic desensitization of these fears. The program helps alleviate such debilitating symptoms as intrusive thoughts, panic attacks, and episodic depression. The manualized treatment programs outlined in Chapter 1 are also examples of ready-made programs. Donald Meichenbaum (1994) published a comprehensive handbook for dealing with PTSD that includes a practical manual.

Counselors add value by helping clients adapt "set" programs to their particular needs. Consider the following cases.

**A prevention program for pedophilia.** While there are many treatment programs for pedophilic clients *after* the fact, prevention programs are much scarcer. Consider this case:

After a couple of rather aimless sessions, the helper said to Ahmed, "We've talked about a lot of things, but I'm still not sure why you came in the first place." This challenged Ahmed to reveal the central issue, though he needed a great deal of help to do so. It turned out that Ahmed was sexually attracted to prepubescent children of both sexes. Although he had never engaged in pedophilic behavior, the temptation to do so was growing.

The counselor adapted a New Zealand program called Kia Marama (Hudson et al., 1995), a comprehensive cognitive-behavioral program for incarcerated child molesters, to Ahmed's situation. The original program includes intensive work in challenging distorted attitudes, reviewing a wide range of sexual issues, seeing the world from the point of view of the victim, developing problem-solving and interpersonal-relationship skills, stress management, and relapse-prevention training. The helper and Ahmed spent some time assessing which parts of the program might be of most help before embarking on an intensive tailored program.

The economics of prevention far outweigh the economics of rehabilitation. Not only did Ahmed stay out of trouble, but much of what he learned from the program—for instance, stress management—applied to other areas of his life.

**A program for helping people on welfare become successful at work.** One community-based mental-health center worked extensively with people on welfare. When new legislation was passed forcing welfare recipients to get work, they searched for programs that helped people on welfare get and keep jobs. They learned a great deal from one program sponsored by a major hotel chain (see Milbank, 1996). The hotel targeted welfare recipients because it made both economic and social sense. Because of the problems with this particular population, however, the hotel's recruiters, trainers, and supervisors had to become paraprofessional helpers, though they never used that term. The people they recruited—battered women, ex-convicts, addicts, homeless people, including those who had been thrown out of shelters, and so forth—had all sorts of problems. In the beginning, the hotel's staff did many things *for* the trainees.

> They drive welfare trainees to work, arrange their day care, negotiate with their landlords, bicker with their case workers, buy them clothes, visit them at home, coach them in everything from banking skills to self-respect, and promise those who stick with it full-time jobs. (Milbank, 1996, A1)

But the trainers also challenged their "clients'" mind-set that they were not responsible for what happened to them, enforced the hotel's code of behavior with equity, and persevered. The hotel program was far from perfect, but it did help many of the participants develop much-needed self-discipline and find new lives both at work and outside.

The counselors from a local mental-health center who acted as consultants to the program learned that some of the new employees benefitted greatly from wholesale upfront involvement of trainers and supervisors in their lives. It kick-started a constructive-change process. They also saw that the recruiters, trainers, and supervisors also benefitted. So they started a volunteer program at the mental-health center, looking for people willing to do the kinds of things that the hotel trainers and supervisors did. They knew that both the clients and the volunteers would benefit.

**General well-being programs: Exercise.** Some programs that contribute to general well-being can be used as adjuncts to all approaches to helping. Exercise programs are probably one of the most underused adjuncts to helping (Burks & Keeley, 1989). McAuley, Mihalko, and Bane (1997) have explored the multidimensional relationship between exercise and self-efficacy. There is evidence showing that exercise programs can help in the treatment of schizophrenia and alcohol dependence. Such programs also help more directly to reduce depression, manage chronic pain, and control anxiety (Tkachuk & Martin, 1999). The self-discipline developed through exercise programs can be a stimulus to increased self-regulation in other areas of life. Kate Hays has done a comprehensive review of the positive psychology possibilities of exercise in *Working It Out: Using Exercise in Psychotherapy* (1999). To end on a personal note regarding exercise: Once, I got my gear together and started out to get some exercise. When I hesitated, I asked myself: "Have you *ever* regretted exercising?" I answered, "Never," and headed out the door.

Finally, not all useful ready-made programs are found in sophisticated manuals. Many are found in the best of the self-help literature. Books like *Thoughts and Feelings* (McKay, Davis, & Fanning, 1997) are filled with systematic strategies for the treatment of a wide variety of psychological problems. The best are realistic, practical translations of some of the best thinking in the field.

## Evaluation Questions for Step III-C

Helpers can ask themselves the following questions as they help clients formulate the kinds of plans that actually drive action:

- To what degree do I prize and practice planning in my own life?
- How effectively have I adopted the hologram mind-set in helping, seeing each session and each intervention in the light of the entire helping process?
- How quickly do I move to planning when I see that it is what the client needs to manage problems and develop opportunities better?
- What do I do to help clients overcome resistance to planning? How effectively do I help them identify the incentives for and the payoff of planning?
- How effectively do I help clients formulate subgoals that lead to the accomplishment of overall preferred-scenario goals?
- How practical am I in helping clients identify the action needed to accomplish subgoals, sequence those actions, and establish realistic time frames for them?
- How well do I adapt the specificity and detail of planning to the needs of each client?
- Even at this planning step, how easily do I move back and forth among the different stages and steps of the helping model as the need arises?
- How readily do clients actually move to action because of my work with them in planning?
- How human is the technology of constructive change in my hands?
- How well do I adapt the constructive-change process to the style of the client?
- How effectively do I help clients tailor generic or ready-made change programs to their specific needs?

# THE ACTION ARROW: MAKING IT ALL HAPPEN

The action arrow of the helping model represents the difference between planning and action. The nine steps of Stages I, II, and III all revolve around planning for change, not change itself. However, the need to incorporate action into planning and planning into action has been emphasized throughout the book. That is, the "little actions" needed to get the change process moving right from the start have been noted and illustrated. We now take a more formal look at results-producing action—both the obstacles to action and the ways to overcome those obstacles.

# "How Do I Make It All Happen?" Helping Clients Get What They Want and Need

**HELPING CLIENTS BECOME EFFECTIVE TACTICIANS**

Help Clients Develop "Implementation Intentions"

Help Clients Avoid Imprudent Action

Help Clients Develop Contingency Plans

Help Clients Overcome Procrastination

Help Clients Identify Possible Obstacles to and Resources for Implementing Plans

  Obstacles

  Facilitating forces

Help Clients Find Incentives and Rewards for Sustained Action

Help Clients Develop Action-Focused Self-Contracts and Agreements

Help Clients Be Resilient After Mistakes and Failures

  Social support

  Cognitive skills

  Psychological resources

**GETTING ALONG WITHOUT A HELPER: DEVELOPING SOCIAL NETWORKS FOR SUPPORTIVE CHALLENGE**

  Challenging relationships

  Feedback from significant others

  An amazing case of getting along without a helper

**THE SHADOW SIDE OF IMPLEMENTING CHANGE**

Helpers as Agents

Client Inertia: Reluctance to Get Started
    Passivity
    Learned helplessness
    Disabling self-talk
    Vicious circles
    Disorganization
Entropy: The Tendency of Things to Fall Apart
Choosing Not to Change

In a book called *True Success* (1994), Tom Morris lays down the conditions for achieving success. They include

- determining what you want—that is, a goal or a set of goals "powerfully imagined";
- focus and concentration in preparation and planning;
- the confidence or belief in oneself to see the goal through—that is, self-efficacy;
- a commitment of emotional energy;
- being consistent, stubborn, and persistent in the pursuit of the goal;
- the kind of integrity that inspires trust and gets people pulling for you;
- a capacity to enjoy the process of getting there.

The role of the counselor is to help clients engage in all these internal and external behaviors in the interest of goal accomplishment.

Some clients, once they have a clear idea of what to do to handle a problem situation or develop some opportunity, go ahead and do it, whether they have a formal plan or not. They need little or no further support and challenge from their helpers. They either find the resources they need within themselves or get support and challenge from the significant others in the social settings of their lives. However, other clients, although able to choose goals and come up with strategies for implementing them, are, for whatever reason, stymied when it comes to action. Most clients fall between these two extremes.

Discipline and self-control play an important part in implementing change programs. Kirschenbaum (1987) found that many things can contribute to not getting started or giving up: low initial commitment to change, weak self-efficacy, poor outcome expectations, the use of self-punishment rather than self-reward, depressive thinking, failure to cope with emotional stress, lack of consistent self-monitoring, failure to use effective habit-change techniques, giving in to social pressure, failure to cope with initial relapse, and paying attention to the wrong things—for instance, focusing on the difficulty of the problem situation rather than the attractiveness of the opportunity.

We have seen that self-determination and self-control are essential for action. Kanfer and Schefft (1988, p. 58) differentiate between two kinds of self-control. In *decisional self-control*, a single choice terminates a conflict. For instance, a couple makes the decision to get a divorce and goes through with it. In *protracted self-control*, continued resistance to temptation is required. For instance, it is not enough for a client to decide that she has to keep her anger under control when disagreements with others arise. Each time a conflict arises, she has to renew her resolve. It helps enormously if she develops the attitude that conflicts are learning opportunities and not just interpersonal struggles. This is a positive way of staying on guard.

Most clients need both kinds of self-control to manage their lives better. A client's choice to give up alcohol completely (decisional self-control) needs to be complemented by the ability to handle inevitable longer-term temptations. Protracted self-control calls for a preventive mentality and a certain degree of street

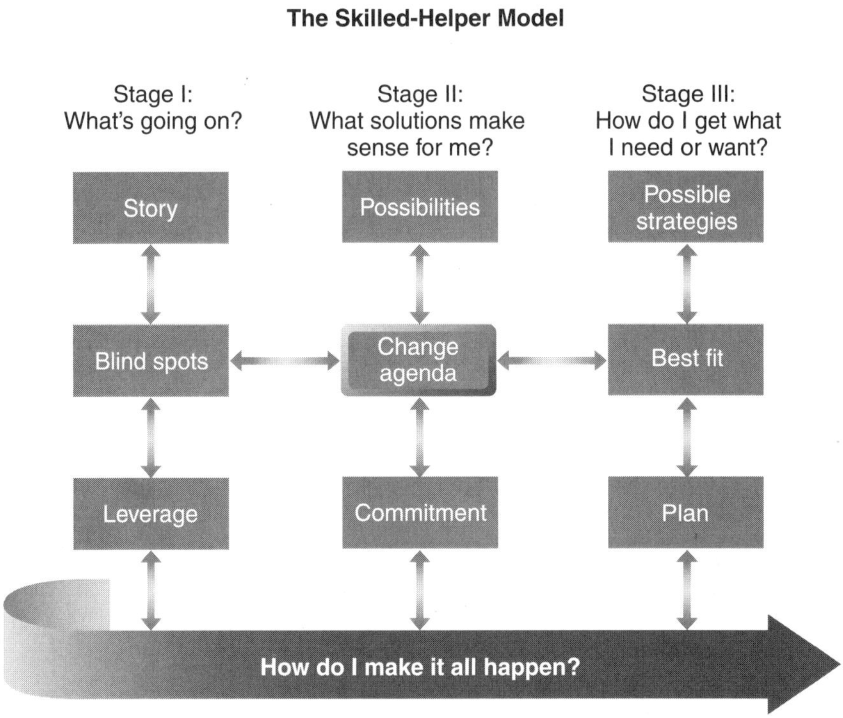

**The Skilled-Helper Model**

FIGURE 21-1
The Helping Model—Complementing Planning with Action

smarts. It is easier for the client who has given up alcohol to turn down an invitation to go to a bar in the first place than to sit in a bar all evening with friends and refrain from drinking.

Figure 21-1 adds the action arrow to the helping model.

## HELPING CLIENTS BECOME EFFECTIVE TACTICIANS

In the implementation phase, strategies for accomplishing goals need to be complemented by tactics and logistics. A strategy is a practical plan to accomplish some objective. Tactics is the art of adapting a plan to the immediate situation. This includes changing the plan on the spot to handle unforeseen complications. Logistics is the art of providing the resources needed to implement a plan in a timely way.

During the summer, Rebecca wanted to take an evening course in statistics so that the first semester of the following school year would be lighter. Having more time would enable her to act in one of the school plays, a high priority for her. But she didn't have the money to pay for the course, and at the university she planned to attend, prepayment for summer courses was the rule. Rebecca had counted on paying for the course from her summer earnings, but she would not have the money until later. Consequently, she did some quick shopping around and found that the same course was being offered by a community college not too far from where she lived. Her tuition there was minimal, since she was a resident of the area the college served.

In this example, Rebecca keeps to her overall plan (strategy). However, she adapts the plan to an unforeseen circumstance, the demand for prepayment (tactics), by locating another resource (logistics).

Since many well-meaning and motivated clients are simply not good tacticians, counselors can add value by using the following principles to help them engage in focused and sustained goal-accomplishing action.

## Help Clients Develop "Implementation Intentions"

Commitment to goals (see Chapter 17) must be followed by commitment to courses of action. Gollwitzer (1999) has researched a simple way to help clients cope with the common problems associated with translating goals into action: failing to get started, becoming distracted, reverting to bad habits, and so forth. Strong commitment to goals is not enough. Equally strong commitment to specific actions to accomplish goals is required. Good intentions, Gollwitzer points out, don't deserve their poor reputation. Strong intentions—"I strongly intend to study for an hour every weekday before dinner"—are "reliably observed to be realized more often than weak intentions" (p. 493).

> Implementation intentions are subordinate to goal intentions and specify the when, where, and how of responses leading to goal attainment. They have the structure of "When situation x arises, I will perform response y!" and thus link anticipated opportunities with goal-directed responses. (p. 494)

Gwendolyn, an aide in a nursing home, may say, "When Enid (a patient) becomes abusive, I will not respond immediately. I'll tell myself that it's her illness that's talking. Then I'll respond with patience and kindness." Her ongoing goal is to control her anger and other negative responses to patients. However, Gwendolyn keeps pursuing this goal by continually refreshing her strong implementation intentions. Since Enid has been a particularly difficult patient, Gwendolyn needs to refresh her intentions frequently. However, her initial strong intention to substitute anger and impatience with kindness and equanimity means that in most cases her responses are more or less automatic. The environmental cue—patient anger, abuse, lack of consideration—"triggers" the appropriate response in Gwendolyn. In a way, poor patient behaviors become "opportunities" for her responses. You can help clients enunciate to themselves strong specific intentions that will help them "automatically" handle many of the obstacles to goal implementation.

## Help Clients Avoid Imprudent Action

For some clients, the problem is not that they refuse to act but that they act imprudently. Rushing off to try the first "strategy" that comes to mind is often imprudent.

Elmer injured his back and underwent a couple of operations. After the second operation, he felt a little better, but then his back began troubling him again. When the doctor told him that further operations would not help, Elmer was faced with the problem of handling chronic pain. It soon became clear that his psychological state affected the level of pain. When he was anxious or depressed, the pain always seemed much worse.

Elmer was talking this through with a counselor. One day, he read about a pain clinic located in a Western state. Without consulting anyone, he signed up for a six-week program. Within ten days he was back, feeling more depressed than ever. He had gone to the program with extremely high expectations because his needs were so great. The program was a holistic one that helped the participants develop a more realistic lifestyle. It included programs dealing with nutrition, stress management, problem solving, and quality of interpersonal life. Group counseling was part of the program, and training was part of the group experience. For instance, the participants were trained in behavioral approaches to the management of pain.

The trouble was that Elmer had arrived at the clinic, which was located on a converted farm, with unrealistic expectations. He had not really studied the materials that the clinic had sent him. He had bought a "packaged" program without studying the package carefully. Since he had expected to find marvels of modern medicine that would magically help him, he was extremely disappointed when he found that the program focused mainly on reducing and managing rather than eliminating pain.

Elmer's goal was to be completely free of pain, but he failed to explore the realism of his goal. A more realistic goal would have centered on the reduction and management of pain. Elmer's counselor failed to help him avoid two mistakes: setting an unrealistic goal and, in desperation, acting on the first strategy that came along. Obviously, action cannot be prudent if it is based on flawed assumptions—in this case, Elmer's assumption that he could be pain free.

## Help Clients Develop Contingency Plans

If counselors help clients brainstorm both possibilities for a better future (goals) and strategies for achieving those goals (courses of action), then clients will have the raw materials, as it were, for developing contingency plans. Contingency plans answer the question, "What will I do if the plan of action I choose is not working?" Contingency plans help make clients more effective tacticians. The formulation of contingency plans is based on the fact that we live in an imperfect world. Often enough, goals have to be fine-tuned or even changed. The same is true for strategies for accomplishing goals.

Jackson, the man dying of cancer, decided to become a resident in the hospice he had visited. The hospice had an entire program in place for helping patients like Jackson die with dignity. Once there, however, he had second thoughts. He felt incarcerated. Fortunately, he had worked out alternative scenarios with his helper. One was living at the home of an aunt he loved and who loved him dearly, with some outreach services from the hospice. He moved out of the hospice into his aunt's home. He spent his final days at the hospice.

Contingency plans are needed especially when clients choose a high-risk program to achieve a critical goal. Having backup plans also helps clients develop more responsibility. If they see that a plan is not working, then they have to decide whether to try the contingency plan. Backup plans need not be complicated. A counselor might merely ask, "If that doesn't work, then what will you do?" As in the case of Jackson, clients can be helped to specify a contingency plan further once it is clear that the first choice is not working out.

## Help Clients Overcome Procrastination

At the other end of the spectrum are clients who keep putting action off. There are many reasons for procrastination. Take the case of Eula:

Eula, disappointed with her relationship with her father in the family business, decided that she wanted to start her own. She thought that she could capitalize on the business skills she had picked up in school and in the family business. Her goal, then, was to establish a small software firm that created products for the family-business market.

But a year went by and she still did not have any products ready for market. A counselor helped her see two things. First, her activities—researching the field, learning more about family dynamics, going to information technology seminars, getting involved for short periods with professionals such as accountants and lawyers who did a great deal of business with family-owned firms, drawing up and redrafting business plans, and creating a brochure—were helpful, but they did not create products. The counselor helped Eula see that at some level of her being, she was afraid of starting a new business. She had a lot of half-finished products. Over-preparation and half-finished products were signs of that fear. So she plowed ahead, finished a product, and brought it to market on the Internet. To her surprise, it was successful. Not a roaring success, but it meant that the cork was out of the bottle. Once she got one product to market, she had little problem developing and marketing others.

Eula certainly was not lazy. She was very active. She did all sorts of useful things. But she avoided the most critical actions—creating and marketing products.

## Help Clients Identify Possible Obstacles to and Resources for Implementing Plans

Years ago, Kurt Lewin (1969) codified common sense by developing what he called "force-field analysis." In ordinary language, this is simply a review by the client of the major obstacles to and the major facilitating forces for implementing action plans. The slogan is "forewarned is forearmed."

**Obstacles.** The identification of possible obstacles to the implementation of a program helps make clients forewarned.

Raul and Maria were a childless couple living in a large Midwestern city. They had been married for about five years and had been unable to have children. They finally decided that they would like to adopt a child, so they consulted a counselor familiar with adoptions. The counselor, in helping Raul and Maria work out a plan of action, helped them examine their motivation, review their suitability to be adoptive parents, contact an agency, and prepare themselves for an interview. After the plan of action had been worked out, Raul and Maria, with the help of the counselor, identified two possible obstacles or pitfalls: the negative feelings that often arise on the part of prospective parents when they are being scrutinized by an adoption agency, and the feelings of helplessness and frustration caused by the length of time and uncertainty involved in the process.

The assumption here is that if clients are aware of some of the "wrinkles" that can accompany any given course of action, they will be less disoriented when they encounter them. Identifying possible obstacles is, at its best, a straightforward census of likely pitfalls rather than a self-defeating search for every possible thing that could go wrong.

Obstacles can come from within the clients themselves, from others, from the social settings of their lives, and from larger environmental forces. Once an obstacle is spotted, ways of coping with it need to be identified. Sometimes simply being aware of pitfalls is enough to help clients mobilize their resources to handle them. At other times, a more explicit coping strategy is needed. For instance, the counselor arranged a couple of role-playing sessions with Raul and Maria in which she assumed the role of the examiner at the adoption agency and took a "hard line" in her questioning. These rehearsals helped them stay calm during the actual interviews. The counselor also helped them locate a mutual-help group of parents working their way through the adoption process. The members of the group shared their hopes and frustrations and provided support for one another. In short, Raul and Maria were trained to cope with the restraining forces they might encounter on the road toward their goal.

**Facilitating forces.** In a more positive vein, counselors can help their clients identify unused resources that can facilitate action.

> Nora found it extremely depressing to go to her weekly dialysis sessions. She knew that without them she would die, but she wondered whether it was worth living if she had to depend on a machine. The counselor helped her see that she was making life more difficult for herself by letting herself think such discouraging thoughts. He helped her learn how to think thoughts that would broaden her vision of the world instead of narrowing it down to herself, her discomfort, and the machine. Nora was a religious person and found in the Bible a rich source of positive thinking. She initiated a new routine: The day before she visited the clinic, she began to prepare herself psychologically by reading from the Bible. Then, as she traveled to the clinic and underwent treatment, she meditated slowly on what she had read.

In this case, the client substituted positive thinking, an underused resource, for poor-me thinking. Brainstorming resources that can counter obstacles to action can be very helpful for some clients. Helping clients brainstorm facilitating forces raises the probability that they will act in their own interests. They can be simple things. George was avoiding an invasive diagnostic procedure. After a brainstorming session, he decided to get a friend to go with him. This meant two things. Once he asked for his friend's help, he "had to go through with it." Second, his friend's very presence distracted him from his fears. Or consider Lucy, who had a history of letting her temper get the better of her. This was especially the case when she returned home after experiencing crises at work. Her mother-in-law and children became the targets of her wrath. After a counseling session, she took two photographs with her to work. One was a wedding-day picture that included her mother-in-law. The second was a recent picture of her three children. When she parked the car at work, she placed the pictures on the driver's seat. Then, when she got in the car in the evening, the first things she saw were the two photographs of her life at its best. This made her think on the way home about how she wanted to enter the house.

## Help Clients Find Incentives and Rewards for Sustained Action

Clients avoid engaging in action programs when the incentives and the rewards for not engaging in the programs outweigh the incentives and the rewards for doing so.

> Miguel, a policeman on trial for use of excessive force with a young offender, had a number of sessions with a counselor from an HMO that handled police health insurance. In the sessions, the counselor learned that, although this was the first time Miguel had run afoul of the law, it was in no way the first expression of a brutal streak within him. He was a bully on the beat and a despot at home, and he had gotten into run-ins with strangers when he visited bars with his friends. Some of this came out during the trial.
>
> Up to the time of his arrest, he had gotten away with his aggressive behavior, even though his friends had often warned him to be more cautious. His badge had become a license to do whatever he wanted. His arrest and now the trial shocked him. Before, he had seen himself as invulnerable; now, he felt very vulnerable. The thought of being a cop in prison understandably horrified him. He was found guilty, was suspended from the force for several months, and received probation on the condition that he continued to see the counselor.

Beginning with his arrest, Miguel had modified his aggressive behavior a great deal, even at home. Of course, fear of the consequences of his aggression was a strong incentive to change his behavior. The next time, the courts would show no sympathy. The counselor took a tough approach to this tough cop. He confronted Miguel for "remaining an adolescent" and for "hiding behind his badge." He called the power Miguel exercised over others "cheap power." He challenged the "decent person" to "come out from behind the screen." He told Miguel point blank that the fear he was experiencing was probably not enough to keep him out of trouble in the future. After probation, the fear would fade and Miguel could easily fall back into his old ways. Even worse, fear was a "weak man's" crutch.

In a more positive vein, the counselor saw in Miguel's expressions of vulnerability the possibility of a much more decent human being, one "hiding" under the tough exterior. The real incentives, he suggested, came from the "decent guy" buried inside. He had Miguel paint a picture of a "tough but decent" cop, family man, and friend. He asked Miguel to come up with "experiments in decency"—at home, on the beat, with his buddies—to get first-hand experience of the rewards associated with decency.

The counselor was not trying to change Miguel's personality. Indeed, the counselor didn't believe in personality transformations. But he pushed Miguel hard to find and bring to the surface a different, more constructive set of incentives to guide his dealings with people. The new incentives had to drive out the old.

The incentives and the rewards that help a client get going on a program of constructive change in the first place may not be the ones that keep the client going.

Dwight, a man in his early thirties who was recovering from an accident at work that had left him partially paralyzed, had begun an arduous physical rehabilitation program with great commitment. Now, months later, he was ready to give up. The counselor asked him to visit the children's ward. Dwight was both shaken by the experience and amazed at the courage of many of the kids. He was especially struck by one teenager who was undergoing chemotherapy. "He seems so positive about everything," Dwight said. The counselor told him that the boy was tempted to give up, too. Dwight and the boy saw each other frequently. Dwight put up with the pain. The boy hung in there. Three months later, the boy died. Dwight's response, besides grief, was, "I can't give up now; that would really be letting him down."

Dwight's partnership with the teenager proved to be an excellent incentive. It helped him renew his resolve. While the counselor joined with Dwight in celebrating his newfound commitment, he also worked with Dwight to find backup incentives for those times when current incentives seem to lose their power. One was the possibility of marrying and starting a family despite residual limitations resulting from the accident.

Constructive-change activities that are not rewarded tend over time to lose their vigor, decrease, and even disappear. This process is called extinction. It was happening with Luigi.

Luigi, a middle-aged man, had been in and out of mental hospitals a number of times. He discovered that one of the best ways of staying out was to use some of his excess energy helping others. He had not returned to the hospital once during the three years he worked at a soup kitchen. However, finding himself becoming more and more manic over the past six months and fearing that he would be rehospitalized, he sought the help of a counselor.

Luigi's discussions with the counselor led to some interesting findings. He discovered that, whereas in the beginning he had worked at the soup kitchen because he wanted to, he was now working there because he thought he should. He felt guilty about leaving and also thought that doing so would lead to a relapse. In sum, he had not lost his interest in helping others, but his current work was no longer interesting or challenging. As a result of his sessions with the counselor, Luigi began to work for a group that provided housing for the homeless and the elderly. He poured his energy into his new work and no longer felt manic.

The lesson here is that incentives cannot be put in place and then be taken for granted. They need tending.

## Help Clients Develop Action-Focused Self-Contracts and Agreements

Earlier we discussed self-contracts as a way of helping clients commit themselves to what they want—that is, their goals. Self-contracts are also useful in helping clients both initiate and sustain problem-managing action and the work involved in developing opportunities. For instance, Feller (1984) developed the following "job-search agreement" to help job seekers persist in their searches. In the agreement, clients respond "true" to all the statements and then act on those "truths." By so doing, clients commit themselves not only to job-seeking behavior but also to sound psychological practices that promote the right mentality for such behavior.

> I agree that no matter how many times I enter the job market, or the level of skills, experiences, or academic success I have, the following appear TRUE:
>
> 1. It takes only one YES to get a job; the number of no's does not affect my next interview.
> 2. The open market lists about 20% of the jobs presently open to me.
> 3. About 80% of the job openings are located by talking to people.
> 4. The more people who know my skills and know that I'm looking for a job, the more I increase the probability that they'll tell me about a job lead.
> 5. The more specifically I can tell people about the problems I can solve or outcomes I can attain, rather than describe the jobs I've had, the more jobs they may think I qualify for.
>
> I agree that regardless of how much I need a job, the following appear TRUE:
>
> 6. If I cut expenses and do more things for myself, I reduce my money problems.
> 7. The more I remain positive, the more people will be interested in me and my job skills.
> 8. If I relax and exercise daily, my attitude and health will appear attractive to potential employers.
> 9. The more I do positive things and the more I talk with enthusiastic people, the more I will gain the attention of new contacts and potential employers.
> 10. Even if things don't go as I would like them to, I choose my own thoughts, feelings, and behaviors each day.

It is easy to see how similar "agreements" could act as drivers of action in many different kinds of problem-managing and opportunity-developing situations. Self-contracts and agreements with others focus clients' energies.

Here is another example. In this case, several parties had to commit themselves to the provisions of the contract.

A boy in the seventh grade was causing a great deal of disturbance by his outbreaks in class. The usual kinds of punishment did not seem to work. After the teacher discussed the situation with the school counselor, the counselor called a meeting of all the stakeholders—the boy, his parents, the teacher, and the principal. The counselor offered a simple contract. When the boy disrupted the class, one and only one thing would happen: He would go home. Once the teacher indicated that his behavior was disruptive, he was to go to the principal's office and check out without receiving any kind of lecture. He was to go immediately home and check in with whichever parent was at home, again without receiving any further punishment. The next day he was to return to school. All agreed to the contract, though both principal and parents said they would find it difficult not to add to the punishment.

The first month, the boy spent a fair number of full or partial days at home. The second month, however, he missed only two partial days, and the third month only one. The truth is that he really wanted to be in school with his classmates. That's where the action was. And so he paid the price of self-control to get what he wanted.

The counselor had suspected that the boy found socializing with his classmates rewarding. But now he had to pay for the privilege of socializing. Reasonable behavior in the classroom was not too high a price.

## Help Clients Be Resilient After Mistakes and Failures

Clients, like the rest of us, stumble and fall as they try to implement their constructive-change programs. However, everyone has some degree of *resilience* within that enables them to get up, pull themselves together, and move on once more. The ability to bounce back is an essential life capability. Holaday and McPhearson (1997) have compiled a list of common factors that influence resilience. Although their study focused on severe-burn victims, what they have to say about resilience applies to all of us and our clients. They distinguish between *outcome* resilience and *process* resilience. While resilience in general is the ability to overcome or adapt to significant stress or adversity, outcome resilience implies a return to a previous state. This is "bounce back" resilience. Dora goes through the trauma of divorce, but within a few months she bounces back. Her friends say to her, "You seem to be your old self now." She replies, "Both older and wiser." Process resilience, on the other hand, represents the continuous effort to cope that is a "normal" part of some people's lives. The sufferers Holaday and McPhearson studied would say such things as, "Resilience? It's my spirit and it's the reason I'm here," and resilience "is deep inside of you, it's already there, but you have to use it," and "To do well takes a lot of determination, courage, and struggling, but it's your choice" (p. 348).

You can encourage both kinds of resilience in clients. Take outcome resilience. Kerry finds himself in a financial mess because of a tendency to be a spendthrift and because of a few poor financial decisions. Although he makes a couple more mistakes, he works his way out of the mess. Once he reaches his goal, he puts himself on a strict budget, and things stabilize. It's not difficult for him to walk the financial straight and narrow because the mess has been too painful to repeat. Now a couple of examples of process resilience. Oscar finds that controlling his anger is a constant struggle. He has to keep finding the resources within himself to keep plugging away. And then there is Nadia. Suffering from chronic fatigue syndrome, she has to dig deep within herself every day to find the will to go on. Like many people suffering from this condition, she wants to do her best and make a good impression (see Albrecht & Wallace, 1998). On the days she's successful in pulling herself together, the people she meets cannot believe that she is ill. Running into this kind of disbelief on the part of otherwise intelligent people is part of the grind.

Holaday and McPhearson (1997, 1999) suggest that the factors that go into resilience are social support, cognitive skills, and psychological resources.

**Social support.** This includes the overall values of a society toward people, especially people in trouble; community support—that is, support in the neighborhood, at work, at church, and so forth; personal support through friends and other special relationships; and familial support—the "affectional ties within a family system." One burn victim said, "My wife made me get out of the hospital bed and learn to walk again."

**Cognitive skills.** It seems that at least average intelligence contributes greatly to resilience. But there are different kinds of intelligence: academic intelligence, social intelligence, street smarts, and so forth. And, as Holaday and McPhearson (1997) point out, "intelligence is also associated with the ability to use fantasy and hope" (p. 350). Cognitive skills also include coping style. For instance, a "belligerent style" (Zimrin, 1986) rather than a passively enduring, accepting, or yielding style often contributes more to resilience: "I don't care what others say, it's *not* over; don't tell me I can't do something." Clients can also cope by discussing feelings. One burn victim said, "Sometimes I still choose to feel sorry for myself and have a bad day, and that's OK." Other useful coping strategies include avoiding self-blame and using the energy of anger to cope with the world rather than damage the self. One client said, "When I was little, I wanted the scars to go away, but now I don't care about them any more. They're part of me." Other cognitive factors in resilience include the degree and the way clients exercise personal control in their lives and how they interpret their experiences. One client who fell off the wagon and got drunk for a couple of days said, "It's a glitch, not a pattern. I can expect a glitch now and again. Glitches can be dealt with. Patterns are damaging."

**Psychological resources.** Certain personality characteristics or dispositions protect people from stress and contribute to bounce back. They include an internal locus of control, empathy, curiosity, a tendency to seek novel experiences, a high activity level, flexibility in new situations, a sense of humor, the ability to elicit positive regard from others, accurate and positive self-appraisal, personal integrity, a sense of self-protectiveness, pride in accomplishments, and a capacity for fun.

However, lest this sound like a recitation of the Boy Scout oath, the point is this: There is a range of "resilience levers" in every client. Your job is to help them discover the levers, pull them, and bounce back. Resilience is "deep inside you" and inside your clients. It's part of their self-healing nature.

## GETTING ALONG WITHOUT A HELPER: DEVELOPING SOCIAL NETWORKS FOR SUPPORTIVE CHALLENGE

In most cases, helping is a relatively short-term process. But even in longer-term therapy, clients must eventually get on with life without their helpers. Ideally, the counseling process not only helps clients deal with specific problem situations and unused opportunities, but also, as outlined in Chapter 1, equips them with the working knowledge and skills needed to manage those situations more effectively on their own.

Because adherence to constructive-change programs is often difficult, social support and challenge in their everyday lives can help them move to action, persevere in action programs, and both consolidate and maintain gains. When it comes to social support and challenge, there are a number of possible scenarios at the implementation stage and beyond:

- Counselors help clients with their plans for constructive change, and then clients, using their own initiative and resources, take responsibility for the plans and pursue them on their own.
- Clients continue to see a helper regularly in the implementation phase.
- Clients see a helper occasionally, either on demand or in scheduled stop-and-check sessions.
- Clients join some kind of self-help group together with one-to-one counseling sessions, which are eventually eliminated.
- Clients develop social relationships that provide both ongoing support and challenge for the changes they are making in their lives.

Support was discussed earlier, and the literature tends to focus on caring *support*, so a few words about caring *challenge* in everyday life are in order.

**Challenging relationships.** It was suggested earlier that support without challenge can be hollow and that challenge without support can be abrasive. Ideally, the people in the lives of clients provide a judicious mixture of support and challenge.

> Harry, a man in his early fifties, was suddenly stricken with a disease that called for immediate and drastic surgery. He came through the operation quite well, getting out of the hospital in record time. For the first few weeks he seemed, within reason, to be his old self. However, he had problems with the drugs he had to take following the operation. He became quite sick and took on many of the mannerisms of a chronic invalid. Even after the right mix of drugs was found, he persisted in invalid-like behavior. Whereas right after the operation he had "walked tall," he now began to shuffle. He also talked constantly about his symptoms and generally used his "state" to excuse himself from normal activities.
> At first Harry's friends were in a quandary. They realized the seriousness of the operation and tried to put themselves in his place. They provided all sorts of support. But gradually they realized that he was adopting a style that would alienate others and keep him out of the mainstream of life. Support was essential, but it was not enough. They used a variety of ways to challenge his behavior, mocking his "invalid" movements, engaging in serious one-to-one talks, turning a deaf ear to his discussions of symptoms, and routinely including him in their plans.

Harry did not always react graciously to his friends' challenges, but in his better moments he admitted that he was fortunate to have such friends. As clients attempt to change their behavior, counselors can help them find people willing to provide a judicious mixture of support and challenge.

**Feedback from significant others.** Gilbert (1978), in his book on human competence, claimed that "improved information has more potential than anything else I can think of for creating more competence in the day-to-day management of performance" (p. 175). Feedback is certainly one way of providing both support and challenge. If clients are to be successful in implementing their action plans, they need adequate information about how well they are performing. Sometimes they know themselves; other times they need a more objective view. The purpose of feedback is not to pass judgment on the performance of clients but rather to provide guidance, support, and challenge. There are two kinds of feedback.

- *Confirmatory feedback.* Through confirmatory feedback, significant others such as helpers, relatives, friends, and colleagues let clients know that they are on course—that is, moving successfully through the steps of their action programs toward their goals.

- *Corrective feedback.* Through corrective feedback, significant others let clients know that they have wandered off course and what they need to do to get back on.

Corrective feedback, whether from helpers or people in the client's everyday life, should incorporate the following principles:

- Give feedback in the spirit of caring.
- Remember that mistakes are opportunities for growth.
- Use a mix of both confirmatory and corrective feedback.
- Be concrete, specific, brief, and to the point.
- Focus on the client's behaviors rather than on more elusive personality characteristics.
- Tie behavior to goals.
- Explore the impact and implications of the behavior
- Avoid name-calling.
- Provide feedback in moderate doses. Overwhelming the client defeats the purpose of the entire exercise.
- Engage the client in dialogue. Invite the client not only to comment on the feedback but also to expand on it. Lectures don't usually help.
- Help the client discover alternative ways of doing things. If necessary, prime the pump.
- Explore the implications of changing over not changing.

The spirit of these "rules" should also govern confirmatory feedback. Very often people give very detailed corrective feedback and then just say "nice job" when a person does something well. All feedback provides an opportunity for learning. Consider the following statement from a father talking to his high school son who stood up for the rights of a friend who was being bullied by some of his classmates:

"Jeb, I'm proud of you. You stood your ground even when they turned on you. They were mean. You weren't. You gave your opinion calmly, but forcefully. You didn't apologize for what you were saying. You were ready to take the consequences. It's easier now that a couple of them have apologized to you, but at the time, you didn't know they would. You were honest to yourself. And now the best of them appreciate it. It made me think of myself. I'm not sure that I would have stood my ground the way you did. . . . But I'm more likely to do so now."

While brief, this is much more than "I'm proud of you, son." Being specific about behavior and pointing out the impact of the behavior turn positive feedback into a learning experience.

Of course, one of the main problems with feedback is finding people in the client's day-to-day life who see the client in action enough to make it meaningful, who care enough to give it, and who have the skills to provide it constructively.

**An amazing case of getting along without a helper.** As indicated earlier, many client problems are coped with and managed, not solved. Consider the following real-life case of a woman who certainly did not choose not to change. Quite the contrary. Her case is a good example of a no-formula approach to developing and implementing a program for constructive change.

> Vickey readily admits that she has never fully "conquered" her illness. Some 20 years ago, she was diagnosed as manic-depressive. The picture looked something like this: She would spend about six weeks on a high; then the crash would come, and for about six weeks she'd be in the pits. After that she'd be normal for about eight weeks. This cycle meant many trips to the hospital. Some seven years into her illness, during a period in which she was in and out of the hospital, she made a decision. "I'm not going back into the hospital again. I will so manage my life that hospitalization will never be necessary." This nonnegotiable goal was her manifesto.
>
> Starting with this declaration of intent, Vickey moved on, in terms of Step II-B, to spell out what she wanted: (a) She would channel the energy of her "highs"; (b) she would consistently manage or at least endure the depression and agony of her "lows"; (c) she would not disrupt the lives of others by her behavior; (d) she would not make important decisions when either high or low. Vickey, with some help from a rather nontraditional counselor, began to do things to turn those goals into reality. She used her broad goals to provide direction for everything she did.
>
> Vickey learned as much as she could about her illness, including cues for crisis times and how to deal with both highs and lows. To manage her highs, she learned to channel her excess energy into useful, or at least nondestructive, activity. Some of her strategies for controlling her highs centered on the telephone. She knew instinctively that controlling her illness meant not just managing problems but also developing opportunities. During her free time, she would spend long hours on the phone with a host of friends, being careful not to overburden any one person. Phone marathons became part of her lifestyle. She made the point that a big phone bill was infinitely better than a stay in the hospital. She called the telephone her "safety valve." She even set up her own phone-answering business and worked very hard to make it a success.
>
> At the time of her highs, she would do whatever she had to do to tire herself out and get some sleep, for she had learned that sleep was essential if she was to stay out of the hospital. This included working longer shifts at the business. She developed a cadre of supportive people, including her husband. She took special care not to overburden him. She made occasional use of a drop-in crisis center but preferred avoiding any course of action that reminded her of the hospital.

It must be noted that the central driving force in this case was Vickey's decision to stay out of the hospital. Her determination drove everything else. This case also exemplifies the spirit of action that ideally characterizes the implementation stage of the helping process. Here is a woman who, with occasional help from a counselor, took charge of her life. She set some simple goals and devised a set of simple strategies for accomplishing them. She never looked back. And she was never hospitalized again. Some will say that she was not "cured" by this process. But her goal was not to be cured but to lead as normal a life as possible in the real world. Some would say that her approach lacked elegance. But it certainly did not lack results.

**Box 21-1    Questions on Implementing Plans**

- Now that I have a plan, how do I move into action?
- What kind of self-starter am I? How can I improve?
- What obstacles lie in my way? Which are critical?
- How can I manage these obstacles?
- How do I keep my efforts from flagging?
- What do I do when I feel like giving up?
- What kind of support will help me to keep going?

Box 21-1 outlines the kinds of questions you can help clients ask themselves as they implement change programs.

## THE SHADOW SIDE OF IMPLEMENTING CHANGE

There are many reasons why clients fail to act in their own behalf. Four are discussed here: helpers who do not have an action mentality, client inertia, choosing not to choose, and client entropy. As you read about these common phenomena, recall what was said about "implementation intentions" earlier. They can play an important role in managing the shadow-side obstacles outlined here.

### Helpers as Agents

Driscoll (1984, pp. 91–97) discusses the temptation of helpers to respond to the passivity of their clients with a kind of passivity of their own, a "sorry, it's up to you" stance. This, he claims, is a mistake.

> A client who refuses to accept responsibility thereby invites the therapist to take over. In remaining passive, the therapist foils the invitation, thus forcing the client to take some initiative or to endure the silence. A passive stance is therefore a means to avoid accepting the wrong sorts of responsibility. It is generally ineffective, however, as a long-run approach. Passivity by a therapist leaves the client feeling unsupported and thus further impairs the already fragile therapeutic alliance. Troubled clients, furthermore, are not merely unwilling but generally and in important ways unable to take appropriate responsibility. A passive countermove is therefore counterproductive, for neither therapist nor client generates solutions, and both are stranded together in a muddle of entangling inactivity. (p. 91)

To help others act, helpers must be agents and doers in the helping process, not mere listeners and responders. The best helpers are active in the helping sessions. They keep looking for ways to enter the worlds of their clients, to get them to become more active in the sessions, to get them to own more of the helping process, and to help them see the need for action—action in their heads and action outside their heads in their everyday lives. And they do all this while espousing the client-centered values outlined in Chapter 3. Although they don't push reluctant clients too hard, thus turning reluctance into resistance, neither do they sit around waiting for reluctant clients to act.

## Client Inertia: Reluctance to Get Started

Inertia is the human tendency to put off problem-managing action. With respect to inertia, I sometimes say something like this to clients who, I suspect, are reluctant to act: "The action program you've come up with seems to be a sound one. The main reason that sound action programs don't work, however, is that they are never tried. Don't be surprised if you feel reluctant to act or are tempted to put off the first steps. This is quite natural. Ask yourself what you can do to get by that initial barrier." The sources of inertia are many, ranging from pure sloth to paralyzing fear. Understanding what inertia is like is easy. We need only look at our own behavior. The list of ways in which we avoid taking responsibility is endless. We'll examine several of them here: passivity, learned helplessness, disabling self-talk, and getting trapped in vicious circles.

**Passivity.**   One of the most important ingredients in the generation and perpetuation of the "psychopathology of the average" is passivity, the failure of people to take responsibility for themselves in one or more developmental areas of life or in various life situations that call for action. Passivity takes many forms: doing nothing—that is, not responding to problems and options; uncritically accepting the goals and solutions suggested by others; acting aimlessly; and becoming paralyzed—that is, shutting down or becoming violent, blowing up (see Schiff, 1975).

> When Zelda and Jerzy first noticed small signs that things were not going right in their relationship, they did nothing. They noticed certain incidents, mused on them for a while, and then forgot about them. They lacked the communication skills to engage each other immediately and to explore what was happening. Zelda and Jerzy had both learned to remain passive before the little crises of life, not realizing how much their passivity would ultimately contribute to their downfall. Endless unmanaged problems led to major blowups until they decided to end their marriage.

Passivity in dealing with little things can prove very costly. The little things have a way of turning into big things.

**Learned helplessness.**   Seligman's (1975, 1991) concept of "learned helplessness" and its relationship to depression has received a great deal of attention since he first introduced it (Garber & Seligman, 1980; Peterson, Maier, & Seligman, 1995). Some clients learn to believe from an early age that there is nothing they can do about certain life situations. There are degrees in feelings of helplessness—from mild forms of "I'm not up to this" to feelings of total helplessness coupled with deep depression. Learned helplessness, then, is a step beyond mere passivity.

Bennett and Bennett (1984) saw the positive side of helplessness. If the problems clients face are indeed out of their control, it is not helpful for them to have an illusory sense of control, unjustly assign themselves responsibility, and indulge in excessive expectations. Somewhat paradoxically, they found that challenging clients' tendency to blame themselves for everything actually fostered realistic hope and change.

The trick is helping clients learn what is and what is not in their control. A man with a physical disability may not be able to do anything about the disability itself, but he does have some control over how he views his disability and the power to pursue certain life goals despite it. The opposite of helplessness is "learned optimism" (Seligman, 1998) and resourcefulness. If helplessness can be learned, so can resourcefulness. Indeed, increased resourcefulness is one of the principal goals of successful helping. Optimism, however, is not an unmixed blessing; nor is pessimism always a disaster (Chang, 2001). While optimists do such things as live longer and enjoy greater success than pessimists, pessimists are better predictors of what is likely to happen. The price of optimism is being wrong a lot of the time. Perhaps we should help our clients be hopeful realists rather than optimists or pessimists.

**Disabling self-talk.** Challenging dysfunctional self-talk on the part of clients was discussed in Chapters 10 and 11. Clients often talk themselves out of things, thus talking themselves into passivity. They say to themselves such things as "I can't do it," "I can't cope," "I don't have what it takes to engage in that program; it's too hard," and "It won't work." Such self-defeating conversations with themselves get people into trouble in the first place and then prevent them from getting out. Helpers can add great value by helping clients challenge the kind of self-talk that interferes with action.

**Vicious circles.** Pyszczynski and Greenberg (1987) developed a theory about self-defeating behavior and depression. They said that people whose actions fail to get them what they want can easily lose a sense of self-worth and become mired in a vicious circle of guilt and depression.

> Consequently, the individual falls into a pattern of virtually constant self-focus, resulting in intensified negative affect, self-derogation, further negative outcomes, and a depressive self-focusing style. Eventually, these factors lead to a negative self-image, which may take on value by providing an explanation for the individual's plight and by helping the individual avoid further disappointments. The depressive self-focusing style then maintains and exacerbates the depressive disorder. (p. 122)

It does sound depressing. One client, Amanda, fit this theory perfectly. She had aspirations of moving up the career ladder where she worked. She was very enthusiastic and dedicated, but she was unaware of the "gentleman's club" politics of the company in which she worked and didn't know how to "work the system." She kept doing the things that she thought should get her ahead. They didn't. Finally, she got down on herself, began making mistakes in the things that she usually did well, and made things worse by constantly talking about how she "was stuck," thus alienating her friends. By the time she saw a counselor, she felt defeated and depressed. She was about to give up. The counselor focused on the entire "circle"—low self-esteem that produced passivity that produced even lower self-esteem—and not only the self-esteem part. Instead of just trying to help her change her inner world of disabling self-talk, he also helped her intervene in her life to become a better problem solver. Small successes in problem solving led to the start of a "benign" circle—success that produced greater self-esteem that led to greater efforts to succeed.

**Disorganization.** Tico lived out of his car. No one knew exactly where he spent the night. The car was chaos, and so was his life. He was always going to get his career, family relations, and love life in order, but he never did. Living in disorganization was his way of putting off life decisions. Ferguson (1987) paints a picture that may well remind us of ourselves, at least at times:

> When we saddle ourselves with innumerable little hassles and problems, they distract us from considering the possibility that we may have chosen the wrong job, the wrong profession, or the wrong mate. If we are drowning in unfinished housework, it becomes much easier to ignore the fact that we have become estranged from family life. Putting off an important project—painting a picture, writing a book, drawing up a business plan—is a way of protecting ourselves from the possibility that the result may not be quite as successful as we had hoped. Setting up our lives to insure a significant level of disorganization allows us to continue to think of ourselves as inadequate or partially-adequate people who don't have to take on the real challenges of adult behavior. (p. 46)

Many things can be behind this unwillingness to get our lives in order, such as defending ourselves against a fear of succeeding.

Driscoll (1984, pp. 112–117) has provided us with a great deal of insight into this problem. He described inertia as a form of control. He says that if we tell some clients to jump into the driver's seat, they will compliantly do so—at least until the journey gets too rough. The most effective strategy, he claims, is to show clients that they have been in the driver's seat right along: "Our task as therapists is not to talk our clients into taking control of their lives, but to confirm the fact that they already are and always will be." That is, inertia, in the form of staying disorganized, is itself a form of control. The client is actually successful, sometimes against great odds, at remaining disorganized and thus preserving inertia. Once clients recognize their power, helpers can help them redirect it.

## Entropy: The Tendency of Things to Fall Apart

Entropy is the tendency to give up action that has been initiated. Kirschenbaum (1987), in a review of the research literature, uses the term *self-regulatory failure*. Programs for constructive change, even those that start strong, often dwindle and disappear. All of us have experienced the problems involved in trying to implement programs. We make plans, and they seem realistic to us. We start the steps of a program with a good deal of enthusiasm. However, we soon run into tedium, obstacles, and complications. What seemed so easy in the planning stage now seems quite difficult. We become discouraged, flounder, recover, flounder again, and finally give up, rationalizing to ourselves that we did not want to accomplish those goals anyway.

Phillips (1987, p. 650) identified what he called the "ubiquitous decay curve" in both helping and in medical delivery situations. Attrition, noncompliance, and relapse are the name of the game. A married couple trying to reinvent their marriage might eventually say to themselves, "We had no idea that it would be so hard to change ingrained ways of interacting with each other. Is it worth the effort?" Their motivation is waning. Wise helpers know that the decay curve is part of life and help clients deal with it. With respect to entropy, a helper might say, "Even sound action programs begun with the best of intentions tend to fall apart over time, so don't be surprised when your initial enthusiasm seems to wane a bit. That's only natural. Rather, ask yourself what you need to do to keep yourself at the task."

Brownell, Marlatt, Lichtenstein, and Wilson (1986) provide a useful caution. They draw a fine line between preparing clients for mistakes and giving them "permission" to make mistakes by implying that they are inevitable. They also make a distinction between "lapse" and "relapse." A slip or a mistake in an action program (a lapse) need not lead to a relapse—that is, giving up the program entirely. Consider Graham, a man who has been trying to change what others see as his "angry interpersonal style." Using a variety of self-monitoring and self-control techniques, he has made great progress in changing his style. On occasion, he loses his temper but never in any extreme way. He makes mistakes, but he does not let an occasional lapse end up in relapse.

## Choosing Not to Change

Some clients who seem to do well in analyzing problems, developing goals, and even identifying reasonable strategies and plans end up by saying—in effect, if not directly—something like this: "Even though I've explored my problems and understand why things are going wrong—that is, I understand myself and my behavior better, and I realize what I need to do to change—right now I don't want to pay the price called for by action. The price of more effective living is too high."

The question of human motivation seems almost as enigmatic now as it must have been at the dawning of the history of the human race. So often we seem to choose our own misery. Worse, we choose to stew in it rather than endure the relatively short-lived pain of behavioral change. Helpers can and should challenge clients to search for incentives and rewards for managing their lives more effectively. They should also help clients understand the consequences of not changing. But in the end, clients make their own choices.

The shadow side of change stands in stark contrast to the case of Vickey. Savvy helpers are not magicians, but they do understand the shadow side of change, learn to see signs of it in each individual case, and, in keeping with the values outlined in Chapter 3, do whatever they can to challenge clients to deal with the shadow side of themselves and the world around them.

## Evaluation Questions for the Action Arrow

How well do I do the following as I try to help this client make the transition to action?

- Understand how widespread both inertia and entropy are and how they are affecting this client
- Help clients become effective tacticians
- Help clients form "implementation intentions" especially when obstacles to goal attainment are foreseen
- Help clients avoid both procrastination and imprudent action
- Help clients develop contingency plans
- Help clients discover and manage obstacles to action
- Help clients discover resources that will enable them to begin acting, to persist, and to accomplish their goals
- Help clients find the incentives and the rewards they need to persevere in action
- Help clients acquire the skills they need to act and to sustain goal-accomplishing action
- Help clients develop a social support-and-challenge system in their day-to-day lives
- Prepare clients to get along without a helper
- Come to grips with what kind of agent of change I am in my own life
- Face up to the fact that not every client wants to change

ABRAMSON, P. R., Cloud, M. Y., Keese, N., & Keese, R. (1994, Spring). How much is too much? Dependency in a psychotherapeutic relationship. *American Journal of Psychotherapy, 48*, 294–301.

ACKOFF, R. (1974). *Redesigning the future.* New York: Wiley.

ADLER, R. B., & TOWNE, N. (1999). *Looking Out/Looking In: Interpersonal Communication.* 9th edition. San Fransisco: Harcourt Brace

ALBANO, A. M. (2000). Treatment of social phobia in adolescents: Cognitive behavioral programs focused on intervention and prevention. *Journal of Cognitive Psychotherapy, 14*(1) 67–76

ALBEE, G. W., & K. D. Ryan–Finn. (1993, November-December). An overview of primary prevention. *Journal of Counseling and Development,* 115–123.

ALBRECHT, F. & Wallace, M. (1998). Detecting chronic fatigue syndrome: The role of counselors. *Journal of Counseling and Development.* [Internet], 76(2), 183(6).

ALFORD, B. A., & Beck, A. T. (1997). Therapeutic interpersonal support in cognitive therapy. *Journal of Psychotherapy Integration, 7*, 105–117.

AMERICAN PSYCHIATRIC ASSOCIATION. (1994). *Diagnostic and statistical manual of mental disorders.* Washington, DC.

*AMERICAN PSYCHOLOGIST.* (1992, December). Ethical principles of psychologists and code of conduct.

*AMERICAN PSYCHOLOGIST.* (1996, October). Special Issue: Outcome assessment of psychotherapy.

ANDERSEN, P. A. (1999). *Nonverbal communication: Forms and functions.* Mountain View, CA: Mayfield Publishing Company.

ANDERSON, T., & Leitner, L. M. (1996). Symptomatology and the use of affect constructs to influence value and behavior constructs. *Journal of Counseling Psychology, 43*, 77–83.

ANKUTA, G. Y., & Abeles, N. (1993, February). Client satisfaction, clinical significance, and meaningful change in psychotherapy. *Professional Psychology: Research and Practice, 24*, 70–74.

ARGYRIS, C. (1982). *Reasoning, learning, and action.* San Francisco: Jossey–Bass.

ARGYRIS, C. (1999). *On organizational learning* (2nd ed.). Cambridge, MA: Blackwell.

ARKIN, R. M., Hermann, A. D. (2000, July). Constructing desirable identities: Self-presentation in psychotherapy and daily life. *Psychological Bulletin, 126*, 501–504

ARKOWITZ, H. (1997). The varieties of support in psychotherapy. *Journal of Psychotherapy Integration, 7*, 151–159.

ARNETT, J. J. (2000, May). Emerging adulthood: A theory of development from the late teens through the twenties. *American Psychologist, 55*, 469–480.

ARNKOFF, D. B. (1995). Two examples of strains in the therapeutic alliance in an integrative cognitive therapy. *Psychotherapy in Practice, 1*, 33–46.

ARNKOFF, D. B. (2000). Two examples of strains in the therapeutic alliance in an integrative cognitive therapy. *Journal of Clinical Psychology, 56* 187–200.

ASAY, T. P., Lambert, M. J. (1999). The empirical case for the common factor in therapy: Quantitative funding. In Hubble, M. A., Duncan, B. L., Miller, S. D. *The heart and soul of change: What works in therapy.* Washington, DC: American Psychological Association.

ATKINSON, D. R., Worthington, R. L., Dana, D. M., & Good, G. E. (1991). Etiology beliefs, preferences for counseling orientations, and counseling effectiveness. *Journal of Counseling Psychology, 38*, 258–264.

ALTMAIER, E. M., Russell, D. W., Kao, C. F., Lehmann, T. R., & Weinstein, J. N. (1993).

Role of self-efficacy in rehabilitation outcome among chronic low back pain patients. *Journal of Counseling Psychology, 40*, 335–339.

AUSTAD. S. A. (1996). *Is long term psychotherapy unethical?* San Francisco: Jossey-Bass.

AXELSON, J. A. (1999). *Counseling and development in a multicultural society* (3rd Ed.). Pacific Grove, CA: Brooks/Cole/Wadsworth.

AZAR, B. (1995). Breaking through barriers to creativity. *APA Monitor (26)* 1.

AZAR, B. (2000, July-August). Psychology's largest prize goes to four extraordinary scientists. *Monitor on Psychology*, 38–40.

BAER, J. S., Krulahan, D. R., & Donovan, D. M. (1999). Integrating skills training and motivational therapies: implications for the treatment of substance dependence. *Journal of Substance Abuse Treatment, 17*, 15–23.

BAILEY, K. G., Wood, H. E., & Nava, G. R. (1992). What do clients want? Role of psychological kinship in professional helping. *Journal of Psychotherapy Integration, 2*, 125–147.

BALDWIN, B. A. (1980). Styles of crisis intervention: Toward a convergent model. *Journal of Professional Psychology, 11*, 113–120.

BALTES, P. B., & Staudinger, U. M. (2000, January). Wisdom: A metaheuristic to orchestrate mind and virtue toward excellence. *American Psychologist, 55*, 122–135.

BANDURA, A. (1977). Self-efficacy: Toward a unifying theory of behavioral change. *Psychological Review, 84*, 191–215.

BANDURA, A. (1980). Gauging the relationship between self-efficacy judgment and action. *Cognitive Therapy and Research, 4*, 263–268.

BANDURA, A. (1982). Self-efficacy mechanism in human agency. *American Psychologist, 37*, 122–147.

BANDURA, A. (1986). *Social foundations of thought and action: A social cognitive theory*. Englewood Cliffs, NJ: Prentice Hall.

BANDURA, A. (1989). Human agency in social cognitive theory. *American Psychologist, 44*,1175–1184.

BANDURA, A. (1990). Foreword. In E. A. Locke & G. P. Latham, *A theory of goal setting and task performance*. Englewood Cliffs, NJ: Prentice Hall.

BANDURA, A. (1991). Human agency: The rhetoric and the reality. *American Psychologist, 46*, 157–161.

BANDURA, A. (Ed.) (1995). *Self-efficacy in changing societies*. New York: Cambridge University Press.

BANDURA, A. (1997). *Self-efficacy: The exercise of control*. New York: Freeman.

BANKOFF, E. A. (1992). The social network of the psychotherapy patient and effective psychotherapeutic process. *Journal of Psychotherapy Integration, 2*, 273–294.

BARRETT-LENNARD, G. T. (1981). The empathy cycle: Refinement of a nuclear concept. *Journal of Counseling Psychology, 28*, 91–100.

Basic Behavioral Science Task Force of the National Advisory Mental Health Council (1996).

Basic behavioral science research for mental health—Family processes and social networks. *American Psychologist, 51*, 622–630.

BAUMEISTER, R.F., Dale, K., Sommer, K.L. (1998). Freudian defense mechanisms and empirical findings in modern social psychology: Reaction formation, projection, displacement, undoing, isolation, sublimation, and denial. *Journal of Personality.* 66 1081–1117.

BEIER, E. G., & Young, D. M. (1984). *The silent language of psychotherapy: Social reinforcement of unconscious processes* (2nd ed.). New York: Aldine.

BENBENISHTY, R., & Schul, Y. (1987). Client-therapist congruence of expectations over the course of therapy. *British Journal of Clinical Psychology, 26*, 17–24.

BENNETT, M. I., & Bennett, M. B. (1984). The uses of hopelessness. *American Journal of Psychiatry, 141*, 559–562.

BERENSON, B. G., & Mitchell, K. M. (1974). *Confrontation: For better or worse*. Amherst, MA: Human Resource Development Press.

BERG, I. K., (1994). *Family-Based services: A solution-focused approach*. New York: W.W. Norton.

BERGER, D. M. (1989). Developing the story in psychotherapy. *American Journal of Psychotherapy, 43*, 248–259.

BERGIN, A. E. (1991). Values and religious issues in psychotherapy and mental health. *American Psychologist, 46*, 394–403.

BERGIN, A. E., & Garfield, S. L. (1994). *Handbook of psychotherapy and behavior change* (4th ed.). New York: Wiley.

BERGIN, A. E., & Richards, P. S. (Eds.). (2000). *Handbook of psychotherapy and religious diversity*. Hyattsville, MD: American Psychological Association.

BERNARD, M. E. (Ed.). (1991). *Using rational-emotive therapy effectively*. New York: Plenum.

BERNARD, M. E., & DiGiuseppe, R. (Eds.). (1989). *Inside rational-emotive therapy*. San Diego, CA: Academic Press.

BERNE, E. (1964). *Games people play*. New York: Grove Press.

BERNHEIM, K. F. (1989). Psychologists and the families of the severely mentally ill: The role of family consultation. *American Psychologist, 44,* 561–564.

BERNSTEIN, R. (1994). *Dictatorship of virtue: Multiculturalism and the battle for America's future.* New York: Knopf.

BERSOFF, D. N. (1995). *Ethical conflicts in psychology.* Hyattsville, MD: American Psychological Association.

BEUTLER, L. E. (1998, February). Identifying empirically supported treatments: What if we didn't? *Journal of Consulting and Clinical Psychology, 66,* 113–120.

BEUTLER, L. E., & Bergan, J. (1991). Values change in counseling and psychotherapy: A search for scientific credibility. *Journal of Counseling Psychology, 38,* 16–24.

BINDER, C. (1990, September). Closing the confidence gap. *Training,* 49–56.

BINDER, J. L., & Strupp, H. H. (1997). "Negative process": A recurrently discovered and underestimated facet of therapeutic process and outcome in the individual psychotherapy of adults. *Clinical Psychology: Science and Practice, 4,* 121–139.

BISCHOFF, M. M., & Tracey, T. J. G. (1995). Client resistance as predicted by therapist behavior: A study of sequential dependence. *Journal of Counseling Psychology, 42,* 487–495.

BLAMPIED, N. M. (2000). Single-case research designs: A neglected alternative. *American Psychologist, 55,* 960

BLOOM, B. L. (1997). *Planned short-termed psychotherapy: A clinical handbook* (2nd ed.). Boston: Allyn & Bacon.

BOHART, A.C., & Greenberg, L. S. (Eds.). (1997, April). *Empathy reconsidered: New directions in psychotherapy.* Washington, DC: American Psychology Association

BOHART, A. C., & Tallman, K. (1999). *How clients make therapy work.* Washington, DC: American Psychological Association.

BORDIN, E. S. (1979). The generalizability of the psychoanalytic concept of the working alliance. *Psychotherapy: Theory, Research and Practice, 16,* 252–260.

BORGEN, F. H. (1992). Expanding scientific paradigms in counseling psychology. In S. D. Brown & R. W. Lent (Eds.), *Handbook of counseling psychology.* pp. 111–139. New York: Wiley.

BORNSTEIN, R. F., & Bowen, R. F. (1995). Dependency in psychotherapy: Toward an integrated treatment approach. *Psychotherapy, 32,* 520–534.

BORSARI, B., & Carey, K. B. (2000). Effects of a brief motivational intervention with college student drinkers. *Journal of Consulting and Clinical Psychology, 68,* 728–733.

BOWMAN, D. (1995). Self-examination therapy: Treatment for anxiety and depression. In Vandecreek, L., Knapp, S., & Jackson, T. (Eds.), *Innovations in clinical practice: A source book.* Sarasota, FL: Professional Resource Press.

BOWMAN, D., Scogin, F., Floyd, M., Patton,E., et al. (1997). Efficacy of self-exanimation therapy in the treatment of generalized anxiety disorder. *Journal of Counseling Psychology.* Jul Vol. 44(3) 367–273

BOWMAN, V., Ward, L., Bowman D., & Scogin, F. (1996). Efficacy of self-examination therapy as an adjunct treatment for depressive symptoms in substance abusers. *Addictive Behaviors, 21,* 129–133.

BRAMMER, L. (1973). *The helping relationship: Process and skills.* Englewood Cliffs, NJ: Prentice Hall.

BRANDSTATTER, H., & Eliasz, A. (2000). *Persons, solutions, and emotions.* New York: Oxford University Press.

BROCK, T. C., Green, M. C., & Reich, D. A. (1998, January). New evidence of flaws in the *Consumers Report* study of psychotherapy. *American Psychologist, 53,* 62–63.

BRODER, M. S. (2000). Making optimal use of homework to enhance your therapeutic effectiveness. *Journal of Rational-Emotive & Cognitive Behavior Therapy, 18,* 3–18.

BRONFENBRENNER, U. (1977). Toward an experimental ecology of human development. *American Psychologist,* 513–531.

BROSKOWSKI, A. T. (1995). The evolution of health care. *Professional Psychology: Research and Practice, 26,* 156–162.

BROWN, C., & O'Brien, K. M. (1998). Understanding stress and burnout in shelter workers. *Professional Psychology: Research and Practice, 29,* 383–385.

BROWNELL, K. D., Marlatt, G. A., Lichtenstein, E., & Wilson, G. T. (1986). Understanding and preventing relapse. *American Psychologist, 41,* 765–782.

BUDMAN, S. H., Hoyt, M. F., & Friedman, S. (Eds.). (1992). *The first session in brief therapy.* NewYork: Guilford.

BUDMAN, S. H., & Steenbarger, B. N. (1997). *The essential guide to group practice in mental health.* NewYork: Guilford Press.

BUGAS, J., & Silberschatz, G. (2000). How patients coach their therapists in psychotherapy.

*Psychotherapy: Theory, Research, Practice, Training, 37,* 64–70.

BURKS, R. J., & Keeley, S. M. (1989). Exercise and diet therapy: Psychotherapists' beliefs and practices. *Professional Psychology: Research & Practice, 20,* 62–64.

BUSHE, G. (1995). Advances in appreciative inquiry as an organizational development intervention. *Organizational Development Journal, 1,* 14–22.

CADE, B., & O'Hanlon, W. H. (1993). *A Brief Guide to Brief Therapy.* New York: W. W. Norton.

CAMERON, J. E. (1999). Social identity and the pursuit of possible selves: Implications for the psychological well-being of university students. *Group Dynamics, 3,* 179–189.

CANTER, M. B., Bennett, B. E., Jones, S. E., & Nagy, T. F. (1994). *Ethics for psychologists: A commentary on the APA ethics code.* Hyattsville, MD: American Psychological Association.

CAPUZZI, D., & Gross, D. R. (1999). *Counseling and Psychotherapy: Theories and Interventions.* Upper Saddle River, NJ: Prentice Hall

CARKHUFF, R. R. (1969a). *Helping and human relations: Vol. 1. Selection and training.* New York: Holt, Rinehart & Winston.

CARKHUFF, R. R. (1969b). *Helping and human relations: Vol. 2. Practice and research.* New York: Holt, Rinehart & Winston.

CARKHUFF, R. R. (1971). Training as a preferred mode of treatment. *Journal of Counseling Psychology, 18,* 123–131.

CARKHUFF, R. R. (1987). *The art of helping* (6th ed.). Amherst, MA: Human Resource Development Press.

CARKHUFF, R. R., & Anthony, W. A. (1979). *The skills of helping: An introduction to counseling.* Amherst, MA: Human Resource Development Press.

CARTER, J. A. (1996). Measuring transference: Can we identify what we have not defined? *Journal of Counseling Psychology, 43,* 257–258.

CARTON, J. S., Kessler, E. A., & Pape, C. L. (1999, Spring). Nonverbal decoding skills and relationship well-being in adults. *Journal of Nonverbal Behavior, 23,* 91–100.

CASTONGUAY, L. G. (1997). Support in psychotherapy: A common factor in need of empirical data, conceptual clarification, and clinical input. *Journal of Psychotherapy Integration, 7,* 99–103

CERVONE, D. (2000). Thinking about self-efficacy. *Behavior modification, 24,* 30–56.

CERVONE, D., and Scott, W. D. (1995). Self-efficacy theory and behavioral change: Foundations, conceptual issues, and therapeutic implica-

tions. In W. O'Donohue & L. Krasner (Eds.), *Theories of behavior therapy: Exploring behavior change* (pp. 349–383). Washington, DC: American Psychological Association.

CHAMBLESS, D. L., & Hollon, S. D. (1998, February). Defining empirically supported therapies. *Journal of Consulting and Clinical Psychology, 66,* 7–18.

CHANG, E. C. (Ed.). (2001). *Optimism and pessimism: Implications for theory, research, and practice.* Washington, DC: American Psychological Association.

CLAIBORN, C. D. (1982). Interpretation and change in counseling. *Journal of Counseling Psychology, 29,* 439–453.

CLAIBORN, C. D., Berberoglu, L. S., Nerison, R. M., & Somberg, D. R. (1994). The client's perspective: Ethical judgments and perceptions of therapist practices. *Professional Psychology: Research and Practice, 25,* 268–274.

CLARK, A. J. (1991). The identification and modification of defense mechanisms in counseling. *Journal of Counseling and Development, 69,* 231–236.

CLARK, A. J. (1998). *Defense mechanisms in the counseling process.* Thousand Oaks, CA: Sage.

COLBY, S. M., et al. (1998). Brief motivational interviewing in a hospital setting for adolescent smoking: A preliminary study. *Journal of Counseling and Clinical Psychology, 66,* 574–578.

COLE, H. P., & Sarnoff, D. (1980). Creativity and counseling. *Personnel and Guidance Journal, 59,* 140–146.

COLE, J. D., et al. (1993). The science of prevention: A conceptual framework and some directions for a national research program. *American Psychologist, 48,* 1013–1022.

COLLINS, J. C., & Porras, J. I. (1994). *Built to last: Successful habits of visionary companies.* New York: Harper Business.

COMPAS, B. E., Haaga, D. A., Keefe, F. J., Leitenberg, H., Williams, D. A. (1998, February). Sampling of empirically supported psychological treatments fom health psychology: Smoking, chronic pain, cancer, and bulimia nervosa. *Journal of Consulting and Clinical Psychology, 66,* 89–112.

CONOLEY, C. W., Padula, M. A., Payton, D. S., & Daniels, J. A. (1994). Predictors of client implementation of counselor recommendations: Match with problem, difficulty level, and building on client strengths. *Journal of Counseling Psychology, 41,* 3–7.

CONSUMER REPORTS. (1994). Annual questionnaire.

CONSUMER REPORTS. (1995, November). Mental health: Does therapy help? 734–739.

COOPER, J. F. (1995). *A Primer of Brief Psychotherapy*. New York: W. W. Norton.

COOPER, S.H. (1998). Changing Notions of Defense within psychoanalytic theory. *Journal of Personality*. 66(6) 947–964

COOPERRIDER, D., & Srivasta, S. (1987). Appreciative inquiry into organizational life. *Organizational Change and Development, 1*, 129–169.

COREY, G. (1996). *Theory and practice of counseling and psychotherapy* (5th ed.). Pacific Grove, CA: Brooks/Cole.

COREY, G., Corey, M. S., & Callanan, P. (1997). *Issues and ethics in the helping professions* (5th ed.). Pacific Grove, CA: Brooks/Cole.

CORRIGAN, P. W. (1995). Wanted: Champions of psychiatric rehabilitation. *American Psychologist, 50*, 514–521.

CORRIGAN, P. W., Lickey, S., Schmoock, A., Virgil, L., & Juricek, M. (1999). Dialogue among stakeholders of severe mental illness. *Psychiatric Rehabilitation Journal, 23*, 62–65.

COSIER, R. A., & Schwenk, C. R. (1990). Agreement and thinking alike: Ingredients for poor decisions. *Academy of Management Executive, 4*, 69–74.

COSTANZO, M. (1992). Training students to decode verbal and nonverbal clues: Effects on confidence and performance. *Journal of Educational Psychology, 84*, 308–313.

COTTONE, R. R., & Claus, R. E. (2000). Ethical decision-making models: A review of the literature. *Journal of Counseling and Development, 78*, 275.

COURSEY, Alford, & Safarjan. (1997). Significant advances in understanding and treating serious mental illness. *Professional Psychology: Research and Practice, 28*, 205–216.

COVEY, S. R. (1989). *The seven habits of highly effective people*. New York: Simon & Schuster (Fireside edition, 1990).

COWEN, E. L. (1982). Help is where you find it. *American Psychologist, 37*, 385–395.

COYNE, J. C., & Racioppo, M. (2000). Never the twain shall meet? Closing the gap between coping research and clinical intervention research. *American Psychologist*. Jun Vol 55(6) 655–664

CRAMER, P. (1998). Coping and defense mechanisms: What's the difference? *Journal of Personality, 66*, 895–918.

CRAMER, P. (2000). Defense mechanisms in psychology today. *American Psychologist, 55*, 637– 646.

CRANO, W. D. (2000, March). Milestones in the psychological analysis of social influence. *Group Dynamics, 4*, 68–80.

CROSS, J. G., & Guyer, M. J. (1980). *Social traps*. Ann Arbor: University of Michigan Press.

CROSS, S. E., & Markus, H. R. (1994). Self-schemas, possible selves, and competent performance. *Journal of Educational Psychology, 86*, 423–438.

CUELLAR, I., & Paniagua, F. A. (Eds.). (2000). *Handbook of multicultural mental health*. San Diego, CA: Academic Press.

CUMMINGS, N. A. (1979). Turning bread into stones: Our modern anti-miracle. *American Psychologist, 34*, 1119–1129.

CUMMINGS, N. A., (2000). *The first session with substance abusers*. San Francisco: Jossey-Bass.

CUMMINGS, N. A., Budman, S. H., & Thomas, J. L. (1998, October). Efficient psychotherapy as a viable response to scarce resources and rationing of treatment. *Professional Psychology: Research and Practice, 29*, 460–469.

CUMMINGS, N. A., & Cummings, J. L. (2000). *The essence of psychotherapy: Reinventing the art in the new era of data*. San Diego: Academic Press.

CUMMINGS, N.A., Pallak, M.S., & Cummings, J.L. (1996). *Surviving the demise of Solo practice: Mental health practitioners prospering in the era of managed care*. Madison, CT: Psychological Press.

DAS, A. K. (1995, Fall). Rethinking multicultural counseling: Implications for counselor education. *Journal of Counseling and Development, 74*, 45–52.

DAUSER, P. J., Hedstrom, S. M., & Croteau, J. M. (1995). Effects of disclosure of comprehensive pretherapy information on clients at a university counseling center. *Professional Psychology: Research and Practice, 26*, 190–195.

DE BONO, E. (1992). *Serious creativity: Using the power of lateral thinking to create new ideas*. New York: Harper Business.

DEFFENBACHER, J. L., Oetting, E. R., Huff, M. E., & Thwaites, G. A. (1995). Fifteen-month follow-up of social skills and cognitive-relaxation approaches to general anger reduction. *Journal of Counseling Psychology, 42*, 400–405.

DEFFENBACHER, J. L., Thwaites, G. A., Wallace, T. L., & Oetting, E. R. (1994). Social skills and cognitive-relaxation approaches to general anger reduction. *Journal of Counseling Psychology, 41*, 386–396.

DENCH, S., & Bennett, G. (2000). The impact of brief motivational intervention at the start of an outpatient day programme for alcohol dependance. *Behavioural & Cognitive Psychotherapy, 28,* 121–130.

DEPAULO, B. M. (1991). Nonverbal behavior and self-presentation. *Psychological Bulletin, 11,* 203–243.

de SHAZER, S. (1985). *Keys to solution in brief therapy.* New York: W. W. Norton.

de SHAZER, S. (1994). *Words were originally magic.* New York: W. W. Norton.

DEUTSCH, M. (1954). Field theory in social psychology. In G. Lindzey (Ed.), *The handbook of social psychology* (Vol. 1). Cambridge, MA: Addison-Wesley.

DIENER, E. (January, 2000). Subjective well-being: The science of happiness and a proposal for a national index. *American Psychologist, 55,* 34–43.

DIMOND, R. E., Havens, R. A., & Jones, A. C. (1978). A conceptual framework for the practice of prescriptive eclecticism in psychotherapy. *American Psychologist, 33,* 239–248.

DIVISION 12 TASK FORCE. (1995). Training in and dissemination of empirically-validated psychological treatments. *The Clinical Psychologist, 49,* 3–23.

DONNEY, D. A. C., & Borgen, F. H. (1999). The incremental validity of vocational self-efficacy: An examination of interest, self-efficacy, and occupation.

DORN, F. J. (1984). *Counseling as applied social psychology: An introduction to the social influence model.* Springfield, IL: Chas. C Thomas.

DORN, F. J. (Ed.). (1986). *The social influence process in counseling and psychotherapy.* Springfield, IL: Chas. C Thomas.

DORNER, D. (1996). *The logic of failure: Why things go wrong and what we can do to make them right.* New York: Holt.

DRISCOLL, R. (1984). *Pragmatic psychotherapy.* New York: Van Nostrand Reinhold.

DRAYCOTT, S., & Dabbs, A. (Sept. 1998a). Cognitive dissonance 1: An overview of the literature and its integration into theory and practice of clinical psychology. *British Journal of Clinical Psychology, 37,* 341–353.

DRAYCOTT, S., & Dabbs, A. (1998b). Cognitive dissonance 2: A theoretical grounding of motivational interviewing. *British Journal of Clinical Psychology, 37,* 355–364.

DRYDEN, W. (1995). *Brief rational emotive behavior therapy.* New York: Wiley.

DRYDEN, W., Neenan, M., Yankura, J., & Ellis, A. (1999). *Counseling Individuals.* London: Whurr Publishers.

DUAN, C., & Hill, C. E. (1996). The current state of empathy research. *Journal of Counseling Psychology, 43,* 261–274.

DURLAK, J. A. (1979). Comparative effectiveness of paraprofessional and professional helpers. *Psychological Bulletin, 86,* 80–92.

DWORKIN, S. H., & Kerr, B. A. (1987). Comparison of interventions for women experiencing body image problems. *Journal of Counseling Psychology, 34,* 136–140.

ECONOMIST. (1995, July). Come feel the noise.

EDWARDS, C. E., & Murdock, N. L. (1994). Characteristics of therapist self-disclosure in the counseling process. *Journal of Counseling and Development, 72,* 384–389.

EGAN, G. In press *Conversations for the 21st Century: The values and skills of dialogue.*

EGAN, G. (1970). *Encounter: Group processes for interpersonal growth.* Pacific Grove, CA: Brooks/Cole.

EGAN, G. (1998). *The skilled helper: A problem-management approach to helping.* (6th ed.) Pacific Grove, CA: Brooks/Cole.

EGAN, G., & Cowan, M. A. (1979). *People in systems: A model for development in the human-service professions and education.* Pacific Grove, CA: Brooks/Cole.

EISENBERG, W., & Strayer, J. (Eds.). (1987). *Empathy and its development.* New York: Cambridge University Press.

EKMAN, P. (1992). *Telling lies: Clues to deceit in the marketplace, politics, and marriage.* New York: W. W. Norton.

EKMAN, P. (1993). Facial expression and emotion. *American Psychologist, 48,* 384–392.

EKMAN, P., & Davidson, R. J. (1994). *The nature of emotion: Fundamental questions.* New York: Oxford University Press.

EKMAN, P., & Friesen, W. V. (1975). *Unmasking the human face: A guide to recognizing emotions from facial cues.* Englewood Cliffs, NJ: Prentice-Hall.

EKMAN, P., & Rosenberg, E. L., (Eds.). (1998). *What the face reveals: Basic and applied studies of spontaneous expression using the facial action coding system (FACS).* New York: Oxford University Press.

ELIAS, M. J., & Clabby, J. F. (1992). *Building social problem-solving skills.* San Francisco: Jossey-Bass.

ELKIN, I. (1999, Spring). A major dilemma in psychotherapy outcome research: Disentangling therapists from therapies. *Clinical Psychology: Science and Practice, 6,* 10–32.

ELLIOTT, C. (1999). *Locating the energy for change: An introduction to appreciative inquiry.* Winnepeg: IISD.

ELLIOTT, R. (1985). Helpful and nonhelpful events in brief counseling interviews: An empirical taxonomy. *Journal of Counseling Psychology*, *32*, 307–322.

ELLIS, A. (1984). Must most psychotherapists remain as incompetent as they are now? In J. Hariman (Ed.), *Does psychotherapy really help people?* Springfield, IL.: Chas. C Thomas.

ELLIS, A. (1985). *Overcoming resistance: Rational-emotive therapy with difficult clients.* New York: Springer.

ELLIS, A. (1987a). The evolution of rational-emotive therapy (RET) and cognitive behavior therapy (CBT). In J. K. Zeig (Ed.), *The evolution of psychotherapy.* New York: Brunner/Mazel.

ELLIS, A. (1987b). Integrative developments in rational-emotive therapy (RET). *Journal of Integrative and Eclectic Psychotherapy*, *6*, 470–479.

ELLIS, A. (1991). The revised ABCs of rational-emotive therapy (RET). *Journal of Rational-Emotive and Cognitive-Behavior Therapy*, *9*, 139–172.

ELLIS, A. (1997). Extending the goals of behavior therapy and cognitive behavior therapy. *Behavior Therapy*, *28*, 333–339.

ELLIS, A. (1999a). Early theories and practices of rational emotive behavior therapy and how they have been augmented and revised during the last three decades. *Journal of Rational-Emotice and Cognitive Behavior Therapy*, *17*, 69–93.

ELLIS, A. (1999b). *How to make yourself happy and remarkably less disturbable.* San Luis Obispo, CA: Impact.

ELLIS, A. (2000). Can rational emotive behavior therapy (REBT) be effectively used with people who have devout beliefs in God and religion? *Professional Psychology: Research and Practice*, *31*, 29–33.

ELLIS, A., & Dryden, W. (1987). *The practice of rational-emotive therapy.* New York: Springer.

ELLIS, A., Dryden, W., Neenan, M., & Yankura, J. (1999). *Counseling individuals: A rational emotive behavioral handbook.* London: Whurr.

ELLIS, A., & Powers. M. (1998). *A guide to rational living.* North Hollywood, CA: Wilshire Books.

ELLIS, D. B. (1999). *Creating your future.* Boston, MA: Houghton Mifflin.

ELLIS, K. (1998). *The magic lamp: Goal setting for people who hate setting goals.* New York: Crown.

ELSON, M. L., & Neufeldt, S. A. (1996, Summer). Building on an empirical foundation: Strategies to enhance good practice. *Journal of Counseling and Development*, *74*, 609–615.

ERICKSON, R.C., Post, R.D., & Paige, A.B. (Apr. 1975). Hope as a psychiatric variable. *Journal of Clinical Psychology*, *31*, 324–330

ETZIONI, A. (1989, July-August). Humble decision making. *Harvard Business Review*, 120–126.

EYSENCK, H.J. (1952). The effects of psychotherapy: An evaluation. *Journal of Consulting Psychology*, *16*, 319–324.

EYSENCK, H. J. (1984). The battle over psychotherapeutic effectiveness. In J. Hariman (Ed.), *Does psychotherapy really help people?* Springfield, IL: Chas. C Thomas.

EYSENCK, H. J. (1994). The outcome problem in psychotherapy: What have we learned? *Behavior Research and Therapy*, *32*, 477–495.

FARRELLY, F., & Brandsma, J. (1974). *Provocative therapy.* Cupertino, CA: Meta Publications.

FELDMAN, R. S., Rime, B. (Eds.) (1991). *Fundamentals of nonverbal behavior.* New York: Cambridge University Press.

FELLER, R. (1984). *Job-search agreements.* Colorado State University, Fort Collins.

FERGUSON, M. (1980). *The aquarian conspiracy: Personal and social transformation in the 1980s.* Los Angeles: J. P. Tarcher.

FERGUSON, T. (1987, January-February). Agreements with yourself. *Medical Self-Care*, 44–47.

FESTINGER, S. (1957). *A theory of cognitive dissonance.* New York: Harper & Row.

FISH, J. M. (1995, Spring). Does problem behavior just happen? Does it matter? *Behavior and Social Issues*, *5*, 3–12.

FISH, J.M. (1995). Solution–focused therapy in global perspective. *World Psychology*, *1*, 43–67.

FISH, J.M. (1997). Paradox for complaints? Stategic thoughts about solution-focused therapy. *Journal of Systematic Therapies*, *16*(3), 266–273.

FISHER, C. B., & Younggren, J. N. (1997) The value and unity of the 1992 ethics code. *Professional Psychology: Research and Practice*, *28*, 582–592.

FISHER, R., & Ury, W. (1981). *Getting to yes: Negotiating agreement without giving in.* Boston: Houghton Mifflin.

FLACK, W. F., & Laird, J. D. (Eds.). (1998). *Emotions and psycopathology.* New York: Oxford University Press.

FOLKMAN, S., & Moskowitz, J. T. (2000). Positive affect and the other side of coping. *American Psychologist*, *55*, 647–664.

FOX, R. E. (1995). The rape of psychotherapy. *Professional Psychology: Research and Practice*, *26*, 147–155.

FOXHALL, K. (2000, July-August). Research for the real world. *Monitor on Psychology*, 28–36.

FRANCES, A., Clarkin, J., & Perry, S. (1984). *Differential therapeutics in psychiatry*. New York: Brunner/Mazel.

FRASER, J. S. (1996). All that glitters is not always gold: Medical offset effects and managed behavioral health care. *Professional Psychology: Research and Practice, 27*, 335–344.

FREIRE, P. (1970). *Pedagogy of the oppressed*. New York: Seabury.

FREMONT, S. K., & Anderson, W. (1986). What client behaviors make counselors angry? An exploratory study. *Journal of Counseling and Development, 65*, 67–70.

FRIEDLANDER, M. L., & Schwartz, G. S. (1985). Toward a theory of strategic self-presentation in counseling and psychotherapy. *Journal of Counseling Psychology, 32*, 483–501.

FRIEMAN, S. (1997). *Time effective psychotherapy: Maximizing outcomes in an era of minimized resources*. Boston: Allyn & Bacon.

FRUEH, B. C., de Arellano, M. A. Turner, S. M.(1997). Systematic desensitization as an alternative exposure strategy for PTSD. *American Journal of Psychiatry, 154*, 287–288.

GALASSI, J. P., & Bruch, M. A. (1992). Counseling with social interaction problems: Assertion and social anxiety. In S. D. Brown & R. W. Lent (Eds.), *Handbook of counseling psychology*. 753–791. New York: Wiley.

GARBER, J. & Seligman, M. (Eds.). (1980). *Human helplessness: Theory and applications*. New York: W.H. Freeman.

GARFIELD, S. L. (1996). Some problems associated with "validated" forms of psychotherapy. *Clinical Psychology: Science and Practice, 3*, 218–229.

GARFIELD S. L. (1997). The therapist as a neglected variable in psychotherapy research. *Clinical Psychology: Research and Practice, 4*, 40–43.

GARTNER, A., & Riessman, F. (Eds.). (1984). *The self-help revolution*. New York: Human Sciences Press.

GASTON, L., et al. (1995). The therapeutic alliance in psychodynamic, cognitive-behavioral, and experimental therapies. *Journal of Psychotherapy Integration, 5*, 1–26.

GATI, I., Krausz, M., & Osipow, S. H. (1996). A taxonomy of difficulties in career decision making. *Journal of Counseling Psychology, 43*, 510–526.

GELATT, H. B. (1989). Positive uncertainty: A new decision-making framework for counseling. *Journal of Counseling Psychology, 36*, 252–256.

GELATT, H. B., Varenhorst, B., & Carey, R. (1972). *Deciding: A leader's guide*. Princeton, NJ: College Entrance Examination Board.

GELSO, C. J., & Carter, J. A. (1994). Components of the psychotherapy relationship: Their interaction and unfolding during treatment. *Journal of Counseling Psychology, 41*, 296–306.

GELSO, C. J., Hill, C. E., Mohr, J. (2000). Client concealment and self-presentation in therapy: Comment on Kelly (2000). *Psychological Bulletin, 126*, 495–500.

GELSO, C. J., Hill, C. E., Mohr, J., Rochlen, A. B., & Zack, J. (1999). The face of transference in successful, long-term therapy: A qualitative analysis. *Journal of Counseling Psychology, 46*, 257–267.

GELSO, C. J., Kivlighan, D. M., Wise, B., Jones, A., & Friedman, S. C. (1997). Tranference, insight, and the course of time-limited therapy. *Journal of Counseling Psychology, 44*, 209–217.

GEORGES, J. C. (1988, April). Why soft-skills training doesn't take. *Training*, 40–47.

GIANNETTI, E. (1997). *Lies we live by: The art of self-deception*. New York and London: Bloomsbury.

GIBB, J. R. (1968). The counselor as a role-free person. In C. A. Parker (Ed.), *Counseling theories and counselor education*. Boston: Houghton Mifflin.

GIBB, J. R. (1978). *Trust: A new view of personal and organizational development*. Los Angeles: The Guild of Tutors Press.

GILBERT, P. & Andrews, B. (Eds.)(1998). *Shame: Interpersonal behavior, psychotherapy, and culture*. New York: Oxford University Press.

GILBERT, T. F. (1978). *Human competence: Engineering worthy performance*. New York: McGraw-Hill.

GILLIAND, B. E., & James, R. K. (1997). *Theory and strategies in counseling and psychotherapy* (4th ed.). Needham Heights, MA: Allyn & Bacon.

GILOVICH, T. (1991). *How we know what isn't so: The fallibility of human reason in everyday life*. New York: The Free Press.

GLASSER, W. (2000). *Reality therapy in action*. New York: HarperCollins.

GODWIN, G. (1985). *The finishing school*. New York: Viking.

GOLEMAN, D. (1985). *Simple Truths: The psychology of Self-deception*. New York: Simon and Schuster.

GOLEMAN, D. (1995). *Emotional intelligence*. New York: Bantam Books.

GOLEMAN, D. (1998). *Working with emotional intelligence*. New York: Bantam Books.

GOLLWITZER, P. M. (1999). Implementation intentions: Strong effects of simple plans. *American Psychologist, 54*, 493–503.

GOODYEAR, R. K., & Shumate, J. L. (1996). Perceived effects of therapist self-disclosure of attraction to clients. *Professional Psychology: Research and Practice, 27*, 613–616.

GONCALVES, O. F., & Machado, P. P. P. (1999, October). Cognitive narrative psychotherapy: Research foundations. *Journal of Clinical Psychology, 55*, 1179–1191.

GOSLIN, D. A. (1985). Decision making and the social fabric. *Society, 22*, 7–11.

GRACE, M., Kivlighan, D. M., Jr., & Kunce, J. (1995, Summer). The effect of nonverbal skills training on counselor trainee nonverbal sensitivity and responsiveness and on session impact and working alliance ratings. *Journal of Counseling and Development, 73*, 547–552.

GREENBERG, L. S. (1986). Change process research. *Journal of Consulting and Clinical Psychology, 54*, 4–9.

GREENBERG, L. S. (1994). What is "real" in the relationship? Comment on Gelso and Carter (1994). *Journal of Counseling Psychology, 41*, 307–309.

GREENBERG, L. S., & Paivio, S. C. (1997). *Working with the emotions in psychotherapy*. New York: Guilford Press.

GREENSON, R. R. (1967). *The technique and practice of psychoanalysis*. New York: International Universities Press.

GUERRERO, L. K., Devito, J. A., & Hecht, M. L. (Eds.) (1999). *The Nonverbal Communication Reader: Classic and Contemporary Readings* (2nd ed.). Prospect Heights, IL: Waveland Press.

GUISINGER, S., & Blatt, S. J. (1994). Individuality and relatedness: Evolution of a fundamental dialectic. *American Psychologist, 49*, 104–111.

HALEY, J. (1976). *Problem solving therapy*. San Francisco: Jossey-Bass.

HALL, E. T. (1977). *Beyond culture*. Garden City, NJ: Anchor Press.

HALLECK, S. L. (1988). Which patients are responsible for their illnesses? *American Journal of Psychotherapy, 42*, 338–353.

HAMMOND, J. S., Keeny, R. L., & Raiffa, H. (1998). The hidden traps in decision making. *Harvard Business Review*, 47–58.

HAMMOND, J. S., Keeny, R. L., & Raiffa, H. (1999). *Smart choices: A practical guide to making better decisions*. Boston, MA: Harvard Business School Press.

HAMMOND, S. (1996). *The thin book of appreciative inquiry*. Plano, TX: Thin Book Publishing.

HAMMOND, S., & Royal, C. (Eds.). (1998). *Lessons from the field: Applying appreciative inquiry*. Plano, TX: Practical Press.

HANDELSMAN, M. M., & Galvin, M. D. (1988). Facilitating informed consent for outpatient psychotherapy: A suggested written format. *Professional Psychology: Research and Practice, 19*, 223–225.

HANNA, F. J. (1994). A dialectic of experience: A radical empiricist approach to conflicting theories in psychotherapy. *Psychotherapy, 31*, 124–136.

HANNA, F. J., Hanna, C. A., & Keys, S. G. (1999). Fifty strategies for counseling defiant, aggressive adolescents: Reaching, accepting, and relating. *Journal of Counseling and Development, 77*, 395–404.

HANNA, F. J., & Ottens, A. J. (1995). The role of wisdom in psychotherapy. *Journal of Psychotherapy Integration, 5*, 195–219.

HARE–MUSTIN, R., & Marecek, J. (1986). Autonomy and gender: Some questions for therapists. *Psychotherapy, 23*, 205–212.

HARRIS, C., et al. (1991). The will and the ways: Development and validation of an individual differences measure of hope. *Journal of Personality and Social Psychology, 60*, 570–585.

HARRIS, G. A. (1995). *Overcoming resistance: Success in counseling men*. Alexandria, VA: American Counseling Association.

HARVARD MENTAL HEALTH LETTER. (1993a, March). Self-help groups. Part I.

HARVARD MENTAL HEALTH LETTER. (1993b, April). Self-help groups. Part II.

HATTIE, J. A., Sharpley, C. E., & Rogers, H. J. (1984). Comparative effectiveness of professional and paraprofessional helpers. *Psychological Bulletin, 95*, 534–541.

HAYS, K. F. (1999). *Working it out: Using exercise in psychotherapy*. Washington, DC: American Psychological Association.

HEADLEE, R., & Kalogjera, I. J. (1988). The psychotherapy of choice. *American Journal of Psychotherapy, 42*, 532–542.

HEINSSEN, R. K. (1994, June). Therapeutic contracting with schizophrenic patients: A collaborative approach to cognitive-behavioral treatment. Paper presented at the 21st International Symposium for the Psychotherapy of Schizophrenia, Washington, DC.

HEINSSEN, R. K., Levendusky, P. G., & Hunter, R. H. (1995). Client as colleague: Therapeutic contracting with the seriously mentally ill. *American Psychologist, 50*, 522–532.

HELLER, K. (1993, November-December). Prevention activities for older adults: Social structures and personal competencies that maintain useful social roles. *Journal of Counseling and Development*, 124–130.

HELLSTROEM, O. (1998). Dialogue medicine: A health-liberating attitude in general practice. *Patient Education & Counseling*, 35, 221–231.

HENDERSON, S. J. (2000). "Follow your bliss": A process for career happiness. *Journal of Counseling and Development*, 78, 305–315

HENDRICK, S. S. (1988). Counselor self-disclosure. *Journal of Counseling and Development*, 66, 419–424.

HENDRICK, S. S. (1990). A client perspective on counselor disclosure. *Journal of Counseling and Development*, 69, 184.

HENGGELER, S. W., Schoenwald, S. K., & Pickrel, S. G. (1995, Fall). Multisystemic therapy: Bridging the gap between university and community-based treatment. *Journal of Consulting and Clinical Psychology*, 63, 709–717.

HEPPNER, P. P. (1989). Identifying the complexities within clients' thinking and decision making. *Journal of Counseling Psychology*, 36, 257–259.

HEPPNER, P. P., & Claiborn, C. D. (1989). Social influence research in counseling: A review and critique [Monograph]. *Journal of Counseling Psychology*, 36, 365–387.

HEPPNER, P. P., & Frazier, P. A. (1992). Social psychological processes in psychotherapy: Extrapolating basic research to counseling psychology. In S. D. Brown & R. W. Lent (Eds.), *Handbook of counseling psychology*. New York: Wiley.

HERMAN, H. J. M., & Kempen, H. J. G. (1998). Moving cultures: Perilous problems of cultural dichotomies in a globalizing society. *American Psychologist*, 53, 1111–1120.

HERINK, R. (Ed.) (1980). *The Psychotherapy Handbook*. New York: Meridian.

HICKSON, M. L. & Stacks, D. W. (1993). Nonverbal communication: Studies and applications (2nd ed.). Madison, WI: Brown & Benchmark.

HIGGINSON, J. G. (1999). Defining, excusing, and justifying deviance: Teen mothers' accounts for statutory rape. *Symbolic Interaction*, 22, 25–44.

HIGHLEN, P. S., & Hill, C. E. (1984). Factors affecting client change in counseling. In S. D. Brown & R. W. Lent (Eds.), *Handbook of counseling psychology* (pp. 334–396). New York: Wiley.

HIGHT, T. L., Worthington, E. L., Ripley, J. S., & Perrone, K. M. (1997). Strategic hope-focused relationship-enrichment counseling with individual couples. *Journal of Counseling Psychology*, 44, 381–389.

HILL, C. E. (1994). What is the therapeutic relationship? A reaction to Sexton and Whiston. *The Counseling Psychologist*, 22, 90–97.

HILL, C. E., & Corbett, M. M. (1993, January). A perspective on the history of process and outcome research in counseling psychology. *Journal of Counseling Psychology*, 40, 3–24.

HILL, C. E., Gelso, C. J., & Mohr, J .J. (2000). Client concealment and self-presentation therapy: Comment on Kelly (2000). *Psychological Bulletin*, 126(4) 495–500.

HILL, C. E., Nutt-Williams, E., Heaton, K. J., Thompson, B. J., & Rhodes, R. H. (1996). Therapist retrospective recall of impasses in long-term psychotherapy: A qualitative analysis. *Journal of Counseling Psychology*, 43, 207–217.

HILL, C. E., & O'Brien, K.M. (1999). *Helping Skills: facilitating exploration, insight, and action*. American Psychological Association, Washington, DC.

HILL, C. E., Thompson, B. J., Cogar, M. C., & Denmann, D. W., III. (1993). Beneath the surface of long-term therapy: Therapist and client report of their own and each other's covert processes. *Journal of Counseling Psychology*, 40, 278–287.

HILL, C. E., & Williams, E.N. (2000). The process of individual counseling. In R. Lent & S. Brown (Eds.), *Handbook of Counseling Psychology* (3rd ed.). New York: Wiley.

HILLIARD, R. B. (1993, June). Single-case methodology in psychotherapy process and outcome research. *Journal of Consulting and Clinical Psychology*, 61, 373–380.

HILLS, M. D. (1984). Improving the learning of parents' communication skills by providing for the discovery of personal meaning. Unpublished doctoral dissertation, University of Victoria, British Columbia, Canada.

HOGAN-GARCIA, M. (1999). *Four Skills of Cultural Diversity Competence: A Process for Understanding and Practice*. CA: Wadsworth.

HOLADAY, M., & McPhearson, R. W. (1999, May-June). Resilience and severe burns. *Journal of Counseling and Development*, 75, 346–356.

HOLADAY, M., & McPhearson, R. W. (1997). Resilience and severe burns. *Journal of Counseling and Development*, 75, 346–356.

HOLMES, G. R., Offen, L., & Waller, G. (1997). See no evil, hear no evil, speak no evil: Why do relatively few male victims of childhood sexual abuse receive help for abuse-related issues in adulthood? *Clinical Psychology Review*, 27, 69–88.

HOOKER, K., Fiese, B. H., Jenkins, L., Morfei, M. Z., & Schwagler, J. (1996) Possible selves among parents of infants and preschoolers. *Developmental Psychology*, 32, 542–550.

HORVATH, A. O. (2000). The therapeutic relationship: From transference to alliance. *Journal of Clinical Psychology, 56*, 163–173.

HORVATH, A. O., & Symonds, B. D. (1991). Relation between working alliance and outcome in psychotherapy: A meta-analysis. *Journal of Counseling Psychology, 38*, 139–149.

HOUSER, R. F., Feldman, M., Williams, K., & Fierstien, J. (1998, July). Persuasion and social influence tactics used by mental health counselors. *Journal of Mental Health Counseling, 20*, 238–249.

HOWARD, G. S. (1991). Culture tales: A narrative approach to thinking, cross–cultural psychology, and psychotherapy. *American Psychologist, 46*, 187–197.

HOWELL, W. S. (1982). *The empathic communicator*. Belmont, CA: Wadsworth.

HOYT, M. F. (1995a). Brief Psychotherapies. In A. S. Gurman & S. B. Messer, (Eds.), *Essential Psychotherapies* (pp. 441–487). New York: Guilford Press.

HOYT, M. F. (1995b). *Brief Therapy and managed care: readings for contemporary practice*. San Francisco: Jossey-Bass.

HOYT, W. T. (1996). Antecedents and effects of perceived therapist credibility: A Meta- analysis. *Journal of Counseling Psychology, 43*, 430–447.

HUBER, C. H., & Baruth, L. G. (1989). *Rational-emotive family therapy*. New York: Springer.

HUDSON, S. M., Marshall, W. L., Ward, T., Johnston, P. W., et al. (1995). Kia Marama: A cognitive–behavioral program for incarcerated child molesters. *Behavior Change, 12*, 69–80.

HUMPHREYS, K. (1996). Clinical psychologists as therapists: History, future, and alternatives. *American Psychologist, 51*, 190–197.

HUMPHREYS, K., & Rappaport, J. (1994). Researching self-help mutual aid groups and organizations: Many roads, one journey. *Applied and Preventive Psychology, 3*.

HUNTER, R. H. (1995). Benefits of competency-based treatment programs. *American Psychologist, 50*, 509–513.

HURVITZ, N. (1970). Peer self-help psychotherapy groups and their implications for psychotherapy. *Psychotherapy: Theory, Research, and Practice, 7*, 41–49.

HURVITZ, N. (1974). Similarities and differences between conventional psychotherapy and peer self-help psychotherapy groups. In P. S. Roman & H. M. Trice (Eds.), *The sociology of psychotherapy*. New York: Jason Aronson.

ICKES, W. (1993). Empathic accuracy. *Journal of Personality, 61*, 587–610.

ICKES, W. (1997). Introduction. In W. Ickes (Ed.), *Empathic Accuracy* (pp. 1–16). New York: Guilford Press.

ISHIYAMA, F. I. (1990). A Japanese perspective on client inaction: Removing attitudinal blocks through Morita therapy. *Journal of Counseling and Development, 68*, 566–570.

IVEY, A. E., & Ivey, M. B. (1999). *Intentional interviewing and counseling* (4th ed.). Pacific Grove, CA: Brooks/Cole.

IVEY, A. E., Ivey, M. B., & Simek-Morgan, L. (1997). *Counseling and psychotherapy: A multicultural perspective* (4th ed.). Needham Heights, MA: Allyn & Bacon.

JACOBSON, N. S. & Christiensen, A. (1996, October). Studying the effectiveness of psychotherapy: How well can clinical trials do the job? *American Psychologist, 51*, 1031–1039.

JANIS, I. L., & Mann, L. (1977). *Decision making: A psychological analysis of conflict, choice, and commitment*. New York: Free Press.

JANOSIK, E. H. (Ed.). (1984). *Crisis counseling: A contemporary approach*. Belmont, CA: Wadsworth.

JENSEN, J. P., Bergin, A. E., & Greaves, D. W. (1990). The meaning of eclecticism: New survey and analysis of components. *Professional Psychology: Research and Practice, 21*, 124–130.

JOHNSON, W. B., Ridley, C.R., & Nielson, S.L. (2000). Religiously sensitive rational emotive behavior therapy: Elegant solutions and ethical risks. *Professional Psychology: Research and Practice, 31*, 14–20.

JONES, A. S., & Gelso, C. J. (1988). Differential effects of style of interpretation: Another look. *Journal of Counseling Psy-chology, 35*, 363–369.

JONES, B. F., Rasmussen, C. M., Moffitt, M. C. (1997) *Real-Life Problem Solving: A Collaborative Approach to Interdisciplinary Learning*. Washington, DC: American Psychological Association.

KAGAN, J. (1996). Three pleasing ideas. *American Psychologist, 51*, 901–908.

KAGAN, N. (1973). Can technology help us toward reliability in influencing human interaction? *Educational Technology, 13*, 44–51.

KAHN, J. H., Kelly, A. E., & Coulter, R. G. (1996). Client self-presentations at intake. *Journal of Counseling Psychology, 43*, 300–309.

KAHN, M. (1990). *Between therapist and client*. New York: W. H. Freeman.

KAMINER, W. (1992). *I'm Dysfunctional, You're Dysfunctional*. Reading, MA: Addison-Wesley.

KANFER, F. H., & Schefft, B. K. (1988). *Guiding therapeutic change*. Champaign, IL: Research Press.

KARASU, T.B., (1986). The psychotherapies: Benefits and limitations. *American Journal of Psychotherapy.* Jul Vol 40(3) 324–342.

KAROLY, P. (1995). Self-control theory. In W. O'Donohue & L. Krasner (Eds.), *Theories of behavior therapy: Exploring behavior change* (pp. 259–285). Washington, DC: American Psychological Association.

KAROLY, P. (1999). A goal systems self-regulatory perspective on personality, psychopathology, and change. *Review of General Psychology, 3,* 264–291.

KATZ, A. H., & Bender, E. I. (1990). *Helping one another: Self-help groups in a changing world.* Oakland, CA: Third Party Publishing Company.

KAUFMAN, G. (1989). *The psychology of shame.* New York: Springer.

KAYE, H. (1992). *Decision power.* Upper Saddle River, NJ: Prentice Hall.

KAZANTIS, N. (2000). Power to detect homework effects in psychotherapy outcome research. *Journal of Consulting and Clinical Psychology, 68,* 166–170.

KAZANTIS, N. & Deane, F. P. (1999). Psychologists' use of homework assignments in clinical practice. *Professional Psychology: Research & Practice, 30,* 581–585.

KAZDIN, A. E. (1996). Validated treatments: Multiple perspectives and issues: Introduction to the series. *Clinical Psychology: Science and Practice, 3,* 216–217.

KEITH–SPIEGEL, P. (1994, November). The 1992 ethics code: Boon or bane? *Professional Psychology: Research and Practice, 25,* 315–316.

KELLY, A. E. (2000a, July). Helping construct desirable identities: A self-presentational view of psychotherapy. *Psychological Bulletin, 126,* 475–496.

KELLY, A. E. (2000b, July). A self-presentational view of psychotherapy: Reply to Hill, Gelso and Mohr (2000) and to Arkin and Hermann (2000). *Psychological Bulletin, 126,* 505–511.

KELLY, A.E., Kahn, J.H., & Coulter, R.G. (1996). Client self-presentations at intake. *Journal of Counseling Psychology, 43(3)* 300–309.

KELLY, E. W., Jr. (1994). *Relationship-centered counseling: An integration of art and science.* New York: Springer.

KELLY, E. W., Jr. (1997, May/June). Relationship-centered counseling: A humanistic model of integration. *Journal of Counseling and Development, 75,* 337–345.

KENDALL, P. C. (1992). Healthy thinking. *Behavior Therapy, 23,* 1–11.

KENDALL, P. C. (1998). Empirically supported psychological therapies. *Journal of Consulting and Clinical Psychology, 66,* 3–6.

KERR, B. & Erb, C. (1991). Career counseling with academically talented students: Effects of a value-based intervention. *Journal of Counseling Psychology, 38,* 309–314.

KIERULFF, S. (1988). Sheep in the midst of wolves: Person-responsibility therapy with criminals. *Professional Psychology: Research and Practice, 19,* 436–440.

KIESLER, D. J. (1988). *Therapeutic Meta-communication.* Palo Alto, CA: Consulting Psychologists Press.

KIRSCH, I. (Ed.). (1999). *How expectancies shape experience.* Washington, DC: American Psychological Association.

KIRSCHENBAUM, D. S. (1985). Proximity and specificity of planning: A position paper. *Cognitive Therapy and Research, 9,* 489–506.

KIRSCHENBAUM, D. S. (1987). Self-regulatory failure: A review with clinical implications. *Clinical Psychological Review, 7,* 77–104.

KIVLIGHAN, D. M., Jr. (1990). Relation between counselors' use of intentions and clients' perception of working alliance. *Journal of Counseling Psychology, 37,* 27–32.

KIVLIGHAN, D. M., Jr., & Arthur, E. G. (2000). Convergence in client and counselor recall of important session events. *Journal of counseling psychology, 47,* 79–84.

KIVLIGHAN, D. M., Jr., & Schmitz, P. J. (1992). Counselor technical activity in cases with improving work alliances and continuing-poor working alliances. *Journal of Counseling Psychology, 39,* 32–38.

KIVLIGHAN, D. M., Jr., & Shaughnessy, P. (2000). Patterns of working alliance development: A typology of client's working alliance ratings. *Journal of Counseling Psychology, 47,* 362–371.

KNAPP, M. L., & Hall, J. A. (1996). *Nonverbal communication in human interaction* (4th edition). San Diego: Harcourt, Brace, Jovanovich.

KNOX, S., Hess. S. A., Petersen, D. A., & Hill, C. E. (1997). A qualitative analysis of client perceptions of the effects of helpful therapist self-disclosure in long-term therapy. *Journal of Counseling Psychology, 44,* 274–283.

KOHUT, H. (1978). The psychoanalyst in the community of scholars. In P. H. Ornstein (Ed.), *The search for self: Selected writings of H. Kohut.* New York: International Universities Press.

KOTTLER, J. A. (1992). *Compassionate therapy: Working with difficult clients.* San Francisco: Jossey-Bass.

KOTTLER, J. A. (1993). *On being a therapist* (2nd ed.). San Francisco: Jossey-Bass.

KOTTLER, J. A. (1997). *Finding your way as a counselor*. Alexandria, VA: American Counseling Association.

KOTTLER, J. A. (2000). *Doing good: Passion and commitment for helping others*. Philadelphia: Brunner/Routledge.

KOTTLER, J. A., & Blau, D. A. (1989). *The imperfect therapist: Learning from failure in therapeutic practice*. San Francisco: Jossey-Bass.

KUSHNER, M. G., & Sher, K. J. (1989). Fear of psychological treatment and its relation to mental health service avoidance. *Professional Psychology: Research and Practice*, 20, 251–257.

LAMBERT, E. W., Salzer, M., & Bickman, L. (1998). Outcome, consumer satisfaction, and ad hoc ratings of improvement. *Journal of Consulting and Clinical Psychology*, 66, 270–279.

LAMBERT, M. J., & Bergin, A. E. (1994). The effectiveness of psychotherapy. In A. E. Bergin & S. L. Garfield, *Handbook of psychotherapy and behavior change* (4th ed. pp. 143–189). New York: Wiley.

LAMBERT, M. J., & Cattani–Thompson, K. (1996, Summer). Current findings regarding the effectiveness of counseling: Implications for practice. *Journal of Counseling and Development*, 74, 601–608.

LAMBERT, M. J., Okiishi, J. C., Finch, A. E., & Johnson, L. D. (1998, February). Outcome assessment: From conceptualization to implementation. *Professional Psychology: Research and Practice*, 29, 63–70.

LANDRETH, G. L. (1984). Encountering Carl Rogers: His views on facilitating groups. *Personnel and Guidance Journal*, 62, 323–326.

LANDSMAN, M. S. (1994). Needed: Metaphors for the prevention model of mental health. *American Psychologist*, 49, 1086–1087.

LANG, P. J. (1995, May). The emotion probe: Studies of motivation and attention. *American Psychologist*, 50, 372–385.

LARSON, L. M. (1998, March). The social cognitive model of counselor training. *The Counseling Psychologist*, 26, 219–273.

LARSON, L. M., & Daniels, J. A. (1998, March). Review of the counseling self-efficacy literature. *The Counseling Psychologist*, 26, 179–218.

LARSON, R. W. (2000, January). Toward a psychology of positive youth development. *American Psychologist*, 55, 170–183.

LATTIN, D. (1992, November 21). Challenging organized religion: Spiritual small-group movement. *San Francisco Chronicle*, pp. A1, A9.

LAVOIE, F., Gidron, B., & Borkman, T. (Eds.) (1995). *Self-help and mutual aid groups: International and multicultural perspectives*. Binghamton, NY: Haworth Press.

LAZARUS, A. A. (1976). *Multi-modal behavior therapy*. New York: Springer.

LAZARUS, A. A. (1981). *The practice of multimodal therapy*. New York: McGraw-Hill.

LAZARUS, A. A. (1993, Fall). Tailoring the therapeutic relationship, or being an authentic chameleon. *Psychotherapy*, 30, 404–407.

LAZARUS, A. A. (1997). *Brief but comprehensible psychotherapy: The multi-modal way (Springer series on behavior therapy and behavioral medicine)*. New York: Springer.

LAZARUS, A. A., Beutler, L. E., & Norcross, J. C. (1992). The future of technical eclecticism. *Psychotherapy*, 29, 11–20.

LAZARUS, A. A., Lazarus, C. N., & Fey, A. (1993). *Don't believe it for a minute!: Forty toxic ideas that are driving you crazy*. San Luis Obispo, CA: Impact.

LAZARUS, R. S. (2000). Toward better research on stressing and coping. *American Psychologist*, 55, 665–673.

LEAHEY, M., & Wallace, E. (1988). Strategic groups: One perspective on integrating strategic and group therapies. *Journal for Specialists in Group Work*, 13, 209–217.

LEAHY, R. L. (1999). Strategic self-limitation. *Journal of Cognitive Psychotherapy*, 13, 275–293.

LEATHERS, D. S. (1996). *Successful nonverbal communication: principles and applications* (3rd ed.). Needham Heights, MA: Allyn & Bacon.

LEE, C. C. (1997). *Multicultural issues in counseling: New approaches to diversity*. (2nd Ed.). Alexandria, VA: American Counseling Association.

LEVIN, L. S., & Shepherd, I. L. (1974). The role of the therapist in Gestalt therapy. *The Counseling Psychologist*, 4, 27–30.

LEWIN, K. (1969). Quasi-stationary social equilibria and the problem of permanent change. In W. G. Bennis, K. D. Benne, & R. Chin (Eds.), *The planning of change*. New York: Holt, Rinehart & Winston.

LIEBERMAN, L. R. (1997). Psychologists as psychotherapists. *American Psychologist*, 52, 181.

LIGHTSEY, O. R. Jr. (1996, October). What leads to wellness? The role of psychological resources in well-being. *The Counseling Psychologist*, 24, 589–735.

LIPSEY, M. W., & Wilson, D. B. (1993). The efficacy of psychological, educational, and behavioral treatment: Confirmation from meta-analysis. *American Psychologist*, 48, 1181–1209.

LLEWELYN, S. P. (1988). Psychological therapy as viewed by clients and therapists. *British Journal of Clinical Psychology*, 27, 223–237.

LOCKE, E. A., & Latham, G. P. (1984). *Goal setting: A motivational technique that works*. Englewood Cliffs, NJ: Prentice Hall.

LOCKE, E. A., & Latham, G. P. (1990). *A theory of goal setting and task performance*. Englewood Cliffs, NJ: Prentice Hall.

LOCKE, E.A., & Latham, G.P. (1994). Goal Setting Theory. In H. F. O'Neil, Jr., & M. Drillings, (Eds.), *Motivation: Theory and Research*. (pp. 13–29). Hillsdale, NJ: Lawrence Erlbaum Associates, Inc.

LOCKWOOD, P., & Kunda, Z. (1999). Increasing the salience of one's best selves can undermine inspiration by outstanding role models. *Journal of Personality and Social Psychology, 76*, 214–228.

LOPEZ, F. G., Lent, R. W., Brown, S. D., & Gore, P. A. (1997). Role of social-cognitive expectations in high school students' mathematics-related interest and performance. *Journal of Counseling Psychology, 1*, 44–52.

LOWENSTEIN, L. (1993). Treatment through traumatic confrontation approaches: The story of S. *Education Today, 43*, 198–201.

LOWMAN, R. L. (Ed.). (1998). *The ethical practice of psychology in organizations*. Washington, DC: American Psychology Association.

LUBORSKY, L. (1993, Fall). The promise of new psychosocial treatments or the inevitability of nonsignificant differences—A poll of the experts. *Psychotherapy and Rehabilitation Research Bulletin*, 6–8.

LUBORSKY, L., et al. (1986). The nonspecific hypothesis of therapeutic effectiveness: A current assessment. *American Journal of Orthopsychiatry, 56*, 501–512.

LUDEMA, J., Wilmot, T., & Srivasta, S. (1997). Organizational hope: Reaffirming the constructive task of social and organizational inquiry. *Human Relations, 50*, 1015–1051.

LUNDERVOLD, D. A., & Belwood, D. A. (2000, Winter). The best kept secret in counseling: Single-case experimental designs. *Journal of Counseling and Development, 78*, 92–102.

LYNCH, R. T., & Gussel, L. (1996). Disclosure and self-advocacy regarding disability-related needs: Strategies to maximize integration in post-secondary education. *Journal of Counseling and Development, 74*, 352–357.

LYND, H. M. (1958). *On shame and the search for identity*. New York: Science Editions.

MacDONALD, G. (1996). Inferences in therapy: Process and hazards. *Professional Psychology: Research and Practice, 27*, 600–603.

MacLAREN, C., & Ellis, A. (1998). *Rational emotive behavior therapy: A therapist's guide*. San Luis Obispo, CA: Impact.

MACHADO, P. P. P., Beutler, L. E., & Greenberg, L. S. (1999). Emotion recognition in psychotherapy: Impact of therapist level of experience and emotional awareness. *Journal of Clinical Psychology, 55*, 39–57.

MADDUX, J. E. (Ed.). (1995). *Self-efficacy, adaptation, and adjustment: Theory, research, and application*. New York: Plenum.

MAGER, R. F. (1992, April). No self-efficacy, no performance. *Training*, 32–36.

MAHALIK, J. R. (1994). Development of the client resistance scale. *Journal of Counseling Psychology, 41*, 58–68.

MAHONEY, M. J. (1991). *Human change processes*. New York: Basic Books.

MAHONEY, M. J., & Patterson, K. M. (1992). Changing theories of change: Recent developments in counseling. In S. D. Brown & R. W. Lent (Eds.), *Handbook of counseling psychology*, pp. 665–689. New York: Wiley.

MAHRER, A. R. (1993, Fall). The experiential relationship: Is it all-purpose or is it tailored to the individual client? *Psychotherapy, 30*, 413–416.

MAHRER, A. R. (1996). *The complete guide to experiential psychotherapy*. New York: Wiley.

MAHRER, A. R., Gagnon, R., Fairweather, D. R., Boulet, D. B., & Herring, C. B. (1994). Client commitment and resolve to carry out postsession behaviors. *Journal of Counseling Psychology, 41*, 407–414.

MALLINCKRODT, B. (1996). Change in working alliance, social support, and psychological symptoms in brief therapy. *Journal of Counseling Psychology, 43*, 448–455.

MANTHEI, R. (1998). *Counseling: The skills of finding solutions to problems*. New York: Routledge.

MANTHEI, R., & Miller, J. (2000). *Good counseling: A guide for clients*. Auckland: Addison Wesley Longman New Zealand.

MARCH, J. G. (1982, November-December). Theories of choice and making decisions. *Society*, 29–39.

MARCH, J. G. (1994). *A primer on decision making: How decisions happen*. New York: The Free Press.

MARKUS, H., & Nurius, P. (1986). Possible selves. *American Psychologist, 41*, 954–969.

MARSH, D.T., & Johnson, D.L. (1997). The family experience of mental illness: Implications for intervention. *Professional Psychology: Research and Practice, 28*, 229–237.

MARTIN, J. (1994). *The construction and understanding of psychotherapeutic change*. New York: Teachers College Press.

MASH, E. J., & Hunsley, J. (1993). Assessment considerations in the identification of failing psychotherapy: Bringing the negatives out of the darkroom. *Psychological Assessment, 5*, 292–301.

MASLOW, A. H. (1968). *Toward a psychology of being* (2nd ed.). New York: Van Nostrand Reinhold.

MASSIMINI, F., & Delle Fave, A. (2000, January). Individual development in a bio-cultural perspective. *American Psychologist, 55*, 24–33.

MASSON, J. F. (1988). *Against therapy: Emotional tyranny and the myth of psychological healing.* New York: Atheneum.

MATHEWS, B. (1988). The role of therapist self-disclosure in psychotherapy: A survey of therapists. *American Journal of Psychotherapy, 42*, 521–531.

MATT, G. E., & Navarro, A. M. (1997). What meta-analyses have and have not taught us about psychotherapy effects: A review and future directions. *Clinical Psychology Review, 17*, 1–32.

MATTHEWS, W., & Edgette, J. (Eds.). (1999). *Current thinking and research in brief therapy: Solutions, strategies, narratives.* New York: Brunner/Mazel.

McAULEY, E., Mihalko, S. L., & Bane, S. M. (1997). Exercise and self-esteem in middle-aged adults: Multidimensional relationships and physical fitness and self-efficacy influences. *Journal of Behavioral Medicine, 20*, 63–87.

McAULIFFE, G. J., & Eriksen, K. P. (1999). Toward a constructivist and developmental identity for the counseling profession: The context-phase-stage-style model. *Journal of Counseling and Development, 77*, 267–279.

McCARTHY, W. C., & Frieze, I. H. (1999, Spring). *Journal of Social Issues, 55*, 33–50.

McCRAE, R. R., & Costa, P. T., Jr. (1997, May). Personality trait structure as a human universal. *American Psychologist, 52*, 509–516.

McCROSKEY. J. C. (1993). *An introduction to rhetorical communication* (5th ed.). Englewood Cliffs, NJ: Prentice-Hall.

McFADDEN, J. (1993). *Transcultural counseling.* Alexandria, VA: American Counseling Association.

McFADDEN, J. (1996, Spring). A transcultural perspective: Reaction to C. H. Patterson's "Multicultural counseling: From diversity to universality." *Journal of Counseling and Development, 74*, 232–235.

McFADDEN, L., Seidman, E., & Rappaport, J. (1992). A comparison of espoused theories of self- and mutual-help groups: Implications for mental health professionals. *Professional Psychology: Research and Practice, 23*, 515–520.

McKAY, M., Davis, M., & Fanning, P. (1997). *Thoughts and feelings: Taking control of your moods and your life.* Oakland, CA: New Harbinger Publications.

McKAY, G. D., & Dinkmeyer, R, D. (1994). *How you feel is up to you.* San Luis Obispo, CA: Impact.

McMILLEN, C., Zuravin, S., & Rideout, G. (1995). Perceived benefit from child sexual abuse. *Journal of Consulting and Clinical Psychology, 63*, 1037–1043.

McNEILL, B., & Stolenberg, C. D. (1989). Reconceptualizing social influence in counseling: The elaboration likelihood model. *Journal of Counseling Psychology, 36*, 24–33.

McWHIRTER, E. H. (1996). *Counseling for empowerment.* Alexandria, VA: American Counseling Association.

MEHRABIAN, A. (1971). *Silent messages.* Belmont, CA: Wadsworth.

MEHRABIAN, A. (1972). *Nonverbal communication.* Chicago: Aldine-Atherton.

MEHRABIAN, A. (1981). *Silent messages: Implicit Communication of Emotions and Attitudes* Belmont, CA: Wadsworth

MEHRABIAN, A., & Reed, H. (1969). Factors influencing judgments of psychopathology. *Psychological Reports, 24*, 323–330.

MEICHENBAUM, D. (1994). *A clinical handbook/practical therapist manual for assessing and treating adults with post-traumatic stress disorder (PTSD).* Waterloo, Ontario: Institute Press.

MEIER, S. T., & Letsch, E. A. (2000). What is necessary and sufficient information for outcome assessment? *Professional Psychology: Research and Practice, 31*, 409–411.

METCALF, L. (1998). *Counseling towards solutions: A practical solution-focused program for working with students, teachers, and parents.* New York: Center for Applied Research in Education.

METCALF, L. (1998). *Solution focused group therapy: Ideas for groups in private practice, schools, agencies, and treatment programs.* New York: The Free Press.

MILAN, M. A., Montgomery, R. W., & Rogers, E. C. (1994). Theoretical orientation revolution in clinical psychology: Fact or fiction? *Professional Psychology: Research and Practice, 4*, 398–402.

MILBANK, D. (1996, October 31). Hiring welfare people, hotel chain finds, is tough but rewarding. *Wall Street Journal*, pp. A1, A14.

MILLER, G. A., Galanter, E., & Pribram, K. H. (1960). *Plans and the structure of behavior.* New York: Holt, Rinehart & Winston.

MILLER, I. J. (1996). Managed care is harmful to outpatient mental health services: A call for accountability. *Professional Psychology: Research and Practice, 27*, 349–363.

MILLER, I. P. (1996a). Some "short-term therapy values" are a formula for invisible rationing. *Professional Psychology: Research and Practice, 27,* 577–582.

MILLER, I. P. (1996b). Time-limited brief therapy has gone too far: The result is invisible rationing. *Professional Psychology: Research and Practice, 27,* 567–576.

MILLER, L. M. (1984). *American spirit: Visions of a new corporate culture.* New York: Morrow.

MILLER, S. D., Hubble, M.A.,& Duncan, B.L. (Eds.). (1996). *Handbook of solution-focused brief therapy.* San Francisco: Jossey-Bass

MILLER, W. C. (1986). *The creative edge: Fostering innovation where you work.* Reading, MA: Addison-Wesley.

MILLER, W. R. (2000, March). Rediscovering fire: Small interventions, large effects. *Psychology of Addictive Behaviors, 14,* 6–18.

MILLER, W. R., & Rollnick, S. (1991). Motivational interviewing: *Behavior and cognitive Psychotherapy, 23,* 325–334.

MOHR, D. C. (1995, Spring). Negative outcome in psychotherapy: A critical review. *Clinical Psychology: Science and Practice, 2,* 1–27.

MORRIS, T. (1994). *True success: A new philosophy of excellence.* New York: Grosset/ Putman.

MULTON, K. D., Brown, S. D., & Lent, R. W. (1991). Relation of self-efficacy beliefs to academic outcomes: A meta-analytic investigation. *Journal of Counseling Psychology, 38,* 30–38.

MULTON, K. D., Patton, M. J., & Kivlighan, D. M., Jr. (1996a). Development of the Missouri Identifying Transference scale. *Journal of Counseling Psychology, 43,* 243–252.

MULTON, K. D., Patton, M. J., & Kivlighan, D. M., Jr. (1996b). Counselor recognition of transference reactions: Reply to Mallinckrodt (1996) and Carter (1996). *Journal of Counseling Psychology, 43,* 259–260.

MURPHY, J. J. (1997). *Solution-focused counseling in middle and high schools.* Alexandria, VA: American Counseling Association.

MURPHY, K. C., & Strong, S. R. (1972). Some effects of similarity self-disclosure. *Journal of Counseling Psychology, 19,* 121–124.

MYERS, D. G. (2000, January). The funds, friends, and faith of happy people. *American Psychologist, 55,* 56–67.

NAJAVITS, L. M., Weiss, R. D., Shaw, S. R., & Dierberger, A. E. (2000). Psychotherapists views of treatment manuals. *Professional Psychology: Research and Practice, 31,* 404–408.

NATHAN, P. E. (1998, March). Practice guidelines: Not yet ideal. *American Psychologist, 53,* 290–299.

NEIMEYER, G. (Ed.). (1993). *Constructivist assessment, a casebook.* Newbury Park, CA: Sage.

NEIMEYER, R. A., & Mahoney, M. J. (1995). *Constructivism in psychotherapy.* Hyattsville, MD: American Psychological Association.

NEIMEYER, R. A., & Raskin, J. D. (2000). *Constructions of disorder: Meaning-making frameworks for psychotherapy.* Washington, DC: American Psychological Association.

NIELSEN, S. J., Johnson, W. B., & Ridley, C. R., (2000). Religiously sensitive rational emotive behavior therapy: Theory, techniques, and brief excerpts from a case. *Professional Psychology: Research and Practice, 31.*

NORCROSS, J. C., & Kobayashi. (1999). Treating anger in psychtherapy: Introduction and cases. *Journal of Clinical Psychology. 55* 275–282.

NORCROSS, J. C., & Wogan, M. (1987). Values in psychotherapy: A survey of practitioners' beliefs. *Professional Psychology: Research and Practice, 18,* 5–7.

NORTON, R. (1983). *Communicator style: Implicit communication of emotions and attitudes.* Beverly Hills, CA: Sage.

O'CONNELL, B. (1998). *Solution-focused Therapy.* London: Sage Publications.

O'HANLON, W. H., & Weiner–Davis, M. (1989). *In search of solutions: A new direction in psychotherapy.* New York: W. W. Norton.

OKUN, B. F., Fried, J., & Okun, M. L. (1999). Diversity as a learning-as-practice primer. Pacific Grove, CA: Brooks/Cole/Wadsworth.

O'LEARY, A. (1985). Self-efficacy and health. *Behavior Research and Therapy, 23,* 437–451.

OMER, H. (2000). Troubles in the therapeutic relationship: A pluralistic perspective. *Journal of clinical psychology, 56,* 201–210.

ORLINSKY, D. E., & Howard, K. I. (1987, Spring). A generic model of psychotherapy. *Journal of Integrative and Eclectic Psychotherapy, 6,* 6–27.

OTANI, A. (1989). Client resistance in counseling: Its theoretical rationale and taxonomic classification. *Journal of Counseling and Development, 67,* 458–461.

PATRICELLI, R. E., & Lee, F. C. (1996). Employer–based innovations in behavioral health benefits. *Professional Psychology: Research and Practice, 27,* 325–334.

PATTERSON, C. H. (1985). *The therapeutic relationship: Foundations for an eclectic psychotherapy.* Pacific Grove, CA: Brooks/Cole.

PATTERSON, C. H. (1996, Spring). Multicultural counseling: From diversity to universality. *Journal of Counseling and Development, 74,* 227–231.

PATTERSON, C. H., & Watkins, C. E. (1996). *Theories of psychotherapy*. Reading, MA: Addison-Wesley.

PAUL, G. L., (1967). Strategy of outcome research in psycotherapy. *Journal of Consulting Psychology*, 31, 109–118.

PAULHUS, D. L., Fridhandler, B. & Hayes, S. (1997). Psychological defense: Contemporary theory and research. In J. Johnson, R. Hogan & S. R. Briggs (Eds.). *Handbook of Personality* (pp. 544–580), New York: Academic Press.

PAYNE, E. C., Robbins, S. B., & Dougherty, L. (1991). Goal directedness and older-adult adjustment. *Journal of Counseling Psychology*, 38, 302–308.

PEDERSEN, P. (1994). A handbook for developing multicultural awareness (2nd ed.). Alexandria, VA.: American Counseling Association.

PEDERSEN, P. B. (1997). *Culture-Centered Counseling Interventions: Striving for Accuracy*. London: Sage.

PEKARIK, G., & Guidry, L. L. (1999). Relationship of satisfaction to symptom change, follow-up adjustment, and clinical significance in private practice. *Professional Psychology: Research and Practice*, 5, 474–478.

PEKARIK, G., & Wolff, C. B. (1996). Relationship of satisfaction to symptom change, follow-up adjustment, and clinical significance. *Professional Psychology: Research and Practice*, 27, 202–208.

PENNEBAKER, J. W. (1995a). *Emotion, disclosure, and health*. Washington, DC: American Psychological Association.

PENNEBAKER, J. W. (1995b). Emotion, disclosure, and health: An overview. In J. W. Pennebaker (Ed.), (pp. 3–10). *Emotion, disclosure, and health*, Washiington, DC: American Psychological Association.

PENNEBAKER, J. W. & Seagal, J. D. (1999, October). Forming a story: The health benefits of narrative. *Journal of Clinical Psychology*, 55, 1243–1254.

PERROTT, L. A. (1998, April). When will it be coming to the large discount chain stores? Psychotherapy as commodity. *Professional Psychology: Research and Practice*, 29, 168–173.

PERRY, M. J., & Albee, G. W. (1994). On "the science of prevention." *American Psychologist*, 49, 1087–1088.

PERSONS, J. B. (1991, February) Psychotherapy outcome studies do not accurately represent current models of psychotherapy: A proposed remedy. *American Psychologist*, 46, 99–106.

PETERSON, C. (2000, January). The future of optimism. *American Psychologist*, 55, 44–55.

PETERSON, C., Maier, S. F., & Seligman, M. E. P. (1995). *Learned helplessness: A theory for the age of personal control*. New York: Oxford University Press.

PETERSON, C., Seligman, M. E. P., & Vaillant, G. E. (1988). Pessimistic explanatory style as a risk factor for physical illness: A thirty-five-year longitudinal study. *Journal of Personality and Social Psychology*, 55, 23–27.

PFEIFFER, A. M., Martin, J. M., & Whelan, J. P., (2000). Decision-making bias in psychotherapy: Effects of hypothesis source and accountability. *Journal of Counseling Psychology*, 47, 429–436.

PHILLIPS, E. L. (1987). The ubiquitous decay curve: Service delivery similarities in psychotherapy, medicine, and addiction. *Professional Psychology: Research and Practice*, 18, 650–652.

PLUTCHIK, R. (2001). *Emotions in the practice of psychotherapy: Clinical implications affect theories*. Washington, DC: American Psychological Association.

PONTEROTTO, J. G., Fuertes, J. N., & Chen, E. C. (2000). Models of multicultural counseling. In S. D. Brown & R. W. Lent (Eds.), *Handbook of counseling psychology*. (3rd edition. pp. 639–669). New York: Wiley.

POPE, K. S., & Tabachnick, B. G. (1994). Therapists as patients: A national survey of psychologists' experiences, problems, and beliefs. *Professional Psychology: Research and Practice*, 25, 247–258.

PRESTON, J. (1998). *Integrative brief therapy: Cognitive, psychodynamic, humanistic & neurobehavioral approaches*. San Luis Obispo, CA: Impact.

PRESTON, J., Varzos, N., & Liebert, D. (1995). *Every session counts: Making the most of your brief therapy*. San Luis Obispo, CA: Impact.

PROCHASKA, J. O., & Norcross, J. C. (1998). *Systems of psychotherapy: A transtheoretical analysis* (4th ed.). Pacific Grove, CA: Brooks/Cole.

PROCTOR, E. K., & Rosen, A. (1983). Structure in therapy: A conceptual analysis. *Psychotherapy: Theory, Research, and Practice*, 20, 202–207.

PUTNAM, R. D. (2000). *Bowling alone*. NY: Simon & Schuster.

PYSZCZYNSKI, T., & Greenberg, J. (1987). Self-regulatory preservation and the depressive self-focusing style: A self-awareness theory of depression. *Psychological Bulletin*, 102, 122–138.

QUALLS, S.H., & Abeles N. (Eds.). (2000). *Psychology and the aging revolution: How we adapt to longer life*. Washington, DC: American Psychological Association.

RATNER, H. (1998). Solution-focused brief therapy: From hierarchy to collaboration. In R. Bayne, P. Nicholson, & I. Horton, (Eds.), *Counseling and communication skills for medical and health practitioners*. London: BPS Books.

RAUP, J. L., & Myers, J. E. (1989). The empty nest syndrome: Myth or reality? *Journal of Counseling and Development, 68,* 180–183.

REANDEAU, S. G., & Wampold, B. E. (1991). Relationship of power and involvement to working alliance: A multiple-case sequential analysis of brief therapy. *Journal of Counseling Psychology, 38,* 107–114.

RENNIE, D. L. (1994). Clients' deference in psychotherapy. *Journal of Counseling Psychology, 41,* 427–437.

RHODES, R. H., Hill, C. E., Thompson, B. J., & Elliot, R. (1994). Client retrospective recall of resolved and unresolved misunderstanding events. *Journal of Counseling Psychology, 4,* 473–483.

RICHARDS, P. S., & Bergin, A. E. (2000). *Handbook of psychotherapy and religious diversity*. Washington, DC: American Psychological Association.

RICHMOND, V. P., & McCroskey, J. C. (2000). *Nonverbal behavior in interpersonal relations*. (4th ed.). Needham Heights, MA: Allyn & Bacon.

RIDLEY, C. R. (2000). Religiously sensitive rational emotive behavior therapy: Theory, techniques, and brief excerpts from a case. *Professional Psychology: Research and Practice, 31,* 21–28.

RIDLEY, C. R., Nielsen, S. L., & Johnson, W. B. (2000). Religiously sensitive rational emotive behavior therapy: Theory, techniques, and brief excerpts from a case. *Professional Psychology: Research and Practice, 31,* 21–28.

RIESSMAN, F. (1985). New dimensions in self-help. *Social Policy, 15,* 2–4.

RIESSMAN, F. (1990, Summer/Fall). Bashing self-help. *Self-Help Reporter,* 1–2.

RIESSMAN, F., & Carroll, D. (1995). *Redefining self-help*. San Francisco: Jossey–Bass.

ROBBINS, S.B., & Kliewer, W. (2000). Advances in theory and research on subjective well-being. In S. B. Brown & R. W. Lent (Eds.), *Handbook of Counseling Psychology* (3rd edition) pp. 310–345. New York: Wiley.

ROBERTS, M. (1998). A thing called therapy: Therapist-client co-constructions. *Journal of Systematic Therapies, 17,* 14–26.

ROBERTSHAW, J. E., Mecca, S. J., & Rerick, M. N. (1978). *Problem-solving: A systems approach*. New York: Petrocelli Books.

ROBINS, R. W., Gosling, S. D., & Craik, K. H. (1999, February). An empirical analysis of trends in psychology. *American Psychologist, 54,* 117–128.

ROBINSON, B. E., & Bacon J. G. (1996). The "If only I were thin..." treatment program: Decreasing the stigmatizing effects of fatness. *Professional Psychology: Research and Practice, 27,* 175–183.

ROBITSCHEK, C. G., & McCarthy, P. A. (1991). Prevalence of counselor self-reliance in the therapeutic dyad. *Journal of Counseling and Development, 69,* 218–221.

ROGERS, C. R. (1951). *Client-centered therapy*. Boston: Houghton Mifflin.

ROGERS, C. R. (1957). The necessary and sufficient conditions of therapeutic personality change. *Journal of Consulting Psychology, 21,* 95–103.

ROGERS, C. R. (1965). *Client-centered therapy: Its current practice, implications and theory*. Boston: Houghton Mifflin.

ROGERS, C. R. (1975). Empathy: An unappreciated way of being. *Counseling Psychologist, 21,* 95–103.

ROGERS, C. R. (1980). *A way of being*. Boston: Houghton Mifflin.

ROGERS, C. R., Perls, F., & Ellis, A. (1965). *Three approaches to psychotherapy* [Film]. Orange, CA: Psychological Films, Inc.

ROGERS, C. R., Shostrom, E., & Lazarus, A. (1977). *Three approaches to psychotherapy* [Film]. Orange, CA: Psychological Films, Inc.

ROLLNICK, S., & Miller, W.R. (1995). What is motivational interviewing? *Behavioral & Cognitive Psychotherapy*. Vol 23 325–334

ROSEN, G. M. (1993). Self-help or hype? Comments on psychology's failure to advance self-care. *Professional Psychology: Research and Practice, 24,* 340–345.

ROSEN, S., & Tesser, A. (1970). On the reluctance to communicate undesirable information: The MUM effect. *Sociometry, 33,* 253–263.

ROSEN, S., & Tesser, A. (1971). Fear of negative evaluation and the reluctance to transmit bad news. *Proceedings of the 79th Annual Convention of the American Psychological Association, 6,* 301–302.

ROSENBERG, S. (1996). Health maintenance organization penetration and general hospital psychiatric services: Expenditure and utilization trends. *Professional Psychology: Research and Practice, 27,* 345–348.

ROSSI, P. H., & Wright, J. D. (1984). Evaluation research: An assessment. *Annual Review of Sociology, 10,* 331–352.

ROWAN, T., O'Hanlon, W. H., & O'Hanlon, B. (1999). *Solution-oriented therapy for chronic and severe mental illness*. New York: Wiley

RUSSELL, J. A. (1995). Facial expressions of emotion: What lies beyond minimal universality? *Psychological Bulletin, 118*, 379–391.

RUSSELL, J. A., Fernandez–Dols, J–M., & Mandler, G. (Eds.). (1997). *The psychology of facial expression*. New York: Cambridge University Press.

RUSSO, J. E., & Schoemaker, P. (1990). *Decision traps: The ten barriers to brilliant decision-making and how to overcome them*. New York: Fireside.

RUSSO, J. E., & Schoemaker, P. J. H. (1992, Winter). Managing overconfidence. *Sloan Management Review*, 7–17.

RYAN, R. M., & Deci, E. L. (2000, January). Self-determination theory and the facilitation of intrinsic motivation, social development, and well-being. *American Psychologist, 55*, 68–78.

RYFF, C. D. (1991). Possible selves in adulthood and old age: A tale of shifting horizons. *Psychology & Aging, 6*, 286–295.

SAFRAN, J. D., & Muran, J. C. (1995a). Resolving therapeutic alliance ruptures: Diversity and integration. *In Session: Psychotherapy in Practice, 1*, 81–92.

SAFRAN, J. D., & Muran, J. C. (Eds.). (1995b, Spring). The therapeutic alliance [Special issue]. *In Session: Psychotherapy in Practice, 1*.

SALOVEY, P., Rothman, A. J., Detweiler, J. B., & Steward, W. T. (2000, January). Emotional states and physical health. *American Psychologist, 55*, 110–121.

SATCHER, D. (2000). Mental health: A report of the surgeon general—Executive summary. *Professional Psychology: Research and Practice, 31*, 5–13.

SATEL, S. (1996, May 8). Psychiatric apartheid. *Wall Street Journal*, pp. A14.

SCHEIN, E. H. (1990, Spring). A general philosophy of helping: Process consultation. *Sloan Management Review*, 57–64.

SCHIFF, J. L. (1975). *Cathexis reader: Transactional analysis treatment of psychosis*. New York: Harper & Row.

SCHMIDT, F. L. (1992). What do data really mean? Research findings, meta-analysis, and cumulative knowledge in psychology. *American Psychologist, 47*, 1173–1181.

SCHMIDT, F. L., & Hunter, J. E. (1993). Tacit knowledge, practical intelligence, general mental ability, and job knowledge. *Current Directions in Psychological Science, 1*, 8–9.

SCHNEIDER, K. J. (1999). Clients deserve relationships, not merely "treatments." *American Psychologist, 54*, 205–207.

SCHOEMAKER, P. J. H., & Russo, J. E. (1990). *Decision traps*. New York: Doubleday.

SCHWARTZ, B. (2000, January). Self-determination: The tyranny of freedom. *American Psychologist, 55*, 79–88.

SCHWARTZ, R. S. (1993). Managing closeness in psychotherapy. *Psychotherapy, 30*, 601–607.

SCHWARZER, R. (Ed.). (1992). *Self-efficacy: Thought control of action*. Bristol, PA: Taylor & Francis.

SCHWARZER, R., & Fuchs, R. (1995). Changing risk behaviors and adopting health behaviors: The role of self-efficacy beliefs. In A. Bandura (Ed.), *Self-efficacy in changing societies*. (pp. 259–288). New York: Cambridge University Press.

SCOTT, B. (2000). *Being real: An ongoing decision*. Berkeley, CA: Frog.

SCOTT, N. E., & Borodovsky, L. G. (1990). Effective use of cultural role taking. *Professional Psychology: Research and Practice, 21*, 167–170.

SECUNDA, A. (1999). *The 15 second principle: Short, simple steps to achieving long–term goals*. New York: HarperCollins.

*SELF-HELP REPORTER*. (1985, Vol. 7, No. 2). The self-help ethos.

*SELF-HELP REPORTER*. (1992, Summer). From self-help to social action.

SELIGMAN, M.. (1975). *Helplessness: On depression, development, and death*. San Francisco: W. H. Freeman.

SELIGMAN, M. (1991). *Learned optimism*. New York: Knopf.

SELIGMAN, M. (1994). *What you can change and what you can't*. New York: Knopf.

SELIGMAN, M. (1995). The effectiveness of psychotherapy: The Consumer Reports study. *American Psychologist, 50*, 965–974.

SELIGMAN, M. (1998). *Learned Optimism, 2nd ed*. New York: Pocket Books (Simond and Schuster.

SELIGMAN, M., & Csikszentmihalyi, M. (2000). Positive psychology: An introduction. *American Psychologist, 55*, 5–14.

SEXTON, G. (1999, June). Viewpoint. *Training*, 88.

SEXTON, T. L., & Whiston, S. C. (1994). The status of the counseling relationship: An empirical review, theoretical implications, and research directions. *The Counseling Psychologist, 22*, 6–78.

SHADISH, W. R., & Sweeney, R. B. (1991). Mediators and moderators in meta-analysis:

There's a reason we don't let dodo birds tell us which psychotherapies should have prizes. *Journal of Consulting and Clinical Psychology, 59,* 883–893.

SHAPIRO, D., & Shapiro, D. (1982). Meta-analysis of comparative therapy outcome studies: A replication and refinement. *Psychological Bulletin, 92,* 581–604.

SHARF, R. S. (1999). *Theories of psychotherapy and counseling: Concepts and cases.* 2nd edition. Belmont, CA: Wadsworth.

SIMON, J. C. (1988). Criteria for therapist self-disclosure. *American Journal of Psychotherapy, 42,* 404–415.

SIMONTON, D. K. (2000, January). Creativity: Cognitive, personal, developmental, and social aspects. *American Psychologist, 55,* 151–158.

SMABY, M., & Tamminen, A. W. (1979). Can we help belligerent counselees? *Personnel and Guidance Journal, 57,* 506–512.

SMITH, E. R., & Mackie, D. M. (2000). *Social psychology.* (2nd ed.) Philadelphia: Psychology Press.

SMITH, M., Glass, G., & Miller, T. (1980). *The benefit of psychotherapy.* Baltimore: Johns Hopkins University Press.

SMITH, P. L., & Fouad, N. A. (1999). Subject-matter specificity of self-efficacy, outcome expectancies, interests, and goals: Implications for the social-cognitive model. *Journal of Counseling Psychology, 4,* 461–471.

SNYDER, C. R. (1984, September). Excuses, excuses. *Psychology Today,* 50–55

SNYDER, C. R. (1988). Reality negotiation: From excuses to hope and beyond. *Journal of Social and Clinical Psychology, 8,* 130–157.

SNYDER, C. R. (1994, December). *The psychology of hope: You can get there from here.* New York: Free Press.

SNYDER, C. R. (1994). Hope and optimism. In V.S. Ramachandran (Ed.), *Encyclopedia of human behavior, Vol. 2* (pp. 535–542). San Diego, CA: Academic Press.

SNYDER, C. R. (1995). Conceptualizing, measuring, and nurturing hope. *Journal of Counseling and Development, 73,* 355–360.

SNYDER, C. R. (1998). A case for hope in pain, loss and suffering. In J. H. Harvey, J. Omarzy, & E. Miller (Eds.), *Perspectives on loss: A sourcebook.* Washington, DC: Taylor & Francis, Ltd.

SNYDER, C. R. (Ed.). (1999). *Coping: The psychology of what works.* New York: Oxford University Press.

SNYDER, C. R. (Ed.). (2000). *Handbook of hope: Theory, measures, and applications.* San Diego, CA: Academic Press.

SNYDER, C. R., Cheavens, J., & Sympson, S. C. (1997). Hope: An individual motive for social commerce. *Group Dynamics, 1,* 107–118.

SNYDER, C. R., Harris, C., Anderson, J. R., Holleran, S. A., et al. (1991). The will and the ways: Development and validation of an individual differences measure of hope. *Journal of Personality and Social Psychology, 60,* 570–585

SNYDER, C. R., & Higgins, R. L. (1988). Excuses: Their effective role in the negotiation of reality. *Psychological Bulletin, 104,* 23–35.

SNYDER, C. R., Higgins, R. L., & Stucky, R. J. (1983). *Excuses: Masquerades in search of grace.* New York: Wiley.

SNYDER, C. R., & McDermott, D., & (1999). *Making hope happen.* Oakland/San Fransisco: New Harbinger Press.

SNYDER, C. R., McDermott, D., Cook, W., & Rapoff, M. (1997). *Hope for the journey: Helping children through the good times and the bad.* Boulder, CO: Westview/Basic Books.

SNYDER, C. R., Michael, S. T., & Cheavens, J. S. (1999). Hope as a psychotherapeutic foundation for nonspecific factors, placebos, and expectancies. In M. A. Huble, B. Duncan, & S. Miller (Eds.), *Heart and soul of change: What works in therapy.* (pp. 179–200) Washington, DC: American Psychological Association.

SNYDER, C. R., et al. (1996). Development and validation of the State Hope Scale. *Journal of Personality and Social Psychology, 2,* 321–335.

SOLDZ, S., & McCullough, L. (Eds.). (2000). *Reconciling empirical knowledge and clinical experience: The art and science of psychotherapy.* Washington, DC: American Psychology Association.

SOMMERS-FLANAGAN, J., & Sommers-Flanagan, R. (1995). Psychotherapeutic techniques with treatment-resistant adolescents. *Psychotherapy, 32,* 131–140.

SOMBERG, D. R., Stone, G. L., & Claiborn, C. D. (1993). Informed consent: Therapists' beliefs and practices. *Professional Psychology: Research and Practice, 24,* 153–159. Revised edition.

SRIVASTA, S, & Cooperrider, D. (1999). *Appreciative management and leadership.* Revised edition. Euclid, OH: Williams Custom Publishing.

STEENBARGER, B. N., & Smith, H. B. (1996, November-December). Assessing the quality of counseling services. *Journal of Counseling and Development,* 145–150.

STERNBERG, R. J. (1990). Wisdom and its relations to intelligence and creativity. In R. J. Sternberg (Ed.), *Wisdom: Its nature, origins, and development* (pp. 124–159). New York: Cambridge University Press.

STERNBERG, R. J., (1990). *Wisdom: Its Nature, Origins, and development*. Cambridge, UK: Cambridge University Press.

STERNBERG, R. J. & Lubart, T. I. (1996). Investing in creativity. *American Psychologist, 51,* 677–688.

STERNBERG, R. J., Wagner, R. K., Williams, W. M., & Horvath, J. A. (1995, Winter). Testing common sense. *American Psychologist, 50,* 912–927.

STERNBERG, R.J. (1998). A balance theory of wisdom. *Review of General Psychology, 2* 347–365.

STILES, W. B. (1994, Fall). Drugs, recipes, babies, bathwater, and psychotherapy process–outcome relations. *Journal of Consulting and Clinical Psychology, 62,* 955–959.

STILES, W. B., & Shapiro, D. A. (1994, Fall). Disabuse of the drug metaphor: Psychotherapy process-outcome correlations. *Journal of Consulting and Clinical Psychology, 62,* 942–948.

STOTLAND, E. (1969). *The psychology of hope*. San Francisco: Jossey-Bass.

STRAUSS. R., & Goldberg, W. A. (1999). Self and possible selves during the transition to fatherhood. *Journal of Family Psychology, 13,* 244–259.

STRICKER, G. (1995). Failures in psychotherapy. *Journal of Psychotherapy Integration, 5(2),* 91–93.

STRICKER, G., & Fisher, A. (Eds.). (1990). *Self-disclosure in the therapeutic relationship*. New York: Plenum.

STROH, P., & Miller, W. W. (1993, May). HR professionals should thrive on paradox. *Personnel Journal, 132–135.*

STRONG, S. R. (1968). Counseling: An interpersonal influence process. *Journal of Counseling Psychology, 15,* 215–224.

STRONG, S. R. (1991). Social influence and change in therapeutic relationships. In C. R. Snyder & D. R. Forsyth (Eds.), *Handbook of social and clinical psychology: The health perspective* (pp. 540–562). New York: Pergamon Press.

STRONG, S. R., & Claiborn, C. D. (1982). *Change through interaction: Social psychological processes of counseling and psychotherapy*. New York: Wiley.

STRONG, S., Yoder, B., & Corcoran, J. (1995). Counseling: A social process for encouraging personal powers. *The Counseling Psychologist, 23,* 374–384.

STRUPP, H. H., Hadley, S. W., & Gomes-Schwartz, B. (1977). *Psychotherapy for better or worse: The problem of negative effects*. New York: Jason Aronson.

SUE, D. W., (1990). Culture-specific strategies in counseling: A conceptual framework. *Professional Psychology: Research and Practice, 21(6),* 424–433.

SUE, D. W., Carter, J. M., Casas, J. M., & Fouad, N. A. (1998). Multicultural counseling competencies: Individual and organizational development. *Multicultural Aspects of Counseling Series, 11*. London: Sage.

SUE, D. W., Ivey, A. E., & Pedersen, P. B. (1996). *Theory of Multicultural Counseling and Therapy*. CA: Wadsworth.

SUE, D. W., & Sue, D. (1999). *Counseling the culturally different: Theory and practice*. (3rd ed.). New York: Wiley.

SULLIVAN, T., Martin, W., Jr., & Handelsman, M. (1993). Practical benefits of an informed-consent procedure: An empirical investigation. *Professional Psychology: Research and Practice, 24,* 160–163.

SWINDLE, R., Heller, K., Pescosolido, B., & Kikuzawa, S. (2000). Responses to nervous breakdowns in America over a 40-year period: Mental health policy implications. *American Psychological Association, 55,* 740–747.

SYKES, C. J. (1992). *A nation of victims*. New York: St. Martin's Press.

TAYLOR, S. E., Kemeny, M. E., Reed, G. M., Bower, J. E., & Gruenewald, T. L. (2000, January). Psychological resources: Positive illusions and health. *American Psychologist, 55.* 99–109.

TAYLOR, S. E., Pham, L. B., Rivkin, I. D., & Armor, D. A. (1998). Harnessing the imagination: Mental stimulation, self-regulation, and coping. *American Psychologist, 53,* 429–439.

TAUSSIG, I. M. (1987). Comparative responses of Mexican Americans and Anglo-Americans to early goal setting in a public mental health clinic. *Journal of Counseling Psychology, 34,* 214–217.

TESSER, A., & Rosen, S. (1972). Similarity of objective fate as a determinant of the reluctance to transmit unpleasant information: The MUM effect. *Journal of Personality and Social Psychology, 23,* 46–53.

TESSER, A., Rosen, S., & Batchelor, T. (1972). On the reluctance to communicate bad news (the MUM effect): A role play extension. *Journal of Personality, 40,* 88–103.

TESSER, A., Rosen, S., & Tesser, M. (1971). On the reluctance to communicate undesirable messages (the MUM effect): A field study. *Psychological Reports, 29,* 651–654.

THOMPSON, K. R., Hochwater, W. A., & Mathys, N. J. (1997). Stretch targets: What makes them effective? *Academy of Management Executive, 11,* 48–60.

TINSLEY, H. E. A., Bowman, S. L., & Barich, A. W. (1993). Counseling psychologists' perceptions of the occurrence and effects of unrealistic expectations about counseling and psychotherapy among their clients. *Journal of Counseling Psychology, 40,* 46–52.

TKACHUK, G. A., & Martin, G. L. (1999, June). Exercise therapy for patients with psychiatric disorders: Research and clinical implications. *Professional Psychology: Research and Practice, 30,* 275–282.

TRACEY, T. J. (1991). The structure of control and influence in counseling and psychotherapy: A comparison of several definitions and measures. *Journal of Counseling Psychology, 38,* 265–278.

*TRAINING.* (1993). Take STRIDES in needs analysis. pp. 14–17.

TREVINO, J. G. (1996, April). Worldview and change in cross-cultural counseling. *Counseling Psychologist, 24,* 198–215.

TRIGEORGIS, L. (1998). *Real Options: Managerial Flexibilityand Strategy in Resource Allocation.* Cambridge: The MIT Press.

TRIGEORGIS, L. (1999). *Real Options and Business Strategy: Applications to Decision Making.* London: Risk Books.

TRYON, G. S., & Kane, A. S. (1993). Relationship of working alliance to mutual and unilateral termination. *Journal of Counseling Psychology, 40,* 33–36.

TYLER, F. B., Pargament, K. I., & Gatz, M. (1983). The resource collaborator role: A model for interactions involving psychologists. *American Psychologist, 38,* 388–398.

VACHON, D. O., & Agresti, A. A. (1992). A training proposal to help mental health professionals clarify and manage implicit values in the counseling process. *Professional Psychology: Research and Practice, 23,* 509–514.

VAILLANT, G. E. (2000, January). Adaptive mental mechanisms: Their role in a positive psychology. *American Psychologist, 55.* 89–98.

VAILLANT, G. E., Davis, J. T. (2000). Social/emotional intelligence and midlife resilience in schoolboys with low tested intelligence. *American Journal ofOrthopsychiatry, 70,* 215– 222.

VALSINER, J. (1986). Where is the individual subject in scientific psychology. In J. Valsiner (Ed.), *The individual subject and scientific psychology* (pp. 1–16). New York: Plenum.

WACHTEL, P. L. (1989, August 6). Isn't insight everything? [Review of the book *How does treatment help: On the modes of therapeutic action of psychoanalytic psychotherapy*]. *New York Times,* p. 18.

WACHTEL, P. L., & Messer, S. B. (Eds.). (1997). *Theories of psychotherapy: Origins and evolution.* Washington, DC: American Psychological Association.

WADDINGTON, L. (1997). Clinical judgement and case formulation. *British Journal of Clinical Psychology, 36,* 309–311.

WAEHLER, C. A., Kalodner, C. R., Wampold, B. E., & Lichtemberg. J. W. (2000, September). Empirically supported treatments (ESTs) in perspective: Implications for counseling training. *The Counseling Psychologist, 28,* 657–671.

WAEHLER, C. A., & Lenox, R. (1994). A concurrent (versus stage) model for conceptualizing and representing the counseling process. *Journal of Counseling and Development, 73,* 17–22.

WAHLSTEN, D. (1991). Nonverbal behavior and self-presentation. *Psychological Bulletin, 110,* 587–595.

WALTER, J., & Peller, J. (1992). *Becoming solution-focused in brief therapy.* New York: Brunner-Mazel.

WAMPOLD, B. E., et al. (1997, November). A meta-analysis of outcome studies comparing bona fide psychotherapies: Empirically, "All must have prizes." *Psychological Bulletin, 122,* 203–215.

WATKINS, C. E., Jr. (1990). The effects of counselor self-disclosure: A research review. *The Counseling Psychologist, 18,* 477–500.

WATKINS, C. E., Jr., & Schneider, L. J. (1989). Self-involving versus self-disclosing counselor statements during an initial interview. *Journal of Counseling and Development, 67,* 345–349.

WATSON, J. C., Greenberg, L. S. (2000). Alliance ruptures and repairs in experimental therapy. *Journal of Clinical Psychology, 56,* 175–186.

WATSON, D. L., & Tharp, R. G. (1997). *Self-directed behavior* (7th ed.). Pacific Grove, CA: Brooks/Cole.

WEICK, K. E. (1979). *The social psychology of organizing* (2nd ed.). Reading, MA: Addison-Wesley.

WEINBERGER, J. (1995, Spring). Common factors aren't so common: The common factors dilemma. *Clinical Psychology: Science and Practice, 2,* 45–69.

WEINER, M. F. (1983). *Therapist disclosure: The use of self in psychotherapy* (2nd ed.). Baltimore: University Park Press.

WEINRACH, S. G. (1989). Guidelines for clients of private practitioners: Committing the structure to print. *Journal of Counseling and Development, 67,* 299–300.

WEINRACH, S. G. (1995). Rational emotive behavior therapy: A tough-minded therapy for a tender-minded profession. *Journal of Counseling and Development, 73,* 296–300.

WEINRACH, S. G. (1996). Nine experts describe the essence of rational emotive therapy while standing on one foot. *Journal of Counseling and Development, 74,* 326–331.

WEINRACH, S. G., & Thomas, K. R. (1996, Summer). The counseling profession's commitment to diversity-sensitive counseling: A critical reassessment. *Journal of Counseling and Development, 74,* 472–477.

WEINRACH, S. G., & Thomas, K. R. (1998). Diversity-sensitive counseling today: A postmodern clash of values. *Journal of Counseling and Development, 76,* 115–122

WEISZ, J. R., Donenberg, G. R., Han, S. S., & Weisz, B. (1995, Fall). Bridging the gap between laboratory and clinic in child and adolescent psychotherapy. *Journal of Consulting and Clinical Psychology, 63,* 688–701.

WEISZ, J. R., Rothbaum, F. M., & Blackburn, T. C. (1984). Standing out and standing in: The psychology of control in America and Japan. *American Psychologist, 39,* 955–969.

WELFEL, E. R., Danzinger, P. R., & Santoro, S. (2000). Mandated reporting of abuse/maltreatment of older adults: A primer for counselors. *Journal of Counseling and Development, 78,* 284.

WELLENKAMP, J. (1995). Cultural similarities and differences regarding emotional disclosure: Some examples from Indonesia and the Pacific. In J. W. Pennebaker (Ed.), *Emotion, Disclosure, and Health,* 293–311.

WESTERMAN, M. A. (Ed.). (1989). Putting insight to work [Special section]. *Journal of Integrative and Eclectic Psychotherapy, 8,* 195–250.

WHEELER, D. D., & Janis, I. L. (1980). *A practical guide for making decisions.* New York: Free Press.

WHISTON, S. C., & Sexton, T. L. (1993, February). An overview of psychotherapy outcome research: Implications for practice. *Professional Psychology: Research and Practice, 24,* 43–51.

WHYTE, G. (1991). Decision failures: Why they occur and how to prevent them. *Academy of Management Executive, 5,* 23–31.

WILLIAMS, R. (1989, January-February). The trusting heart. *Psychology Today,* 36–42.

WILSON, G. T. (1998). Manual-based treatment and clinical practice. *Clinical Psychology: Science & Practice, 5,* 363–375.

WINBORN, B. (1977). Honest labeling and other procedures for the protection of consumers of counseling. *Personnel and Guidance Journal, 56,* 206–209.

WING, R. R. & Jeffery, R. W. (1999). Benefits of recruiting participants with friends and increasing social support for weight loss and maintenance. *Journal of Consulting & Clinical Psychology, 67,* 132–138.

WOOD, B. J., Klein, S., Cross, H. J., Lammers, C. J., & Elliot, J. K. (1985). Impaired practitioners: psychologists' opinions about prevalence, and proposals for intervention. *Professional Psychology: Research and Practice, 16,* 843–850.

WOODY, R. H. (1991). *Quality care in mental health.* San Francisco: Jossey-Bass.

WORSLEY, A. (1981). In the eye of the beholder: Social and personal characteristics of teenagers and their impressions of themselves and fat and slim people. *British Journal of Medical Psychology, 54,* 231–242.

YALOM, I. D. (1989). *Love's executioner and other tales of psychotherapy.* Scranton, PA: Basic Books.

YALOM, V. & Bugental, F. T. (1997). Support in existential-humanistic psychotherapy. *Journal of Psychotherapy Integration, 7,* 119–128.

YANKELOVICH, D. (1992, October 5). How public opinion really works. *Fortune,* 102–108.

YUN, K. A. (1998). Relational closeness and production of excuses in events of failure. *Psychological Reports, 83,* 1059–1066.

ZAYAS, L. H., Torres, L. R., Malcom, J., & DesRosiers, F. S. (1996). Clinicians' definitions of ethnically sensitive therapy. *Professional Psychology: Research and Practice, 27,* 78–82.

ZEMKE, R. (1999, June). Don't fix that company. *Training,* 26–33.

ZIMMERMAN, B. J. (1996). Enhancing student academic and health functioning: A self-regulatory perspective. *School Psychology Quarterly, 11,* 47–66.

ZIMMERMAN, T. S., Prest, L. A., & Wetzel, B. E. (1997). Solution-focused couples therapy group: An experimental study. *Journal of Family Therapy, 19,* 125–144.

ZIMRIN, H. (1986). A profile of survival. *Child Abuse and Neglect, 10,* 339–349.

# N A M E   I N D E X

Abeles, 11, 86
Abramson, P. R., 55
Ackoff, R., 254
Adler, 224
Agresti, A. A., 44
Albano, 44
Albrecht, 358
Alford, 273, 317
Altmaier, 302
Amkoff, 60
Andersen, 67
Anderson, 163
Ankuta, 11
Anthony, W. A., 209
Argyris, C., 181–182, 194
Aristotle, 7
Arkin, 159
Arkowitz, 317
Armor, 263
Arnett, 86
Arnkoff, D. B., 42
Asay, 249
Atkinson, D. R., 227
Austad, 35
Axelson, 49
Azar, 6, 264

Baer, 216
Bailey, K. G., 42
Baldwin, B. A., 234
Baltes, 18–20, 19
Bandura, A., 221, 301, 305
Bane, 345
Barich, 59
Barrett-Lennard, 48
Batchelor, T., 226
Beck, 317
Beier, E. G., 194
Belwood, 33
Benbenishty, R., 61
Bennett, 216
Bennett, B. E., 364

Bennett, M. B., 364
Bennett, M. I., 59
Berberoglu, L. S., 59
Berenson, B. G., 220
Berg, 243
Bergan, J., 44
Bergin, A. E., 13, 37, 44, 49
Bernard, M. E., 176
Berne, E., 193
Bernheim, 272
Bernstein, R., 49
Bersoff, D. N., 59
Beutler, L. E., 37, 44, 80, 216
Bickman, 11
Binder, C., 60
Bischoff, M. M., 163
Blackburn, 289–290
Blampied, 33
Blatt, S. J., 49–50
Blau, D. A., 227
Bloom, 244
Bohart, 48, 156–157, 227
Bordin, E. S., 43
Borgen, F. H., 176, 302
Bornstein, 55
Borodovsky, L. G., 110–111
Borsari, 216
Bowen, 55
Bowman, 59, 238
Brammer, L., 131
Brandsma, J., 56
Brock, 11
Broder, 156
Bronfenbrenner, U., 305
Brown, S. D., 133, 302
Brownell, K. D., 367
Bruch, M. A., 301
Budman, 35
Bugas, 65

Bugental, 317
Burks, 345
Bushe, 246

Cade, 244
Caliborn, 55
Callanan, P., 59
Cameron, 263
Canter, M. B., 59
Capauzzi, 37
Carey, R., 216, 332
Carkhuff, R. R., 66, 196, 209, 319
Carter, J. A., 49
Carton, 83–85
Casas, J. M., 49
Castonguay, 317
Cattani-Thompson, K., 10
Cervone, D., 301
Chang, E. C., 6, 365
Cheavens, 261
Chen, 49
Christensen, 11
Claiborn, C. D., 59, 61, 176
Clark, A. J., 163, 173
Clarkin, J., 232
Claus, 59
Cloud, M. Y., 55
Cogar, M. C., 89
Colby, 216
Cole, H. P., 263
Collins, 288
Conoley, C. W., 214
Cook, 261
Cooper, 244
Cooperrider, 244
Corbett, C. A., 10
Corcoran, J., 55
Corey, G., 37, 59
Corey, M. S., 59
Corrigan, 65
Cosier, R. A., 254

Costa, 36
Costanzo, 84
Cottone, 59
Coulter, 159
Coursey, 273
Covey, S. R., 48, 154
Cowen, E. L., 9, 86
Coyne, 290
Cramer, 171, 172
Crano, William, 55
Cross, J. G., 227, 263
Croteau, J. M., 61
Cuellar, 49
Cummings, Janet, 35
Cummings, N. A., 214
Cummings, Nick, 35

Dabbs, 224
Dana, 227
Daniels, J. A., 214
Das, A. K., 49
Dauser, P. J., 61
Davis, 80, 83, 346
Deane, 155
de Arellano, 344
De Bono, E., 264
Deffenbacher, J. L., 320
Delle Fave, 45
Dench, 216
Denmann, D. W., III, 89
de Shazer, 43
Detweiler, 80–81
Deutsch, M., 147–148
Diener, 7
Dierberger, 12
Dimond, R. E., 165–167
Dinkmeyer, 80
Donenberg, G. R., 10
Donnay, 302
Donovan, 216
Dorn, F. J., 55, 176
Dorner, D., 249, 337
Dougherty, L., 251

Draycott, 224
Driscoll, R., 8, 58, 106,
    165–167, 195, 314,
    363, 366
Dryden, W., 185
Duan, C., 48
Duncan, 244
Durlak, J. A., 16
Dworkin, S. H., 114

Edgette, 244
Edwards, C. E., 207
Egan, G., 48, 86, 164
Ekman, P., 67
Elliot, 227
Ellis, 13, 163, 185, 186,
    192, 297
Ellis, Albert, 156, 185
Ellis, D., 251
Ellis, K., 251
Erb, C., 44
Erickson, 261
Eriksen, 86
Etzioni, A., 254
Eysenck, H. J., 9

Fall, 12
Fanning, 80, 346
Farrelly, F., 56
Fay, 185
Feldman, 55
Feldstein, S., 67
Feller, R., 357
Ferguson, M., 366
Fernandez-Dols, 67
Festinger, S., 224
Fierstien, 55
Fiese, 263
Fish, J. M., 147, 244
Fisher, A., 59
Fisher, R., 207
Floyd, 238
Folkman, 290
Fouad, 49
Foxhall, 13
Frances, A., 232
Frazier, P. A., 55
Freire, P., 55–56
Fremont, S. K., 163
Friedlander, M. L., 163
Friedman, 42
Frieman, 244
Friesen, 67
Frieze, 55
Frueh, 344
Fuchs, 302
Fuertes, 49

Galanter, E., 8–9
Galassi, J. P., 301
Galvin, M. D., 61

Garber, 364
Garfield, S. L., 12, 13,
    226
Gati, I., 254
Gelatt, H. B., 243, 256,
    332
Gelso, C. J., 42, 159,
    220
Giannetti, Eduardo, 178
Gibb, 53
Gilbert, T. F., 360
Gilland, 37
Gilovich, 254
Gist, 238
Glass, G., 10
Glasser, 147
Godwin, Gail, 262
Goldberg, 263
Goldfried, M. R., 42
Goleman, 48, 49, 178
Gollwitzer, 352
Gomes-Schwartz, B., 13
Good, 227
Goodyear, R. K., 208
Gore, 302
Goslin, D. A., 331–332
Grace, M., 67
Greaves, D. W., 37
Green, 11
Greenberg, J., 48, 60,
    79, 80, 365
Greenberg, L. S., 42,
    285
Greenson, R. R., 43
Gross, 37
Guidry, 11
Guisinger, S., 49–50

Hadley, S. W., 13
Haley, Jay, 197
Hall, E. T., 67, 90
Halleck, S. L., 194
Hammond, 257, 258
Han, S. S., 10
Handelsman, M., 61
Hanna, F. J., 18–20, 19,
    170
Hare-Mustin, R., 55
Harper, 185
Harris, 163
Hattie, J. A., 16
Havens, 165–167
Hays, Kate, 345
Headlee, R., 254
Heaton, K. J., 60
Hedstrom, S. M., 61
Heller, 4–5
Hellstroem, 65
Hendrick, S. S., 207, 209
Henggeler, S. W., 10

Heppner, P. P., 55, 254
Herman, 49
Hermann, 159
Hess, 207
Hicksen, 67
Higgins, R. L., 194, 261
Higginson, 194
Highlen, P. S., 67
Hill, M. D., 10, 42, 43,
    48, 60, 67, 89, 159,
    207, 209
Hilliard, 33
Hogan-Garcia, 49
Holaday, 359
Holmes, Offen, Waller,
    159
Hooker, 263
Horvath, A. O., 42
Horvath, J. A., 18–20,
    42, 43
Houser, 55
Howard, G. S., 37
Howard, K. I., 146
Howell, W. S., 57
Hoyt, W. T., 35, 55, 244
Hubble, 244
Huff, M. E., 320
Hunsley, J., 33
Hunter, R. H., 18–20,
    56, 58

Ickes, 97
Ivey, A. E., 49, 66
Ivey, M. B., 49

Jacobson, 11
James, W., 37, 151
Janis, I. L., 196, 255, 289
Jeffery, 318
Jenkins, 263
Jensen, J. P., 37
Johnson, 186, 273
Jones, A. S., 42, 59,
    165–167, 220
Jones, S. E., 5
Juricek, 65

Kagan, J., 13
Kagan, N., 148
Kahn, M., 159
Kalodner, 12
Kalogjera, I. J., 254
Kane, A. S., 212
Kanfer, F. H., 350
Kao, 302
Karoly, B. E., 19
Karoly, P., 251
Kathn, 42
Kaufman, G., 164
Kaye, H., 254
Kazantzis, 155

Kazdin, 12
Keeley, 345
Keeney, 257, 258
Keese, N., 55
Keese, R., 55
Keith-Spiegel, P., 59
Kelly, E. W., Jr., 42, 159
Kempen, 49
Kendall, P. C., 12
Kerr, B. A., 44, 114
Kessler, D. J., 83–85
Keys, 170
Kiesler, D. J., 184
Kikuzawa, 4–5
Kirsch, 301
Kirschenbaum, D. S.,
    342, 350, 366
Kivlighan, D. M., Jr.,
    42, 67, 212, 216
Klein, 227
Kliewer, 7
Knapp, M. L., 67
Knox, 207
Kobayashi, 79
Kohut, H., 48
Kottler, J. A., 18–20,
    163, 167–168, 227
Krausz, M., 254
Kunce, J., 67
Kunda, 270

Lambert, M. J., 10, 11,
    249
Łammers, 227
Landreth, 212
Lang, P. J., 80
Latham, G. P., 251, 281,
    282, 301
Lazarus, A. A., 37, 44,
    185, 238, 249, 290,
    291
Lazarus, Richard, 6
Leahey, M., 289
Leahy, R., 291
Lee, C. C., 49
Lehmann, 302
Lent, R. W., 302
Letsch, 8
Levendusky, P. G., 56,
    58
Levin, L. S., 19
Lewin, Kurt, 354
Lichtenberg, 12
Lichtenstein, 367
Lickey, 65
Liebert, 36
Lightsey, Jr., 301
Lipsey, M. W., 10
Llewelyn, S. P., 243
Locke, E. A., 251, 281,
    282, 301

Lockwood, 270
Lopez, 302
Lowenstein, L., 215
Lubart, 263
Luborsky, L., 11, 13
Ludema, 247
Lundervold, 33
Lynd, H. M., 164

MacDonald, G., 205
Machado, 80
MacLaren, 185
Maddux, J. E., 301
Mager, R. F., 302
Mahalik, J. R., 163
Mahoney, M. J., 176
Mahrer, A. R., 44, 155
Maier, 364
Mallinckrodt, B., 43
Mandler, 67
Mann, L., 255
Manthei, 7, 43, 56, 58, 244
March, J. G., 254, 257
Marcus, H., 263
Marecek, J., 55
Markus, H., 263
Marlatt, 367
Marsh, 273
Martin, W., Jr., 61, 176, 259, 345
Mash, E. J., 33
Maslow, A. H., 151
Massimini, 45
Masson, J. F., 9–10
Mathews, B., 207
Matt, 10
Matthews, B., 244
McAuley, 345
McAuliffe, 86
McCarthy, P. A., 55, 210
McCrae, 36
McCroskey, 67, 92
McCullough, 12
McDermott, 261
McKay, 80, 346
McMillen, C., 221
McNeill, B., 55
McPhearson, 358, 359
McWhirter, E. H., 55
Mecca, S. J., 263
Mehrabian, A., 67, 146
Meichenbaum, Donald, 344
Meier, 8
Messer, 37
Metcalf, 244
Michael, 261
Mihalko, 345
Milan,, M. A., 37
Milbank, D., 345

Miller, 8–9, 10, 56, 58, 196, 216, 244, 254
Miller, I. P., 35
Miller, William, 153
Mitchell, K. M., 220
Moffitt, 5
Mofrei, 263
Mohr, D. C., 13, 42, 159
Montgomery, R. W., 37
Morfei, 263
Morris, Tom, 350
Moskowitz, 290
Multon, K. D., 302
Murdock, N. L., 207
Murphy, K. C., 209, 244
Myers, 7

Nadya, 302
Nagy, G. R., 59
Najavits, 12
Nathan, 12
Nava, G. R., 42
Navarro, 10
Neenan, 185
Neimeyer, G., 176
Nerison, 59
Nielsen, 186
Norcross, J. C., 37, 44, 79
Norton, 67
Nurius, P., 263
Nutt-Williams, E., 43, 60

O'Brien, 133, 209
O'Connell, 244
Oetting, E. R., 320
O'Hanlon, W. H., 7, 43, 243, 244
Okun, 49
O'Leary, A., 302
Omer, 60
Orlinsky, D. E., 37
Osipow, S. H., 254
Otani, A., 163
Ottens, A. J., 18–20

Padula, M. A., 214
Page, 261
Paivio, 79
Pallak, 35
Paniagua, 49
Pape, 83–85
Patel, 51
Patterson, 42, 49, 176
Patton, M. J., 238
Paul, 12
Payne, E. C., 251
Payton, D. S., 214
Pedersen, K. M., 49
Pekarik, 11

Peller, 244
Pennebaker, J. W., 139
Perls, F., 192, 297
Perry, S., 232
Persons, 33
Pescosolio, 4–5
Petersen, P., 207
Peterson, C., 91, 364
Pfeiffer, 259
Pham, 263
Phillips, E. L., 366
Pickrel, S. G., 10
Plato, 7
Plutchik, 79
Ponterotto, J. G., 49
Pope, K. S., 227
Porras, 288
Post, 261
Prest, 244
Preston, 36, 244
Pribram, K. H., 8–9
Prochaska, J. O., 37
Proctor, E. K., 61
Putnam, Robert, 317–318
Pyszczynski, T., 365

Qualls, 86

Racioppo, 290
Raiffa, 257, 258
Rapoff, 261
Rasmussen, 5
Ratner, 244
Raue, P. J., 42
Reandeau, S. G., 42
Reed, H., 146
Reich, 11
Rennie, D. L., 58
Rerick, M. N., 263
Rhodes, R. H., 60
Richards, 49
Richmond, 67, 92
Rideout, G., 221
Ridley, C. R., 186
Rivkin, 263
Robbins, S. B., 7, 251
Roberts, 65
Robertshaw, J. E., 263
Robitschek, C. G., 210
Rochlen, 42
Rogers, 16, 37, 48–49, 192, 200, 238, 297
Rogers, Carl, 42, 48, 66, 76–77
Rollnick, 216
Rosen, 61, 226
Rosenberg, 67
Rossi, P. H., 10
Rothbaum, 289–290
Rothman, 80–81
Rowan, 7, 43

Russell, J. A., 67, 302
Russo, J. E., 254

Safarjan, 273
Salovey, 80–81
Salzer, 11
Sarnoff, D., 263
Satcher, 10
Schefft, B. K., 350
Scherer, John, 146
Schiff, J. L., 364
Schmidt, 10, 18–20
Schmitz, P. J., 212
Schmook, 65
Schneider, L. J., 212
Schoemaker, P. J., 254
Schoenwald, S. K., 10
Schul, Y., 61
Schwagler, 263
Schwartz, G. S., 59, 163
Schwarzer, 301, 302
Schwenk, C. R., 254
Scogin, 238
Scott, 24, 110–111, 301
Secunda, 251
Seligman, M. E., 11, 91, 364, 365
Sexton, G., 10, 42, 194–195
Sharf, 37
Sharpley, C. E., 16
Shaw, 12
Shazer, 244
Shepherd, I. L., 19
Sheri, 14
Shostrom, E., 238
Shumate, J. L., 208
Siegman, A. W., 67
Silberschatz, 65
Simek-Morgan, L., 49
Simon, J. C., 207
Smaby, M., 169
Smith, M., 10, 302
Smith and Mackie, 55
Snyder, C. R., 194, 261, 262, 290
Snyder, Rick, 261
Soldz, 12
Somberg, D. R., 59, 61
Sommers-Flanagan, 170
Spring, 226
Srivastra, 244, 247
Stacks, 67
Staudinger, 18–20, 19
Steenbarger, 35
Sternberg, R. J., 18–20, 19, 263
Steward, 80–81
Stolenberg, C. D., 55
Stone, G. L., 61
Stotland, 261
Strauss, 263

Stricker, G., 207
Stroh, P., 254
Strong, S. R., 55, 176, 209
Strupp, H. H., 13, 60
Stucky, R. J., 194, 261
Sue, 49, 114
Sullivan, T., 61
Swindle, 4–5
Sykes, C. J., 79
Symonds, B. D., 42

Tabachnick, 227
Tallman, 156–157, 227
Tamminen, A. W., 169
Taussig, I. M., 285, 288
Taylor, 263
Tesser, 226
Tharp, R. G., 168
Thomas, K. R., 35, 49, 52–53
Thompson, B. J., 60, 89
Thwaites, G. A., 320
Tinsley, 59
Tkachuk, 345
Towne, 224
Toynbee, A. J., 149
Tracey, T. J., 55, 163
Trevino, J. G., 176
Trigeorgis, 292
Tryon, G. S., 212
Turner, 344

Vachon, D. O., 44
Vaillant, G. E., 83, 91, 172
Valsiner, 33
Varenhorst, B., 332
Varzos, 36
Virgil, 65

Wachtel, P. L., 37
Waddington, 12
Waehler, C. A., 12
Wagner, R. K., 18–20
Wahlsten, D., 84
Wallace, 289, 320, 358
Walter, 244
Wampold, B. E., 12, 42
Ward, T., 238
Watkins, C. E., Jr., 207, 212
Watson, 42, 60, 168
Weick, K. E., 34
Weinberger, J., 42, 43
Weiner, M. F., 207
Weiner-Davis, M., 7, 43, 243
Weinrach, S. G., 49, 52–53, 177
Weinstein, 302
Weiss, B., 10, 12
Weisz, J. R., 10, 289–290
Wellenkamp, J., 141, 163
Wetzel, 244
Wheeler, D. D., 196
Whelan, 259

Whiston, S. C., 10, 42
Whyte, G., 254
Williams, 18–20, 42, 55
Wilmot, 247
Wilson, D. B., 10, 12, 367
Winborn, B., 61
Wing, 318
Winrach, 52, 61
Wise, 42
Wogan, 44
Wolff, 11
Wood, H. E., 42, 227
Woody, R. H., 61
Worsley, A., 114
Worthington, 227
Wright, J. D., 10

Yalom, I. D., 227, 317
Yankelovich, D., 23
Yankura, 185
Yoder, B., 55
Young, D. M., 194
Younggren, 59
Yun, 194

Zack, 42
Zemke, 244, 245, 247, 248
Zimmerman, 244, 302
Zimrin, 359
Zuravin, S., 221

Accomplishments, helping
  clients state needs and wants
  as, 277–278
Accuracy in responding to key
  experiences and behaviors in
  clients' stories, 102–103
Acting, discrepancies between
  thinking/saying and, 180
Action
  bias for, as metagoal, 292–293
  finding incentives and rewards
    for sustained, 355–357
  goal identification, choice
    and shaping, 304–305
  helping clients avoid
    imprudent, 352–353
  leverage and, 239
  linking best-fit strategies to,
    331
  linking challenge to, 224
  linking new perspectives to,
    182–183
  linking strategies to, 320–321
  storytelling and, 153–157
  values in, 44–46
Action arrow, 31, 106, 347–367
  evaluation questions for, 368
Action-focused self-contracts and
  agreements, helping clients
  develop, 357–358
Action strategies (Step III-A),
  21, 30, 310–322
  brainstorming, 313–315
  developing frameworks for,
    315–317
  evaluation strategies for, 322
  linking, to goals, 320–321
  questions on developing, 317
  support for, 317–319
  working knowledge and skills,
    319–320
Active listening, 75–92. See also
  Listening
Adaptive goals, 288–291

Advice, 115
Affect, 79–81
Agendas
  helping clients deal with com-
    peting, 299–300
  keeping in focus, 47
Agents, helpers as, 363–364
Agreement, 115–116
Aims, helping clients move from
  broad, to clear and specific
  goals, 278–280
Alcoholics Anonymous, 207, 284
Alternatives, satisfactory,
  289–291
Ambiguity, acceptance of, 263
Analysis, in decision making, 253
Anxiety, reducing, in developing
  self-efficacy, 304
Appreciate inquiry, 244
Appreciation of self-healing
  nature of clients, 156–157
Assertiveness, 96
Attraction, immediacy and, 213
Awareness, problem solving
  and, 23

Balance-sheet method, 296
  for choosing strategies,
    328–331
  realism in using, 330–331
Basic Behavioral Science Task
  Force of the National
  Advisory Mental Health
  Council, 318
Behavioral statements, 101
Behaviors, 79
  challenging, within helping
    sessions, 195–198
  challenging self-limiting
    external, 188–190
  challenging self-limiting
    internal, 186–188
  external, 180
  internal, 179–180

responding accurately to key, in
  clients' stories, 102–103
Best-fit strategies (Step III-B),
  21, 30–31, 323–333
  balance-sheet method for
    choosing, 328–331
  evaluation questions for, 333
  helping clients choose,
    325–327
  in keeping clients' values,
    326–327
  linking, to action, 331
  questions on, 327
  shadow side of selecting,
    331–333
  strategy sampling in,
    327–328
Best practice, search for,
  37–38
Bias
  for action as metagoal,
    292–293
  status quo, 257
Blind spots, 139
  breaking through (Step I-B),
    27, 139, 177–181
  nature of, 177–179
  challenging, 51, 177–181
  of helpers, 227–228
  moving to new perspectives,
    181–183
    applications, 184–198
  questions to uncover, 198
  specific targets, 179–181
Bodily behavior in nonverbal
  communication, 67
Brainstorming
  in accomplishing goals,
    313–315
  as tool for divergent thinking,
    266–267
Brief therapy, 35–36, 244–248
  beneficial effects of, 249
Browser, helping model as, 37–38

Cardinal value, 160
Caring
  in dealing with client defensiveness, 223
  lack of, 178–179
Challengers
  discrediting, 225
  persuading, to change their views, 225
Challenging (Step I-B), 21, 27, 176–229
  basic concept, 176–198
    evaluation questions, 217
    specific skills, 199–216
  behaviors within helping sessions, 195–198
  being specific in, 222
  being tentative in, 220–221
  blind spots as targets of, 177–181
  versus confrontation, 177
  confrontation and, 215, 217
  developing social networks for supportive, 359–363
  discrepancies, 190–192
  earning right for, 220
  encouragement and, 215–216, 217
  evaluation questions for, 217, 228–229
  excuses for not, 226–227
  excuses in, 194
  goals of, 184
  guidelines for effective, 219–223
  helper self-disclosure and, 207–209, 217
  immediacy, 209–213, 217
  information sharing and, 205–207, 217
  keeping goals of, in mind, 219
  linking, to action, 224
  message behind the message, 200–205
  mind-sets, 185–186
  predictable dishonesties of everyday life, 192–194
  self-limiting external behavior, 188–190
  self-limiting internal behavior, 186–188
  shadow side of, 224–228
  specific skills in, 199–216
  suggestions and recommendations in, 213–215, 217
  using probes in, 129
  wisdom of, 219–229
Change
  fear of, 165
  humanizing the technology of constructive, 341–344
  misgivings about, 163–165

shadow side of implementing, 363–367
Change agenda, helping clients to commit to, 21, 29–30, 294–307
Change programs, plans in adding value to clients, 336–337
Channel of communication, non-verbal behavior as a, 67
Choices (Step II-B), 21, 29, 276–293
  helping clients shape goals, 276–286
  making, in decision making, 253
Clarification (Stage 1), 21, 26–27
  of key issues, 144–146
  probes in helping clients achieve, 122–123
  Step I-A: storytelling, 21, 27, 138–160
  Step I-B: challenging, 21, 27, 176–229
  Step I-C: leverage, 21, 27–28, 231–240
Clichés, 115
Client-centered approach, 42
Client-helper contract, 58–59
Client Resistance Scale, 163
Clients
  adding value to change programs, 336–337
  appreciating the self-healing nature of, 156–157
  assessing severity of problems, 146–147, 231
  assuming goodwill of, 47
  in avoiding imprudent action, 352–353
  in becoming effective tacticians, 351–359
  in becoming more effective decision makers, 251–259
  in brainstorming strategies to accomplish goals, 313–315
  building on successes of, 221–222
  challenging unused strengths in, 221
  in choosing best-fit strategies, 325–327
  in clarifying key issues, 144–146
  commitment of, 295–300
  communicating understanding, 97, 104
  in dealing with competing agendas, 299–300
  dealing with defensiveness of, 223

in dentifying possible obstacles to and resources for implementing plans, 354–355
in determining outcomes, 28–30, 243–259
in developing action-focused self-contracts and agreements, 357–358
in developing contingency plans, 353
in developing frameworks for stimulating clients' thinking about strategies, 315–317
in developing implementation intentions, 352
in developing self-efficacy, 302–304
in developing strategies for accomplishing goals, 30–31
in developing work ethic, 56–57
in discovering and using their power through goal setting, 249–251
empowerment as an outcome value, 55–58
in exploring problem situations and unexploited opportunities, 141–153
in finding incentives and rewards for sustained action, 355–357
fragility of, 58
helping tell stories, 138–161
human tendencies, 59
in identifying themes in stories, 202–203
in implementing plans, 31–32
inertia, 364–366
intentions or proposals
  communicating understanding of, 104–105
  listening to, 82–83
keeping agenda in focus, 47
listening to decisions of, 82
listening to nonverbal messages and modifiers, 83–85
listening to stories of, 77–81
in making connections they may be missing, 203–204
in making plans, 335–346
in making the implied explicit, 201–202
in overcoming procrastinations, 353–354
points of view
  communicating understanding of, 103–104
  listening to, 81
with problems situations missed or opportunities and unused potential, 4–5

resilience in after mistakes and failures, 358–359
respecting values of, 222–223
responding accurately to feelings of, 99–102
responding accurately to the key experiences and behaviors, 102–103
responding selectively to core messages of, 106–107
searching for resources in storytelling of, 149–151
self-efficacy of, 301–304
in setting goals, 296–299
shadow side of listening to, 89–92
in shaping goals, 276–286
in spotting and developing unused opportunities, 151–152
as storytellers, 158–159
strategies in keeping with values of, 326–327
tailoring ready-made programs to needs of, 344–346
in talking productively about the past, 147–149
tentativeness in challenging, 220–221
understanding, 49–53
using probes
  to challenge, 129
  complete to picture, 123–125
  concreteness and clarity, 122–123
  engage in therapeutic dialogue, 122
  process, 125–127
  in viewing problem situations and opportunities, 125
visibly tuning in to, 66–70
in working on right things, 231–240
Clinically determined therapy, 35
Coercion, reacting to, 165–167
Cognitive skills, 359
Collaborative nature of helping, 43
Commitment (Step II-C), 21, 29–30, 46–47, 294–307
  action in, 304–305
  client self-efficacy in, 301–304
  evaluation questions for, 307
  goal setting and, 305–306
  helping clients with, 295–300
Common sense in helping professions, 18–20
Communication
  internal, 48, 70–72
  mutual talk and, 209–213
  nonverbal behavior as channel of, 67
  proficiency in using, 134–135

shadow side of, 70–72, 135–136
summarizing in, 131–134
tactics for empathic highlights, 112
of understanding to clients, 97, 103–105
Competence, 46–47
Competing agendas, helping clients deal with, 299–300
Complexity, tolerance of, 264
Concerns, identifying, in stories, 138–160
Concreteness, using probes in helping clients achieve, 122–123
Confidence, 263
Confirmatory feedback, 361
Confirming, 84
Confirming-evidence trap, 257–259
Confrontation, 215, 217
  challenge versus, 177
Confusing, 84
Connecting, 65
Constructive change
  adapting process, to client's style, 342–343
  humanizing technology of, 341–344
Context
  responding to, 107–108
  understanding clients through, 86–87
Contingency plans, helping clients develop, 353
Contracts
  client-helper, 58–59
  in structuring process, 298
Controlling, 84–85
Controls, helping clients choose goals that are under their, 283
Coping, 290–291
  problem-focused, 290–291
Corrective feedback, 302, 361
Cost-effectiveness in goal setting, 296
Costs, estimation of, 23
Counseling
  goals, 7
  guidelines for integrating diversity and multiculturalism into, 52–53
  sessions as work sessions, 56–57
Counterdependency, immediacy and, 213
Creativity
  in dealing with client defensiveness, 223
  helping and, 263–264
Crisis, client management of, 234

Culture
  problem management and, 36–37
  putting values into broader context of personal, 44–45
Curiosity, 264

Decisional self-control, 350
Decision engineering, 257
Decision making
  helping clients become more effective in, 251–259
  rational, 252–253
  shadow side of, 251–256
Decisions
  listening to clients', 82
  making smarter, 256–259
  responding with emphatic highlights to clients', 104
  using probes to explore, 127–129
Defense avoidance, 255–256
Defensiveness, avoiding, 54
Deliberation, problem solving and, 23
Denying, 84
Dependency, immediacy and, 213
Dialogue. See Therapeutic dialogue
Differential therapeutics, 232
Direction
  lack of, 212–213
  summarizing in providing, 131–134
Direct self-disclosure, 207
Disabling self-talk, 365
Discord, helpers as sowers of, 176–177
Discrepancies, challenging, 190–192
Discretionary change, nature of, 159–160
Disorganization, 366
  fear of, 164
Dissonance, dealing with, 224–226
Distorted listening, forms of, 90–92
Distracting questions, 114–115
Divergent thinking, 264–265
  brainstorming as tool for, 266–267
Diversity, 49–53
  empathic highlights in bridging gaps, 110–111
  guidelines for integrating, into counseling, 52–53
  immediacy and, 213
  understanding, 50–51
  valuing, 52
Drive, 264
Dying well, 270–271

Economics of helping, 231
Educated hunches, sharing, based on empathic understanding, 204–205
Efficacy studies, 10, 11–13
Emerging goals, 288
Emotions, 79–81
  being sensitive in naming, 100–101
  coming out from under self-defeating, 158
  responding accurately to, 99–102
  sharing emphatic highlights about, 101–102
Empathic highlights, 93–116
  in bridging diversity gaps, 110–111
  in helping process, 105–106
  key building blocks of, 98–105
  message behind the message, 200–205
  nature of, 48–49
  as primary orientation value, 48–53
  principles for sharing, 105–106
  relationship between, and using probes, 129–131
  responding skills and, 95–96
  shadow side of sharing, 113–116
  sharing, 97
  as social-influence process, 108
  in stimulating movement in helping process, 108–109
  tasks for communicating, 112
Empathic listening, 48–49, 76–77
Empathic presence, importance of, 66–70
Empathic relationships, importance of, 112–113
Empathy, advanced, message behind the message, 200–205
Emphasizing, 84
Empirically supported treatments (ESTs), 12
Empirically validated treatments (EVTs), 12
Empowerment, as an outcome value, 55–58
Encouragement, 215–216, 217
  of self-challenge, 219
Entropy, 366–367
Environmental cue, 352
Ethics, in the helping relationship, 59
Evaluation, 38
  of helping process, 32–33
Evaluation questions
  for action arrow, 368
  for Step I-A: storytelling, 161
  for Step I-B: challenging, 217, 228–229

for Step I-C: leverage, 240
for Step II-A: possibilities, 274
for Step II-B: choices, 293
for Step II-C: commitment, 307
for Step III-A action strategies, 322
for Step III-B best-fit strategies, 333
for Step III-C planning, 346
Evaluative listening, 90–91
Event-focused immediacy, 211–212
Everyday life, choices in, 251–256
Evidence, for effectiveness of helping, 10
Excuses, challenge, 194
Experiences, 78–79
  responding accurately to key, in clients' stories, 102–103
External behavior, 180
Extinction, 356
Eye contact in nonverbal communication, 67, 69–70

Facial expressions in nonverbal communication, 67
Fact-centered listening, 91
Fads, 38
Failures, helping clients be resilient after, 358–359
Faking, 116
Fear
  of change, 165
  of disorganization, 164
  of intensity, 164
  reducing, in developing self-efficacy, 304
Feedback
  confirmatory, 361
  corrective, 302, 361
  positive, 303
  from significant others, 360
Feelings, 79–81
  distinguishing between expressed and discussed, 99–100
  identifying key, 85–86
  responding accurately to, 99–102
  sharing empathic highlights about, 101–102
Filtered listening, 90
Flexibility
  helping clients choose goals that have some, 284
  in helping relationships, 44
  in using helping model, 33–34
Focus
  leverage and, 238–239
  summarizing in providing, 131–134
Formal helpers, 3

Framebending, 181
Framebreaking, 181
Frameworks, developing in stimulating clients' thinking about strategies, 315–317
Future
  cases featuring possibilities for a better, 270–274
  helping clients commit to better, 295–300
  helping client talk about past to prepare for, 149
  possibilities for better (Step II-A), 21, 28–29, 261–274
  skills for identifying possibilities for better, 263–270
Future-oriented probes, 268–269

Genuineness as a professional value, 53–54
Gestalt psychology, 91
Goal identification, choosing, and shaping (Stage II), 21, 28, 243
  Step II-A: possibilities, 21, 28, 261–274
  Step II-B: choices, 21, 29, 276–293
  Step II-C: commitment, 21, 29–30, 307
Goals
  adaptive, 288–291
  bias for action as meta-, 292–293
  of challenging, 184, 219
  consistency with values, 284–285
  defined, 276
  developing strategies for accomplishing, 30–31, 309, 313–315
  emerging, 288
  establishing realistic time frames for the accomplishment of, 285–286
  flexibility of, 264, 284
  formulating realistic, 282–283
  of helping, 7–9
  helping clients with setting, 249–251, 296–299
  moving from board aims to clear and specific, 278–280
  prudent, 282
  real-options approach to, 292
  shadow side of setting, 305–306
  shaping, 276–286, 287
  sustainable, 283–284
  that make a difference, 280–281
Goal strategies (Stage III), 21, 30, 309–346
  Step III-A: action strategies, 21, 30, 310–322

Step III-B: best-fit strategies, 21, 30–31, 323–333
Step III-C: planning, 21, 31, 335–346
Goodwill of client, 47

Happiness, 7
Hearing. *See also* Listening
opportunities and resources, 83
processing in, 85–88
slant or spin, 87–88
Helpers
as agents, 363–364
blind spots of, 227–228
challenge and shadow side of, 226–228
employing the client as, 169
formal, 3
human tendencies, 59
identifying good and bad, 13
informal, 3
internal conversation, 88–89
nonverbal behavior of, 67–68
self-disclosure, 207–209, 217
as sowers of discord, 176–177
Helping, 3–20
brief therapy and a hologram approach to, 35–36
clients implement plans, 31–32
collaborative nature of, 43
conclusions on, 13–14
creativity and, 263–264
defined, 3–5
economics of, 231
evidence for effectiveness of, 10
formal, 3
importance of dialogue in, 65–66
informal, 3
messiness of, 17–18
as a natural, two-way influence process, 57
positive psychology and, 6–7
principal goals of, 7–9
problem-management and op-portunity-development ap-proach to, 24–25
shadow side of, 16–20
as a social-influence process, 55
solution-focused, 243–248
usefulness of, 9–14
whole process mentality about, 36
Helping models. *See also* Skilled helper model
action and, 239
as browser, 37–38
building planning mentality into, 341–342
empathic highlights in, 105–106

empathic highlights in stimu-lating movement in, 108–109
evaluation of, 32–33
flexibility in using, 33–34
probes in, 125–127
as problem-management and opportunity-development approach to helping, 24–25
rigid application of, 38–39
shadow side of, 38–39
stages and steps of, 25–31
Helping professions, common sense and wisdom in the, 18–20
Helping relationships, 42–61
client empowerment as out-come value, 55–58
client-helper contract in, 58–59
empathy as primary orientation value, 48–53
ethics in, 59
flexibility in, 44
as forum for relearning, 43–44
genuineness as a professional value, 53–54
as means to an end, 42–43
respect as foundation value, 46–47
shadow-side realities in, 59–61
values in action, 44–46
as working alliance, 43–44
Helping sessions, challenging be-haviors within, 195–198
Helplessness, 365
Honesty, in dealing with client defensiveness, 223
Hope
benefits of, 262
psychology of, 261–262
Humanizing technology of con-structive change, 341–344
Human universal, 36–37
Hunches, sharing educated, 204–205

Immediacy, 209–213, 217
event-focused, 211–212
relationship, 210–211
situations calling for, 212–213
types of, in helping and princi-ples for using them, 209–212
Implementation intentions, help-ing clients develop, 352
Imprudent action, helping clients avoid, 352–353
Inadequate listening, 75–76
Incentives, 297
helping clients find
for committing to change agendas, 21, 29–30, 294–307

for sustained action, 355–357
searching for, in moving be-yond resistance, 169
Independence, 264
Informal helpers, 3
Information gathering, in deci-sion making, 253, 254
Information processing, in deci-sion making, 254–255
Information sharing, 205–207, 217
Innovation, 276
Intensity
adding, 84
fear of, 164
Intentions
listening to clients, 82–83
in moving to clear and specific goals, 278–279
responding with emphatic highlights to clients', 104–105
Internal behavior, 179–180
Interpersonal communication competence, 70–72
skill, 48
Interpretations, 115
Interrupting, 92
Interventions, tailoring, in diversity-sensitive way, 51
Involuntary clients, 165–166

Judgment, rushing to, 47
Just society, two-person, 169

Key building blocks
principles for sharing, 105–111
tactics for communicating, 112
Key issues, helping clients clarify, 144–146
Know-how, 96

Lack of trust, 164
Lateral thinking, 264–265
"Law of the instrument," 231
Lazarus technique in leverage, 238–239
Learned helplessness, 364–365
Learning, focusing on, 57–58
Leverage (Step I-C), 139, 231–240
action and, 239
evaluation questions for, 240
focus and, 238–239
initial search for, 231–233
Lazarus technique in, 238–239
principles of, 233–237
shadow side of, 240
working on issues that make a difference, 233

Listening
active, 75–92
to clients'
decisions, 82
intentions or proposals,
82–83
nonverbal messages and
modifiers, 83–85
points of view, 81
shadow side of, 89–92
stories, 77–81
empathic, 48–49, 76–77
evaluative, 90–91
fact-centered, 91
filtered, 90
hearing and, 83
inadequate, 75–76
to oneself, 88–89
partial, 75–76
person-centered, 91
processing what you hear and,
85–88
rehearsing and, 76
search for meaning in, 85–88
for the slant, 200
stereotype-based, 91
sympathetic, 91
tape-recorder, 76
tough-minded, 87–88
to words, 77–83
Logistics, 351

Meaning, thoughtful search for,
85–88
Messages
identifying key, 85–86
message behind the, 200–205
responding to clients' core,
106–107
Messiness of helping, 17–18
Meta-analysis, 10
Metagoal bias for action,
292–293
Milieu therapy, 325
Mind-sets, 179
challenging, 185–186
Mining, 37
Mistakes, helping clients be re-
silient after, 358–359
Models, 303
Moods, 79–81
responding accurately to,
99–102
Multiculturalism, 49–53
guidelines for integrating, into
counseling, 52–53
MUM effect, 226
Mutual influencing, 65
Mutual talk, 209–213

Needs
helping clients determine
(Stage 2), 21, 243–259

helping clients get, 350–366
stating, as outcomes, 277–278
tailoring ready-made programs
to client, 344–346
versus wants, 286–288
working knowledge and skills
in getting, 319–320
Negative self-involving statements, 212
Nonconformity, 264
Nonlistening, 75
Nonverbal behavior
as channel of communication,
67
of helpers, 67–68
listening to clients', 83–85
myths about, 92
prompts in, 119
responding to feelings and emotion embedded in, 100

Obstacles, helping clients identify possible, and resources for
implementing plans, 354–355
Oneself, listening to, 88–89
Open-ended questions, 121
Open posture, 69
Opportunities
clients with missed, 5
hearing, 83
helping clients spot and develop, 27–28, 139–160,
231–240
Optimism, 263
Ordinary events, infusing, with
positive meaning, 291
Organizing, 38
Outcomes
cocreating, 65
expectations and, 301
final, 285
helping clients determine,
28–30, 243–259
helping clients state needs and
wants as, 277–278
immediate, 285

Parroting, 115
Partial listening, 75–76
Passivity, 364
Past
dealing with, 245–246
helping clients talk produc-
tively about, 147–149
Perceptiveness, 95–96
Persistence, 264
Personal culture, putting values
into the broader context,
44–45
Person-centered listening, 91
Perspectives
developing, 181–182
linking new, to action, 182–183

Phrases that are, in effect, questions or requests, 120–121
Picture, using probes to help
clients complete, 123–125
Piggybacking, 267
Planning (Step III-C), 21, 31,
335–346
in adding value to change programs, 336–337
building mentality for, into
helping process, 341–342
contingency, 353
evaluation questions on, 346
helping clients identify obstacles and resources for implementation of, 354–355
humanizing the technology of
constructive change,
341–344
lack of, 335–336
questions on, 340
questions on implementing,
361
shaping, 338–340
tailoring ready-made programs
to clients' needs, 344–346
Points of view
exploring with probes, 127–129
listening to, 81
responding with emphatic
highlights to clients',
103–104
Positive feedback, 303
Positive psychology
challenging strengths as approach, 221
goals of helping and, 7
helping and, 6–7
leverage mind-set in, 237
Positive reappraisal, 290
Possibilities
exemplars and models as
sources of, 269–270
identifying, for better future
(Step II-A), 21, 28–29,
261–274
moving from, to choices (Step
II-B), 21, 29, 276–293
questions for exploring, 271
Posttraumatic stress disorder
(PTSD), 344–346
Potential, clients with unused, 5
Pragmatics of values, 45–46
Predictable dishonesties of everyday life, challenging the,
192–194
Prejudices, 185
Present, helping client talk
about past make sense of,
148–149
"Prime the pump," 314
Proactivity, importance of,
154–155

Probes
  in challenging clients, 129
  different forms of, 120–121
  in exploring clients' points of
    view, decisions, and propos-
    als, 127–129
  future-oriented, 268–269
  principles in the use of,
    122–129
  relationship between sharing
    empathic highlights using,
    129–131
  suggestions for the use of, 132
Problem-focused coping,
  290–291
Problem management
  culture and, 36–37
  as underlying process, 37
Problem-management and opportunity-development ap-
  proach to helping, 24–25
Problems, 78
  clients with, 4–5
  finding, 143
  helping clients explore,
    141–153
  identifying and exploring op-
    portunities and, 27–28,
    139–160, 231–240
  with manageable subproblem of
    larger, 235
  probes in exploring, 125
  seeing, as opportunities, 153
Problem-severity formula,
  146–147, 231
Procrastination, helping clients
  overcome, 353–354
Productive approaches
  developing, to deal with reluc-
    tance and resistance,
    168–171
  in talking about past, 147–149
Productivity in using time,
  155–156
Proficiency in using communication skills, 134–135
Prompts
  nonverbal, 119
  vocal and verbal, 119
Prompts and fade technique, 314
Proposals
  exploring with probes, 127–129
  listening to clients', 82–83
  responding with emphatic
    highlights to clients',
    104–105
Protracted self-control, 350–351
Prudent, helping clients set goals
  that are, 282
Psychological defenses, 171–173
Psychological resources, 359
Psychology of hope, 261–262
Psychotherapy, 51

Questions. *See also* Evaluation
  questions
  about client commitment, 300
  on best-fit strategies, 327
  on developing strategies, 317
  distracting, 114–115
  effective use of, 121
  for exploring possibilities, 271
  as form of probes, 120
  leverage, 239
  open-ended, 121
  on planning, 340
  for shaping goals, 287
  single words or phrases as,
    120–121
  to uncover blind spots, 198

Rational decision making, 23–24,
  252–253
Rational-Emotional-Behavior
  Therapy (REBT), 185
Rationalization, 225
Rational problem solving, 23–24
Ready-made programs, tailoring,
  to clients' needs, 344–346
Realism in using balance sheet,
  330–331
Realistic goals, helping clients
  formulate, 282–283
Realistic strategies, 326
Real-options approach, 292
REBT, 186
Reframing, 181
Regulating, 84–85
Rehearsing, 76
Reinforcer, success as, 303
Relationships
  challenging, 360
  immediacy, 210–211
Reluctance, 163
  avoiding unhelpful responses,
    167–168
  developing productive ap-
    proaches to dealing with,
    168–171
  misgivings about change,
    163–165
  shadow side of, 171–173
Remedies, initial search for, 23
Repeating, 84
Requests
  as form of probes, 120
  single words or phrases as,
    120–121
Resiliency, in clients after
  mistakes and failures,
    358–359
Resistance, 163
  avoiding unhelpful responses,
    167–168
  developing productive ap-
    proaches to dealing with,
    168–171

reacting to coercion, 165–167
  shadow side of, 171–173
Resource collaborator, 57
Resources
  hearing, 83
  in helping clients formulate re-
    alistic goals, 282–283
  helping clients identify, for im-
    plementing plans, 354–355
  mobilization of, 157–158
  searching for in storytelling,
    149–151
  strengths and unused, 194–195
Respect as the foundation value,
  46–47
Responding skills, 95
  dimensions of, 95–96
Responses, avoiding unhelpful, to
  reluctance and resistance,
    167–168
Results
  getting, 31–32
  importance of, in helping, 7–8
Rewards, helping clients find,
  for sustained action,
    355–357
Robust strategies, 325–326

Satisfactory alternatives,
  289–291
Saying, discrepancies between
  acting and, 180
Screening, initial search for
  leverage and, 231–233
Self-challenge, encouraging,
  219
Self-contracts, 298, 300
Self-control, 18–19
  decisional, 350
  protracted, 350–351
Self-deception, 178
Self-disclosure, 163
  helper, 207–209
Self-efficacy
  beliefs, 301
  client, 301–304
  helping clients develop,
    302–304
Self-examination therapy (SET),
  238–239
Self-healing nature of clients,
  156–157
Self-involving statements, 212
Self-limitation, strategic, 291
Self-limiting beliefs and assump-
  tions, 185–186
Self-limiting external behavior,
  challenging, 188–190
Self-limiting internal behavior,
  challenging, 186–188
Self-regulatory failure, 366
Self-responsibility, norms for,
  55–58

Severity, assessing, for clients' problems, 146–147, 231
Shadow side
of best-fit strategies, 331–333
of challenging, 224–228
of communication skills, 70–72, 135–136
of decision making, 251–256
of goal setting, 305–306
of helping models, 16–20, 38–39
in the helping relationship, 59–61
for implementing change, 363–367
of listening to clients, 89–92
of reluctance and resistance, 171–173
of sharing empathy, 113–116
of storytelling, 158–160
Shaping
of goals, 276–286, 287
in planning, 338–340
Short responses, 112
Significant others, feedback from, 360
Skilled helper model, 351
Stage 1: clarifying issues calling for change, 21, 26–27, 105, 137–240
Step I-A: storytelling, 21, 27, 138–160
Step I-B: challenging, 21, 27, 176–229
Step I-C: leverage, 21, 27–28, 231–240
Stage II: identifying, choosing, and shaping goals, 21, 105, 243–259
Step II-A: possibilities, 21, 28–29, 261–274, 304
Step II-B: choices, 21, 29, 276–293, 304–305
Step II-C: commitment, 21, 29–30, 294–307
Stage III: developing strategies for accomplishing goals, 21, 30, 105–106, 309–346
Step III-A: action strategies, 21, 30, 310–322
Step III-B: best-fit strategies, 21, 30–31, 323–333
Step III-C: planning, 21, 31, 335–346
Social facilitation, 318
Social-influence process
empathic highlights as, 108
helping as, 55
Social networks, developing, for supportive challenge, 359–363
Social support, 359

Solution-focused helping, 7, 243–248
Solutions versus solutions, 248
Space, as nonverbal communication, 67
Specificity, in challenging, 222
Specific strategies, 325
Spontaneity, 54
Stage I. See under Skilled helper model
Stage II. See under Skilled helper model
Stage III. See under Skilled helper model
Starting points in storytelling, 143–144
Statements, as form of probes, 120
Status quo bias, 257
Step I-B. See under Skilled helper model
Step I-C. See under Skilled helper model
Step II-A. See under Skilled helper model
Step II-B. See underSkilled helper model
Step II-C. See under Skilled helper model
Step III-A. See under Skilled helper model
Step III-B. See under Skilled helper model
Step III-C. See under Skilled helper model
Stereotype-based listening, 91
Stories, 139
listening to, 77–81
responding accurately to key experiences and behaviors in clients', 102–103
Storytellers, clients as, 158–159
Storytelling (Step I-A), 21, 27, 138–160
action and, 153–157
evaluation of, 157–158
evaluation questions for, 161
learning to work with styles of, 141–143
searching for resources in, 149–151
shadow side of, 158–160
starting points in, 143–144
Strategic self-limitation, 291
Strategies. See Action strategies (Step III-A); Best-fit strategies (Step III-B); Goal strategies (Stage III)
Strategy sampling, 327–328
Strengthening, 84
Strengths, challenging unused, 221
Subjective well-being (SWB), 7

Substantial distortions, inviting clients to challenge, 192–193
Success as reinforcer, 303
Suggestions and recommendations, using, 213–215
Summarizing, 131–134
Supportive challenge, developing social networks for, 359–363
Sustained, goals that can be, 283–284
Sustained action, helping clients find incentives and rewards for, 355–357
Sympathetic listening, 91
Sympathy, 115–116
Systematic desensitization, 344–346

Tacticians, helping clients become effective, 351–359
Tactics, 351
Tape-recorder listening, 76
Technology, humanizing, of constructive change, 341–344
Templeton Positive Psychology Prize, 6
Tension, immediacy and, 213
Tentativeness, in challenging, 220–221
Therapeutic dialogue, 63–136
active listening in, 75–92
empathic presence in, 66–70
importance of, in helping, 65–66
probing in, 119–131
responding skills in, 95–96
shadow side of communication skills in, 70–72
sharing emphatic highlights, 95–116
summarizing in, 131–134
Thinking
discrepancies between acting and, 180
divergent, 264–267
lateral, 264–265
Time, productive use of, 155–156
Time frames, helping clients establish realistic, for accomplishment of goals, 285–286
Time-sensitive psychotherapies, 35
Tolerance of complexity, 264
Tough-minded listening and processing, 87–88
Training as treatment, 319
Treatment manuals, 11–13
Trust
immediacy and, 213
lack of, 164

Turn taking, 65
Two-person just society, 169

Ubiquitous decay curve,
  366–367
Unawareness, simple, 178
Uncertainty, acceptance of,
  263
Underlying process, problem
  management as, 37
Understanding
  communication of, to clients,
  97
  recovering from inaccurate,
  109–110
Urgency, problem solving and, 23

Values
  in action, 44–46
  adding, to change programs,
  336–337
  choosing goals consistent with,
  284–285
  client empowerment as an
  outcome, 55–58

empathy as a primary orientation, 48–53
genuineness as professional,
  53–54
pragmatics of, 45–46
putting, into the broader
  context of personal
  culture, 44–45
respect as the foundation,
  46–47
respecting clients, 222–223
Verbal fluency, 264
Verbal prompts, 119
Vicious circles, 365–366
Virtuosity, 39
Visibly tuning in to clients,
  66–70
  questions on, 71
Vocal prompts, 119
Voice-related behavior in nonverbal communication, 67

Wants
  helping clients determine
  (Stage 2), 21, 243–259

helping clients get, 350–366
  needs versus, 286–288
  stating, as outcomes,
  277–278
  working knowledge and skills
  in getting, 319–320
Whole process mentality about
  helping, 36
Wisdom in helping professions,
  18–20
Words. *See also* Storytelling
  (Step I-A)
  listening to, 77–83
  single, that are, in effect,
  questions or requests,
  120–121
Work
  counseling sessions as,
  56–57
  helping clients develop ethics
  of, 56–57
  helping relationships as
  alliance, 43–44
  knowledge and skills of,
  319–320